Entomology Monographs

Series Editor

Hideharu Numata, Kyoto University, Kyoto, Japan

Insects are the most diverse group of organisms, and many biological advances have been derived from this group. Although entomology is considered to be a classic science, recent developments in molecular methods, application of modern theoretical concepts and collaboration with related sciences have opened new directions in entomology. Japanese scientists play a significant role in these fields, and this book series will focus on such developments. The book series Entomology Monographs publishes refereed volumes on all aspects of entomology, including ecology, ethology, physiology, taxonomy, systematics, morphology, evolutionary developmental biology, genetics, biochemistry, and molecular biology in insects and related arthropods. Authors are not restricted to Japanese entomologists, and other international experts will also be considered on the basis of their recent contribution to these fields.

Akira Mizoguchi • Naoki Okamoto
Editors

Insect Peptide Hormones and Neuropeptides

Springer

Editors
Akira Mizoguchi
Nagoya, Aichi, Japan

Naoki Okamoto
Life Science Center for Survival Dynamics,
Tsukuba Advanced Research Alliance (TARA)
University of Tsukuba
Tsukuba, Ibaraki, Japan

ISSN 2522-526X ISSN 2522-5278 (electronic)
Entomology Monographs
ISBN 978-981-95-7509-1 ISBN 978-981-95-7510-7 (eBook)
https://doi.org/10.1007/978-981-95-7510-7

This Springer imprint is published by the registered company Springer Nature Singapore Pte Ltd.
The registered company address is: 152 Beach Road, #21-01/04 Gateway East, Singapore 189721, Singapore

Professor Hironori Ishizaki

This book is dedicated to Hironori Ishizaki (1930–2021), an insect endocrinologist who contributed greatly to the advancement of insect peptide hormone research. Ishizaki had an unusual background as a scientist: he initially studied at an art school with the aim of becoming a painter, influenced by his father, a famous Japanese-style painter. However, he also had a strong interest in science. After much consideration, he re-entered the Faculty of Science at Kyoto University after graduating. Ishizaki started his research under Prof. Mamori Ichikawa, a biologist at Kyoto

University, who was passionate about determining the structure of the prothoracicotropic hormone (PTTH) of insects. In response to Ichikawa's earnest wish, Ishizaki began purifying PTTH from the brains of the silkworm Bombyx mori *in 1960. However, purifying PTTH was extremely difficult, and after struggling for 10 years on his own, he almost abandoned it. However, at that time, he received an offer for collaboration on PTTH sequencing from Prof. Saburo Tamura, an organic chemist, of The University of Tokyo. Accepting his offer gladly, Ishizaki restarted PTTH purification and finally succeeded in determining the complete sequence of PTTH in 1990, 30 years after the start of this project. During this joint research, Ishizaki moved to Nagoya University, and Tamura retired and was succeeded by Akinori Suzuki. Through this collaboration, he also succeeded in sequencing* B. mori *insulin-like peptides. This was the first identification of insulin-related peptides in animals other than vertebrates, and this result was highly appreciated not only by entomologists but also by vertebrate endocrinologists. This process is detailed in Chap. 1 of this book. The joint research of Ishizaki and Suzuki nurtured many young insect hormone researchers, who determined the structures of many insect peptide hormones, including MRCH, PBAN, EH, AT, DH, and IGFLP, during their graduate school days and later research lives. Furthermore, his tenacious and sincere approach to research had a strong influence on young researchers. Thus, Ishizaki contributed greatly to advances in insect peptide hormone research in Japan. For his scientific contributions through these studies, he was awarded several prizes, including the Japan*

Academy Prize, the most prestigious academic award in Japan, and the Zoological Society of Japan Award. In addition to his vigorous research activity, he also played important roles in the Zoological Society of Japan and the Asia and Oceania Society for Comparative Endocrinology. Although he did not choose a career as a professional painter, he continued to paint pictures during his busy schedules. On his days off, he often walked in nature, looking for insects to use as the subject for his paintings. After retiring from the university, he devoted himself to creating pictures, mainly painting butterflies and moths. Hironori Ishizaki was undoubtedly an outstanding scientist, teacher, nature lover, and artist. All of us will remember him forever.

Akira Mizoguchi

Foreword

What is the appeal of insects? It could be their diverse shape, vivid body colors and patterns, graceful flight, unique habits of each species, or metamorphosis in which the shape changes completely. Their small size compared to humans may also be a factor in their familiarity. Insects have adapted to a wide variety of environments and live all over the world. Insects far surpass other animal groups in terms of the number of species and the population. Some are useful insects that are directly connected to primary industries such as sericulture and beekeeping, while others are agricultural pests and sanitary pests that transmit diseases, and thus insects have been deeply involved with humans throughout history. We have the opportunity to encounter insect molting and metamorphosis in our daily lives, and many people have viewed it as a mysterious phenomenon. J. C. Fabre (1823–1915), famous for his *Book of Insects*, not only wrote down a huge amount of information about the ecology and unique behavior of insects but also conducted research to clarify the mechanisms, but strangely, there are almost no descriptions of molting or metamorphosis. With such detailed observations, it is likely that it was inevitable that he would not have noticed them. The writer and politician J. W. von Goethe (1749–1832) had a deep knowledge of natural sciences and showed an extraordinary interest in the skeletal systems of vertebrates and insect metamorphosis. He conducted detailed observations of the process of insect metamorphosis through dissection and quantitative analysis, but he still only described the phenomenon and was unable to delve into the mechanism. At that time, the word "hormone" did not exist, and therefore the concept of endocrine system did not exist.

The concept of endocrine secretion was established in the late nineteenth century and early twentieth century. In vertebrates, adrenaline, a lipid-soluble low molecular weight compound, was crystallized in 1901 by J. Takamine and K. Uenaka as the first hormone from the adrenal gland. At almost the same time, in 1902, W. Bayliss and E. Starling identified secretin in the duodenum as the first peptide hormone that promotes the secretion of bicarbonate ions from the pancreas. Meanwhile, in invertebrates, S. Kopeć used experimental morphological techniques in the gypsy moth in 1917 to deduce that a hormone (brain hormone) that promotes molting and metamorphosis is secreted from the brain into the body fluids. E. Schaller and B. Schaller are said to have been the first to discover the phenomenon of neuroendocrine secretion in fish in 1928, but Kopeć's experiments preceded them. Kopeć's experiments were confirmed in 1926 by K. Takewaki of the University of Tokyo. However, it was only after the mid-1970s that insect peptide hormones such as proctolin and adipokinetic hormone were identified as compounds, which is quite late compared to vertebrate hormones.

Regarding the endocrine mechanism of insect molting, C. Williams elucidated the roles of the brain, corpus allata, and prothoracic gland in the 1950s. The chemical structures of the juvenile hormone secreted from the corpora allata and the molting hormone secreted from the prothoracic glands were elucidated by H. Röller et al. and A. Butenandt et al., respectively, in the 1960s, but the chemical identity of the brain hormone remained elusive. After World War II, M. Ichikawa of Kyoto University aimed to elucidate the mechanism of insect molting. Later, H. Ishizaki and J. Nishiitsutsuji-Uwo, who were supervised by Ichikawa, independently began purifying brain hormones. Meanwhile, M. Kobayashi et al. of the Sericultural Experimental Station independently began purifying the brain hormone. Finally, in 1990, Ishizaki et al. succeeded in identifying the brain hormone after about 20 years of collaborative research with A. Suzuki et al. of the University of Tokyo and with the cooperation of Kobayashi. The keys to success were the passion and indomitable spirit of the researchers, the availability of a large amount of silkworm material for extraction thanks to the prosperity of sericultural industry, the development of an efficient and precise method for purifying trace peptides and a breakthrough gas-phase protein sequencer, the development of gene cloning technology, the ability to synthesize recombinant proteins through genetic engineering technology, and the availability of monoclonal antibody production technology. In other words, much of this is due to the dramatic advances in technology in life science research. I myself learned a great deal from participating in this project research. The research process is described in detail in Chap. 1.

In the 1980s, while such research was being carried out, research on insect peptide hormones and neuropeptides gradually became more active worldwide, and various peptide hormones were identified, becoming a central issue in the field of insect physiology. Japanese researchers have made enormous contributions to research in this field. Chap. 2 and the following chapters summarize and describe what has been clarified so far about peptide hormones and neuropeptides with various functions. This book is easy to understand for students and graduate students who are about to start research in this field, and researchers who want to get an overview of this field, and is a must-read. I hope that this publication will provide an opportunity for further development of research in this field and an expansion of knowledge.

Department of Applied Biological Chemistry, The University of Tokyo, Tokyo, Japan
July 2025

Hiromichi Nagasawa

Preface

This book was written to provide graduate students and young researchers engaged in insect endocrinology, particularly in the study of peptide hormones (PHs) and neuropeptides (NPs), with an overview of this research field. It is assumed that readers have already studied general biology, including endocrinology and entomology, and possess basic knowledge in these areas, which can be obtained from standard textbooks.

The book is divided into two parts: Part I offers an overview of insect PHs and NPs, while Part II presents a detailed review of how PHs and NPs are involved in various biological processes. Part I consists of the following five chapters. Chapter 1 outlines the history of research on insect PHs and NPs, as well as the methods and technologies that have supported the research. Chapter 2 classifies the insect PHs and NPs on the basis of the antiquity of their evolutionary origins and describes the history of their discovery, structural characteristics, and known physiological functions. Information on the genes, mRNAs, and receptors for each PH/NP is readily available online, and useful references include *Handbook of Hormones: Comparative Endocrinology for Basic and Clinical Research, 2nd edition* (Ando et al. 2021). Chapter 3 focuses on interspecies differences in the PHs and NPs encoded in insect genomes. Chapter 4 provides an overview of the characteristics of insect endocrine systems and compares them with those of vertebrates. Readers who are new to the insect endocrine system may find it useful to begin with this chapter. Chapter 5 focuses on the distribution of PH- and NP-producing cells within the insect body. Part II reviews latest research on how PHs and NPs regulate insect development, growth, diapause, metabolism, reproduction, behavior, body color, and circadian rhythms. Through this book, readers will gain an understanding of the pivotal roles that PHs and NPs play in insect development, homeostasis, and environmental adaptation. We, the editors, believe that research on insect PHs and NPs will not only deepen our understanding of insect survival strategies but also shed light on mechanisms of animal evolution, especially when findings are compared with those from studies in humans and other vertebrates.

Table 1 One-letter codes for amino acids/amino acid residues

Alanine	A	Leucine	L
Arginine	R	Lysine	K
Asparagine	N	Methionine	M
Aspartic acid	D	Phenylalanine	F
Cysteine	C	Proline	P
Glutamic acid	E	Serine	S
Glutamine	Q	Threonine	T
Glycine	G	Tryptophan	W
Histidine	H	Tyrosine	Y
Isoleucine	I	Valine	V

"X" in the amino acid sequence indicates any amino acid residue
"pE" at the N-terminus of the amino acid sequence indicates pyroglutamic acid

In principle, the names of insect PHs and NPs followed the nomenclature proposed by Coast and Schooley (2011). However, when a widely accepted common name differs from that proposal, the common name takes precedence. Abbreviations for PHs, NPs, and other terms are defined at their first appearance in each chapter. A comprehensive list of abbreviations is also provided for reference.

Amino acid residues in peptide sequences are represented by one-letter codes (alphabetic symbols). A comparison table of amino acid names and symbols is shown in Table 1.

Nagoya, Aichi, Japan Akira Mizoguchi
Tsukuba, Ibaraki, Japan Naoki Okamoto

References

Ando H, Ukena K, Nagata S (eds) (2021) Handbook of hormones: comparative endocrinology for basis and clinical research, 2nd edn. Academic Press

Coast GM, Schooley DA (2011) Toward a consensus nomenclature for insect neuropeptides and peptide hormones. Peptides 32:620–631. https://doi.org/10.1016/j.peptides.2010.11.006

Contents

Contributors

Daiki Fujinaga Department of Entomology, University of California, Riverside, Riverside, CA, USA

Takashi Koyama Department of Biology, University of Copenhagen, Copenhagen, Denmark

Akira Mizoguchi Nagoya, Aichi, Japan

Yosuke Mizuno Degree Programs in Life and Earth Sciences, Graduate School of Science and Technology, University of Tsukuba, Tsukuba, Japan

Shinji Nagata Department of Integrated Biosciences, Graduate School of Frontier Sciences, The University of Tokyo, Kashiwa, Japan

Ryusuke Niwa Life Science Center for Survival Dynamics, Tsukuba Advanced Research Alliance (TARA), University of Tsukuba, Tsukuba, Japan

Naoki Okamoto Life Science Center for Survival Dynamics, Tsukuba Advanced Research Alliance (TARA), University of Tsukuba, Tsukuba, Japan

Hitomi Seike Department of Integrated Biosciences, Graduate School of Frontier Sciences, The University of Tokyo, Kashiwa, Japan

Sakiko Shiga Department of Biological Sciences, Graduate School of Science, The University of Osaka, Osaka, Japan

Kunihiro Shiomi Division of Applied Biology, Faculty of Textile Science & Technology, Shinshu University, Nagano, Japan

Yoshiaki Tanaka Institute of Agrobiological Sciences, NARO, Tsukuba, Japan

Abbreviations

20E	20-Hydroxyecdysone
aa	Amino acid
aaNAT	Arylalkylamine-*N*-acetyltransferase
ACP	AKH/CRZ-related peptide
ADF	Antidiuretic factor
ADH	Antidiuretic hormone
AG/AGs	Abdominal ganglion/abdominal ganglia
aIGF	Arthropod IGF
AKH	Adipokinetic hormone
AKT	Protein kinase B
Alk	Anaplastic lymphoma kinase
ALP	Agatoxin-like peptide
ALS	Acid-labile subunit
aMe	Accessory medulla
AMPK	AMP-activated protein kinase
AST-A	Allatostatin A
AST-B	Allatostatin B
AST-C	Allatostatin C
AST-CC (CCC)	Allatostatin-CC (CCC)
AT	Allatotropin
BBB	Blood-brain barrier
BH	Brain hormone
BIGFLP	*Bombyx* IGF-like peptide
BMP	Bone morphogenetic protein
BNGRs	Neuropeptide G protein-coupled receptors of *B. mori*
Burs/burs	Bursicon alpha subunit
CA	Corpora allata, coupus allatum
CaLexA	Calcium-dependent nuclear import of LexA
cAMP	Cyclic AMP
CAP2b	Cardioacceleratory peptide 2b
CC	Corpora cardiaca, corpus cadiacum

CCAP	Crustacean cardioactive peptide
CCHa	CCHamide
CCK	Cholecystokinin
cDNA	Complementary DNA
CG	Chorionic gonadotropin
CHH	Crustacean hyperglycemic hormone
ClC-a	Chloride channel-a
CNS	Central nervous system
CRISPR	Clustered regulatory interspaced short palindromic repeat
CRF-DH	Corticotropin-releasing factor-like diuretic hormone
CRY	Cryptochrome
CRZ	Corazonin
CT	Circadian time
CT-DH	Calcitonin-like diuretic hormone
Dac	Dachshund
Daw	Dawdle
DDC	Dopa decarboxylase
DH	Diuretic hormone
DH	Diapause hormone
DH31	Diuretic hormone 31
DH44	Diuretic hormone 44
DILP	*Drosophila* ILP
DLGR	*Drosophila* LGR
DN	Dorsal neurons
DN1p	Posterior DN1
DopR	Dopamine receptor
DSB	Double-strand break
Dsk	Drosophila sulfakinin
dsRNA	Double-stranded RNA
EC	Enterocyte
ECD	Ecdysteroid
EcR	Ecdysone receptor
EEC	Enteroendocrine cell
Egr	Eiger
EGFR	Epidermal growth factor receptor
EH	Eclosion hormone
ELH	Egg-laying hormone
EPG	Epiproctodeal gland
ERK	Extracellular signal-regulated kinase
ETH	Ecdysis-triggering hormone
Ey	Eyeless
FaRP	FMRFamide-related peptide
FDS	Fat body-derived signal
FG	Frontal ganglion

FMRFa	FMRFamide
FSH	Follicle-stimulating hormone
GABA	γ-Aminobutyric acid
GBP	Growth-blocking peptide
GFP	Green fluorescent protein
GnRH	Gonadotropin-releasing hormone
GPA2/GPB5	Glycoprotein A2/glycoprotein B5
GPCR	G-protein coupled receptor
GPH	Glycoprotein hormone
Grnd	Grindelwald
GSC	Germline stem cell
Gyc76C	Guanylyl cyclase at 76C
HDR	Homologous direct repair
Hh	Hedgehog
HPLC	High performance liquid chromatography
HPO	Hematopoietic organ
IGF	Insulin-like growth factor
IGF-1R	IGF type-I receptor
IGFBP	IGF-binding protein
IGFLP	Insulin-like growth factor-like peptide
IIS	Insulin/insulin-like growth factor signaling
ILP	Insulin-like peptide
Imp-L2	Imaginal morphogenesis protein-late 2 protein
InR	Insulin-like receptor
INSL	Insulin-like peptide (in vertebrates)
IPC	Insulin-producing cell
iPTH	Insect parathyroid hormone
IRP	Insulin-related peptide
IRS	Insulin receptor substrate
ISN	Interoceptive subesophageal zone neuron
ITP	Ion transport peptide
ITPL	ITP-like
JH	Juvenile hormone
JHAMT	Juvenile hormone acid O-methyl transferase
JHE	Juvenile hormone esterase
LC-MS	MS coupled with liquid chromatography
LD cycle	Light-dark cycle
LGR	Leucine-rich repeat-containing GPCR
LH	Luteinizing hormone
LK	Leucokinin
lLNv	Large type LNv
LN	Lateral neurons
LNv	Ventral LN
LST	Limostatin

LTP	Lipid transfer protein
MALDI	Matrix-assisted laser desorption/ionization
MALDI-TOF	Matrix-assisted laser desorption/ionization-time of flight
MAPK	Mitogen-activated protein kinase
MCO2	Multicopper oxidase 2
MIH	Molt-inhibiting hormone
MIP	Myoinhibitory peptide
mNSC	Medial neurosecretory cell
MOIH	Mandibular organ-inhibiting hormone
MRCH	Melanization and reddish coloration hormone
MS	Mass spectrometry
MS	Myosuppressin
MT	Myotropin
MT	Malpighian tubule
Mth	Methuselah
Mthl10	Methuselah-like 10
mTOR	Mechanistic target of rapamycin
NADA	*N*-acetyldopamine
NB	Neuroblast
NBAD	*N*-β-alanyldopamine
NCA	Nervi corporis allati
NCC	Nervi corporis cardiaci
NGS	Next-generation sequencing
NHEJ	Non-homologous endo-joining
NO	Nitric oxide
NOS	Nitric oxide synthase
NP	Neuropeptide
Npc1	Niemann-Pick disease type C1
NPF	Neuropeptide F
NPY	Neuropeptide Y
NSC	Neurosecretory cell
NTL	Natalisin
OEH	Ovary ecdysteroidogenic hormone
OK	Orcokinin
OMP	Ovary maturating parsin
OT	Oxytocin
PAGE	Polyacrylamide gel electrophoresis
PBAN	Pheromone biosynthesis-activating neuropeptide
Pburs	Partner of bursicon
PC	Prohormone convertase
PC	Principal cell
PDF	Pigment dispersing factor
PDH	Pigment dispersing hormone
PKA	Protein kinase A

PKC	Protein kinase C
PKG	Protein kinase G
PER	Period
PETH	Pre-ecdysis triggering hormone
PG	Prothoracic gland
PH	Peptide hormone
PI	Pars intercerebralis
PI3K	Phosphoinositide 3-kinase
PI-NSC	Neurosecretory cell in the PI region of the brain
PISCF/AST	Allatostatin C
PK	Pyrokinin
PMR	Post-mating response
PO	Phenoloxidase
PO/PSO	Perisympathetic organ
PP	Pancreatic polypeptide
PrRH	Prolactin-releasing hormone
PTH	Parathyroid hormone
PTSP	Prothoracicostatic peptide
PTTH	Prothoracicotropic hormone
PVK	Periviscerokinin
rER	Rough endoplasmic reticulum
rk	Rickets
RLN	Relaxin
RLN-RP	RLN-related peptide
RNAi	RNA interference
RNA-seq	RNA-sequencing
ROS	Reactive oxygen species
RPCH	Red pigment-concentrating hormone
RTK	Receptor tyrosine kinase
RT-PCR	Reverse transcription polymerase chain reaction
RXR	Retinoid X receptor
SC	Secondary (or stellate) cell
scRNA-seq	Single-cell RNA sequencing
SDR	Secreted decoy of InR
SecCl	Secretory chloride channel
SG	Subesophageal ganglion
SGNP	Subesophageal ganglion neuropeptide
SK	Sulfakinin
sLNv	Small type of LNv
sNPF	Short neuropeptide F
SP	Sex peptide
SRP	Signal recognition particle
Sun	Stunted
TAG	Terminal abdominal ganglion

TAG	Triacylglycerol
TALEN	Transcription activator-like effector nuclease
TCA	Tricarboxylic acid
TG/TGs	Thoracic ganglion/thoracic ganglia
TGN	Trans-Golgi network
TH	Tyrosine hydroxylase
TK	Tachykinin
TMOF	Trypsin-modulating oostatic factor
TNF-α	Tumor necrosis factor α
TOR	Target of rapamycin
TRP	TK-related peptide
TRP	Transient receptor potential
TRPA1	Transient receptor potential ankyrin 1
TS	Thyrostimulin
TSH	Thyroid-stimulating hormone
TTFL	Transcription-translation feedback loop
UAS	Gal4-upstream activating sequence
Upd2	Unpaired-2
Urn8-R	Urinate receptor
VIH	Vitellogenesis-inhibiting hormone
VKR	Venus kinase receptor
VNC	Ventral nerve cord
VP	Vasopressin
WGS	Whole genome sequencing
YP	Yolk protein
YPT	Yellow protein of the takeout family

Part I
Peptide Hormones and Neuroendocrine Systems in Insects

Chapter 1
History of the Study on Insect Peptide Hormones: From Experimental Morphological Study to Molecular Genetic Approach

Akira Mizoguchi and Daiki Fujinaga

Abstract Research in insect endocrinology, which began in the early twentieth century, made great progress during the last two decades of the twentieth century with the identification of a wide variety of insect hormones and neuropeptides. Moreover, since 2000, this research field has entered a new stage of development, supported by innovations in research techniques, such as genome sequencing, bioinformatics, genetic manipulation, and imaging. This chapter looks back at the history of the development of insect peptide hormone research, focusing on advances and innovations in research methods. However, it goes without saying that the excellent research ability and constant efforts of researchers are essential for the development of research. Thus, to emphasize the importance of consecutive efforts of individual researchers, we first review the history of a purification study of the prothoracicotropic hormone that extended to three decades.

Keywords Insect · Peptide hormones · Neuropeptides · Purification · History · PTTH · Research techniques

1.1 Thirty Years of Prothoracicotropic Hormone Purification Study

Science develops through technological innovation, although consecutive efforts of scientists obviously form the basis of its development. This is also true for insect endocrinology. In this chapter, we look back on the history of the study on insect peptide hormones with special attention to the progress and innovation of research

A. Mizoguchi (✉)
Nagoya, Aichi, Japan
e-mail: mizoamizo@icloud.com

D. Fujinaga
Department of Entomology, University of California, Riverside, Riverside, CA, USA
e-mail: daikif@ucr.edu

A. Mizoguchi, N. Okamoto (eds.), *Insect Peptide Hormones and Neuropeptides*, Entomology Monographs, https://doi.org/10.1007/978-981-95-7510-7_1

methods. However, to emphasize the importance of consecutive efforts of individual researchers, we first review the history of a purification study of prothoracicotropic hormone (PTTH) that extended to three decades.

1.1.1 Background of PTTH Purification

In 1917, Stefan Kopeć, a Polish biologist, suggested that the brain of the gypsy moth *Lymantria dispar* releases a hormone, named as the brain hormone (BH), necessary for the larva-to-pupa transition of the insect, based on his observation that neck ligation or brain removal of final instar larvae at the right time inhibited pupation (Kopeć 1917, 1922). This was the first report to show the presence of a hormone in insects, although in vertebrates, the presence of many hormones had already been demonstrated. However, his claim that the brain secretes a hormone was not accepted by academia at that time and following 20 years. One of the reasons why his claim was neglected by academia may be the people's understanding in the early twentieth century that the nervous tissue could not secrete hormones. The importance of his discovery was only confirmed in 1940, when Vincent Wigglesworth demonstrated that a hormone secreted by the brain regulates ecdysis and metamorphosis of the kissing bug *Rhodnius prolixus* through his careful experiments consisting of brain removal and re-implantation. In particular, he successfully revealed that ecdysis was induced by implanting a specific part of the brain including a group of large apparent neurosecretory cells, but not other parts of the brain, into brain-removed larvae. This result provided convincing evidence for the secretion of BH responsible for ecdysis (Wigglesworth 1940).

A research approach in endocrinology that utilizes various experimental manipulations, such as ligature, removal and/or transplantation of specific tissues, parabiosis, and hormone injection, to locate or identify endocrine organs/tissues or hormones of interest is called experimental morphology. Such approaches were the primary research style of insect endocrinology in the first half of the twentieth century.

In the modern era, the concept of neurosecretion is widely accepted by biologists. However, this concept was only established in the 1950s, although the presence of neurosecretory cells had been increasingly recognized by biologists through the 1930s and 1940s, thanks to the extensive studies by Ernst and Berta Scharrer on neurosecretory systems of vertebrates and invertebrates, respectively (Scharrer and Scharrer 1939). For establishing the concept of neurosecretion, Wolfgang Bargmann also made significant contributions (Bargmann and Scharrer 1951). Currently, the discovery of BH by Kopeć over a century ago is not only regarded as the first description of a hormone in insects but also appreciated as the first demonstration of neurosecretion in animals.

Apart from the discovery of BH, several researchers independently suggested in the 1930s that a thorax-derived hormone regulates the larva-to-pupa transition in various insects, based on their ligature experiments. Thereafter, Soichi Fukuda found out that the source of the thorax-derived hormone is the prothoracic glands (PGs), based on his ligature and tissue implantation experiments on larvae of the silkworm *Bombyx mori* (Fukuda 1940, 1944).

Thus, the BH and the PG hormone are independently discovered as the hormones regulating metamorphosis. However, the relationship between the two hormones was unknown. This problem was solved by Carroll Williams, who used diapausing pupae of the cecropia moth *Hyalophora cecropia* in his experiments. The diapausing pupae of this insect started pupa-adult development if chilled for a certain period of time, suggesting the release of a hormone responsible for metamorphosis. Therefore, he implanted the brain and the PGs from the chilled or unchilled pupae into the isolated abdomen of diapausing pupae either alone or in combination. The isolated abdomens started adult development only when the brain from the chilled pupa was implanted together with the PGs from either chilled or unchilled pupa, leading to a conclusion that the brain activated by chilling secreted a hormone and stimulated the PGs to release another hormone that directly induces metamorphosis (Williams 1946). He further revealed that larval-larval and larval-pupal ecdyses are regulated by the same two-step hormonal mechanism (Williams 1952).

1.1.2 Start of PTTH Purification by Hironori Ishizaki

In the 1950s, some trials to extract and purify BH began without success. Under these circumstances, Hironori Ishizaki at Kyoto University started his research to purify BH of *B. mori* in 1960, with the aim of determining its structure. In order to purify a hormone, a huge amount of extraction material is necessary. Fortunately, in Japan at that time, a large amount of silkworms was reared at national and private research institutions of sericultural industry, and he could receive without charge the pupae or adults of the silkworm that became unnecessary after use in experiments. For purifying a hormone, it is also critical to establish a suitable assay method to detect the hormone. Masatoshi Kobayashi of the Sericultural Experiment Station (a national research institution of the sericultural industry) was one of the researchers who challenged to purify BH. He and his co-workers developed a bioassay method for BH that uses brain-removed *B. mori* pupae (Kobayashi 1955). When the brains of pupae of a specific *B. mori* strain were surgically removed immediately after ecdysis, the pupae stopped development for a long time, becoming dormant pupae. In contrast, the dormant pupae started development into adults when they were injected with a sample containing BH. Thus, by using this system, the presence of BH in test samples could be detected and further quantified by determining the limit of sample dilution at which the injected sample could evoke adult development. Ishizaki also used a similar but not identical bioassay system to detect BH. He used the pupae of the heterologous moth *Samia cynthia ricini* instead of *B. mori* pupae as the recipients of hormone injection for several reasons. The implantation of the *Bombyx* pupal brain was able to evoke adult development of *Samia* dormant pupa, suggesting that BH acts species non-specifically at least between these two moths. Besides, it was much easier to prepare *Samia* dormant pupae than *Bombyx* one, because the critical period when the brain must be removed after pupal ecdysis was longer (within 1 day) and because the *Samia* dormant pupae could be used in the assay for a longer time (several months after production).

Using this assay method, Ishizaki found that only one *Bombyx* brain treated by heating at 90 °C or immersion overnight in acetone remained active in evoking adult development of *Samia* dormant pupae. This result surprised him, because at the time, it had been believed that a dead brain contains little BH. Then, he tried to extract BH with 2% NaCl, assuming that BH is water-soluble. The brain extracts evoked adult development at a dose equivalent to only one-twentieth the brain. The result also amazed him, because Kobayashi previously reported that BH is extractable with organic solvents and that some hundred brain equivalents of the extract could gave a positive response in his bioassay with *Bombyx* dormant pupae (Kobayashi and Kirimura 1958). Thus, Ishizaki discovered for the first time that BH is water-soluble and is contained in the *Bombyx* pupal brain in much higher amounts than predicted. These results were published in *Nature* in 1961 (Ichikawa and Isihzaki 1961). Around that time, many vertebrate hormones had already been chemically characterized: some were steroids, while others were polypeptides. In particular, the hormones produced by the hypothalamus and pituitary were all polypeptides. Therefore, considering the water-soluble nature of BH, Ishizaki predicted that BH is a polypeptide. As he expected, the brain extract was inactivated by proteases, the activity was recovered by precipitation with ammonium sulfate or acetone, and the active material did not path through the dialysis membrane. The results of these experiments for characterizing BH were published in *Nature* in 1963 (Ichikawa and Isihzaki 1963).

A variety of biochemical techniques to purify vertebrate polypeptide hormones had been developed in the 1960s. In starting the purification of BH, however, a pressing issue was securing huge amounts of *Bombyx* brains. As described above, Ishizaki was able to receive *Bombyx* pupae without charge from the Sericultural Experiment Station. However, the nearest branch of the station was located in the city about 2 h away by the train from the town where he lived. He often visited the station to receive tens of thousands of pupae at one time and sent them to his laboratory. The pupae that arrived at the laboratory were once stored in a cold room and then used over several weeks to collect the brains. The brains could be easily taken out of the pupae by horizontally cutting near the tip of the pupal head with a razor blade and then pinching out the brain with fine forceps. Ishizaki continued this operation as long as his strength lasted every day. The collected brains were stored in the freezer in batches of, for example, 10,000 brains. Another essential and laborious task he had to do before and during his hormone purification work was to prepare a large number of brainless *Samia* dormant pupae for bioassays. Understandably, this task was accompanied by mass breeding of *Samia* larvae as well as cultivation of their host plants, the castor oil plants. As the final instar *Samia* larvae consumed large amounts of leaves of the plant, he had to give up his holidays to feed the larvae continually. Securing the castor oil plant in winter also bothered him, because this plant dies in winter in Kyoto. Thus, he had to make frequent trips to the marine biological laboratory of Kyoto University in Shirahama, about 150 km south of Kyoto, where castor leaves grow even in winter. If this moth could undergo diapause in the winter, it would not need food during that time. However, since it does not actually enter diapause at any stage and under any conditions, it was necessary to maintain the colony during the winter by feeding the larvae. As the colony had been maintained in his laboratory for many years, he could never extinct the colony, because the genetic characteristics of the moth population are more or less different between colonies.

The production of debrained pupae was also a burdensome task. To remove the brain, cut the head of the anesthetized pupa at the appropriate position with microscissors to create a small window, pinch out the brain with forceps, and close the wound with melted paraffin. Although the operation to remove the brain itself was not difficult, he had to do the operation on hundreds of pupae at once. The operation must be done at an appropriate timing after pupal ecdysis. However, ecdysis of individual larvae within a population occurs at various times over a period of 1–2 days. As a result, his work to operate newly ecdysed pupae often continued until midnight.

Thus, overcoming these troublesome issues one by one, he started the work to purify *B. mori* BH. Several years later, through a time-consuming trial and error, he developed a BH purification procedure consisting of several steps and published a progress report of his study (Ishizaki and Ichikawa 1967). According to this report, BH was extracted from 118,000 *Bombyx* brains (fresh weight, 454 g) as a "crude extract." The extract contained 992,000 units of BH, where 1 unit was defined as a minimal amount of BH that gave a positive response in the bioassay. In the *Samia* pupal assay, five pupae were injected with the test sample, and the assay result was considered positive if three or more pupae emerged as adult moths. In fact, the occurrence of wing apolysis was used instead of adult emergence to judge the result of bioassay, because this morphological change occurs within a week after BH injection, much earlier than the emergence of the adult. The BH activity in each sample was quantified by assaying a series of twofold dilution of the sample. The crude extract was purified by heating in boiling water, fractional precipitation with ammonium sulfate, gel filtration on Sephadex G-100, stepwise chromatography on DEAE-cellulose, gradient chromatography on DEAE-cellulose, and second gel filtration on Sephadex G-100. In this procedure, about 8000-fold purification was achieved. Troublingly, however, BH exhibited highly heterogeneous molecular forms on gel filtration and ion-exchange chromatography. Moreover, it was evident that the purified BH still contained a large amount of impurity, because the profile of BH activity on chromatograms did not correspond to that of optical density at 280 nm, which represents the amount of protein. It may be important to note here that a large number of bioassays are required to plot a graph of BH activity in each chromatogram. In gradient chromatography on DEAE-cellulose, for example, eluate was collected in 75 fractions (8 mL each). As the first step of bioassay of all the fractions, a small aliquot of each fraction was diluted 1:20 with water, followed by an injection of 20 μL of the dilute solution into each of the five debrained *Samia* pupae. If the assay result was positive, the fraction was estimated to contain at least

8000 units of BH ($8 \times 20/0.02 = 8000$). On the other hand, if the assay result was negative, the fraction was estimated to contain at most 4000 units of BH, which was negligible compared to the total BH activity in the initial crude extract. Next, as the second step of the bioassay, the positive assay samples were further diluted to 2, 4, and 8 times its volume and assayed. When necessary, the third step of the bioassay was conducted. Thus, a large number of bioassays were required to evaluate the result of chromatography. Indeed, as many as more than 600 debrained pupae were used in only one step of purification.

Although about 8000-fold purification of BH was achieved by his tenacious efforts, Ishizaki realized that it would be difficult to continue purifying BH all alone. He desperately needed the help of a chemist skilled in working with proteins. However, such a chemist did not appear for a long time. In 1972, he was invited to an international symposium on insect hormones held in the United States, as a speaker. Determining to take this opportunity to quit his work on BH purification, he presented all his unpublished data at the meeting. Contrary to his determination, however, this meeting provided a major turning point for him, because he received a long-awaited offer of collaboration from a chemist soon after the meeting. The chemist was Saburo Tamura of the University of Tokyo, a leader in the organic chemistry of bioactive natural products in Japan. Accepting his offer gratefully, Ishizaki started the collaborative research with Tamura laboratory, with Ishizaki concentrating on BH bioassays and Tamura lab concentrating on BH purification. Soon after the start of this collaboration, Ishizaki moved to Nagoya University, where he could establish a system to rear a large number of *Samia* larvae with the support of lab members.

1.1.3 Restart of PTTH Purification as a Joint Study with Chemists

Tamura and his lab member Akinori Suzuki first developed a route to obtain a huge number of silkworms, because it was predicted that millions or tens of millions of *Bombyx* brains would be necessary to isolate BH in an amount sufficient to determine its structure. Fortunately, they found that such a huge number of silkworms could be obtained without charge from silkworm egg production farms, where adult silkworms after egg production were discarded. There was no problem in using the adult brains instead of pupal ones, because adult brains contained more BH than pupal brains. Although using the brains as the starting material for BH extraction was desirable, it was impractical to collect such a huge number of brains when considering the financial and time costs. Thus, they decided to extract BH from the adult heads. During egg-laying season, graduate students in the lab went to a silkworm egg production farm in a suburb of Tokyo, packed adult moths into plastic bags, and immediately froze them in dry ice (Fig. 1.1A). The frozen moths were packed tightly into cardboard boxes and then transported to a nearby cold storage

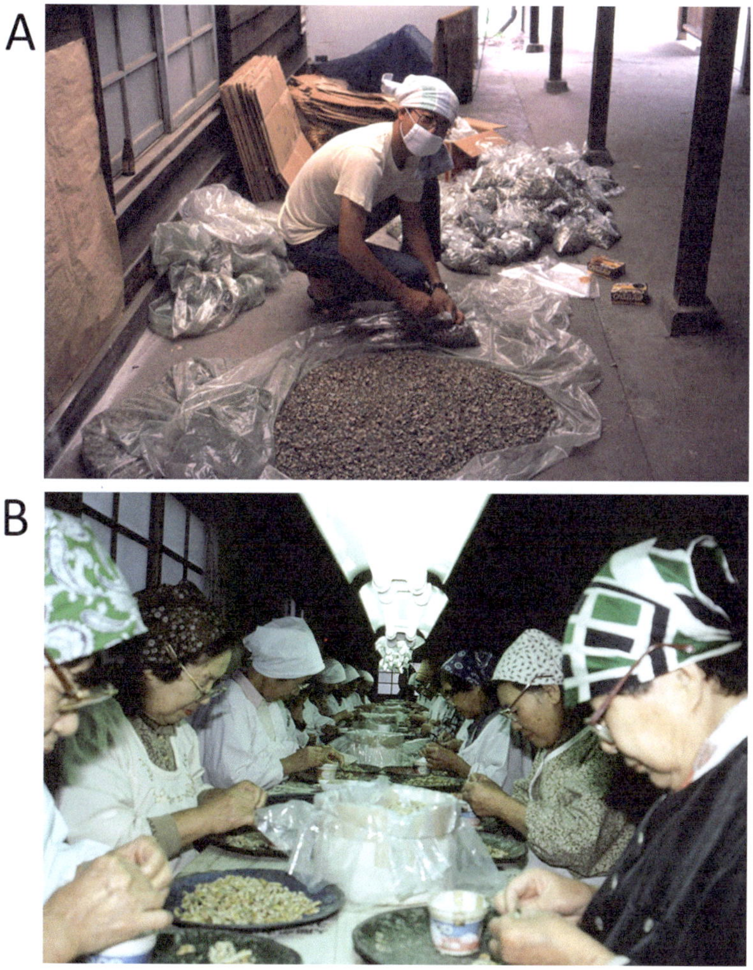

Fig. 1.1 Procurement of adult silkworm heads as a starting material for brain hormone purification. (**A**) Collection and packing of adult silkworms. (**B**) Collection of silkworm heads

warehouse, where a large number of cardboard boxes containing frozen moths were temporarily stored. Forty boxes roughly corresponded to one million moths. At the appropriate time, the heads of the moths were cut off with razor blades and collected by part-time workers (Fig. 1.1B), packed in dry ice, and sent to the lab. In its peak, as many as five million moth heads were harvested per year. Throughout this joint research, which lasted more than 20 years, nearly 30 million *Bombyx* heads were thus procured.

The first 2 years of this joint research was spent establishing an initial process of BH purification suitable for processing a huge amount of *Bombyx* heads. This effort aimed to efficiently extract BH from a large amount of *Bombyx* heads while reducing the total amount of proteins. The established procedure consisted of nine steps of treatments including several kinds of precipitation with organic solvents, extraction with 2% NaCl, and heating (Suzuki et al. 1975). Throughout this procedure, 320 mg protein powder, named crude BH, was obtained per 10,000 *Bombyx* heads (150 g), which served as an important basis for further purification of BH. Using this crude BH, this peptide hormone was chemically characterized as follows: the disulfide bridge(s) in the molecule is essential for its activity; the N- and C-terminus of the polypeptide are both blocked; the sugar chain (if any) is not responsible for its activity; and the tryptophan residue is necessary for its activity, while the sulfhydryl group is not (Ishizaki et al. 1977). In this report, the name of BH was changed to PTTH, because by that time, several other hormones had been discovered from the insect brain one after another and many researchers had begun using the latter name to discriminate BH from other brain-derived hormones. The brain-secretory hormones found in the 1970s include the hypoglycemic hormone, diuretic hormone, ovary-stimulating hormone, and eclosion hormone.

In 1977, Tamura retired from the university, and Suzuki took over the joint research for PTTH purification as a lab leader. One year before that, Hiromichi Nagasawa, one of the graduate students in the Suzuki lab, joined the PTTH purification project. Starting work under the supervision by Suzuki, Nagasawa soon came to play a central role in purifying PTTH in the lab. He tried every type of ion-exchange and adsorption chromatography available one after another.

However, the PTTH activity was distributed across multiple areas in any chromatography, as previously experienced by Ishizaki. To overcome this challenging problem, he repeated a huge number of experiments. For example, he conducted more than 100 preliminary experiments to establish only one step of purification. Needless to say, such numerous chromatography runs were accompanied by an equally large number of bioassays by Ishizaki in Nagoya University. In order to meet the demand for a large number of bioassays, Ishizaki had to prepare a much larger number of debrained *Samia* pupae than ever. The bottleneck in mass production of debrained pupae was microsurgery to remove the brains from newly ecdysed pupae, which was difficult to do for part-time workers. Ishizaki asked his wife to assist with his surgery, and she gladly accepted his request. By this joint work, they could produce 400 *Samia* debrained pupae per hour, four times faster than doing alone.

Using the "crude PTTH (BH)" as the starting material, Nagasawa purified PTTH through four steps of chromatography consisting of gel filtration on Sephadex G-50 (fine), stepwise chromatography on DEAE-Sepharose CL-6B, stepwise chromatography on SP Sephadex C-25, and gel filtration on Sephadex G-50 (superfine), to obtain "highly purified PTTH" (Nagasawa et al. 1979). Based on gel filtration chromatography, the molecular weight (MW) of PTTH was estimated to be approximately 4400. With this procedure, 3 mg of highly purified PTTH was gained from 2.88 g of crude PTTH that was derived from 96,000 adult *Bombyx* heads. This preparation of PTTH was able to elicit adult development of a debrained *Samia* pupa with as little as 6 ng. At this stage of purification, a new difficult problem occurred. It was the loss of hormone activity during chromatography, which is often experienced during a high level of purification. Such a phenomenon can occur due to various reasons including the adsorption of hormone molecules to the surface of glassware.

Then, Nagasawa took every possible countermeasure without success for over a year. However, this problem was quickly solved with the kind advice of Okitsugu Yamashita, who had long been engaged in purifying diapause hormone of *B. mori*. The loss of hormone activity was completely eliminated by adding 1% bovine serum albumin to the samples.

The "highly purified PTTH" derived from 384,000 *Bombyx* heads was next subjected to gradient chromatography on DEAE-Sepharose CL-6B, resulting in the division of the PTTH activity into 2 clear peaks. Then, Nagasawa and co-researchers decided to further purify the fractions containing each PTTH activity peak separately. When the fraction corresponding to one of the two PTTH activity peaks was purified by three steps of ion exchange chromatography, a chromatogram was obtained in which the PTTH activity and OD280 peaks were well overlapped, suggesting that PTTH had been almost isolated. Contrary to expectations, however, a verification experiment by polyacrylamide gel electrophoresis (PAGE) and subsequent bioassay of protein bands proved that PTTH is still a minor component of proteins in the PTTH peak fractions.

It seemed difficult to find a method to further purify PTTH with conventional chromatography. Fortunately, however, this impasse was broken all at once by the use of a newly developed high-performance liquid chromatography (HPLC), which was becoming available in the late 1970s. The HPLC not only greatly shortened the time for chromatography but also dramatically improved the ability to separate substances. This new technology was first developed to purify low MW substances but was gradually improved to be applicable to the purification of larger molecules like polypeptides. As described above, two PTTH activity peaks occurred after gradient chromatography on DEAE-Sepharose CL-6B of highly purified PTTH. When only three fractions corresponding to the center of the first PTTH activity peak were combined, gel-filtrated, and subjected to reverse-phase HPLC on Partisil-10 ODS-3 column, polypeptides in the sample were resolved into many sharp peaks on the chromatogram, and the PTTH activity only resided in one peak (Nagasawa et al. 1984a). In this way, one molecular form of PTTH was finally isolated, 12 years after the start of their collaboration and 24 years after the start of its purification by Ishizaki. The established procedure of PTTH purification consisting of 15 steps is shown in Fig. 1.2. With this procedure, 50 μg of PTTH was isolated from 650,000 *Bombyx* heads. The specific activity increased two million times, the recovery of the PTTH activity was 8%, and only 0.1 ng of isolated PTTH was able to elicit adult development of a debrained *Samia* pupa. Subsequently, two other molecular forms of PTTH were isolated in the same way from other PTTH-active fractions of the abovementioned DEAE-Sepharose chromatography.

1.1.4 Structure Determination of 4K-PTTH

Soon after the isolation of PTTH, Nagasawa began to manually analyze the amino acid (aa) sequence of PTTH by Edman degradation, while having a concern that the isolated PTTH may be insufficient in amount to determine the entire sequence of

Fig. 1.2 Procedure of PTTH isolation from *Bombyx* heads

648,000 male adult *Bombyx* heads (4.86 kg)
↓
Acetone powder
↓
Washing with 80% EtOH
↓
Extraction with 2% NaCl
↓
Heat treatment
↓
Precipitation with 80% saturated $(NH_4)_2SO_4$
↓
Precipitation with 50-75% acetone
↓
Precipitation with 90% saturated picric acid
↓
Precipitation with 80% acetone **"crude PTTH"**
↓
Sephadex G-50 (fine)
↓
DEAE-Sepharose CL-6B (stepwise elution)
↓
SP-Sephadex C-25 (stepwise elution)
↓
Sephadex G-50 (superfine) **"highly purified PTTH"**
↓
DEAE-Sepharose CL-6B (gradient elution)
↓
Sephadex G-50 (superfine)
↓
Reversed-phase HPLC (gradient elution)
↓
PTTH (50 μg)

PTTH. Although he was able to determine the sequence of several N-terminal aa residues of the polypeptide, the progress in the sequence analysis was slow. Fortunately, however, an automatic gas-phase protein sequencer, which enabled the sequencing of extremely small amounts of proteins at picomolar levels, had been developed in the late 1970s and was put on the market in the early 1980s. Nagasawa got a chance to use that sequencer a few years after its release. Using the gas-phase protein sequencer, he determined the N-terminal 19 aa sequences of the abovementioned 3 molecular forms of PTTH. These PTTHs were named 4K-PTTHs after its MW. Surprisingly, these polypeptides (4K-PTTH-I, 4K-PTTH-II, 4K-PTTH-III) had high homology with vertebrate insulins (Nagasawa et al. 1984b). Currently, it is well known that insects and vertebrates share a lot of homologous peptide hormones and neuropeptides. However, this was the first to definitely show the evolutionary relationship between insect and vertebrate hormones by structure determination of polypeptides. Shortly thereafter, Nagasawa succeeded in determining the entire aa sequence of 4K-PTTH-II and its two-chain structure, using his abundant knowledge and skills as an organic chemist. One chain (A-chain) consisted of 20 aa residues, while the other chain (B-chain) consisted of 28 residues. The N-terminus of the B-chain was pyroglutamylated (Nagasawa et al. 1986). His analysis also revealed the location of two interchain disulfide bridges and one intrachain disulfide bridge within the A-chain. The mode of distribution of these disulfide bridges in the 4K-PTTH-II molecule was the same as that in insulin (Nagasawa et al. 1988), confirming that this polypeptide is a homolog of insulin in *B. mori*.

1.1.5 Presence of Two Types of PTTHs in the **Bombyx** *Brain*

There is a reason why PTTH was called "4K-PTTH" in the early 1980s. When PTTH was about to be isolated, Ishizaki conducted a bioassay of the PTTH preparation using debrained *Bombyx* pupae. Surprisingly, the result was negative. Ishizaki immediately began investigating why the purified PTTH showed negative results in the assay with *B. mori*. It was no doubt that the *Bombyx* brain extract was effective in either assay with *Bombyx* and *Samia* debrained pupae. Those results led him to conclude that PTTH is species non-specific at least between *B. mori* and *S. cynthia ricini*. Faced with the above assay results, however, he was forced to doubt this conclusion. Hypothesizing that *Bombyx* brain or head extract contains two types of PTTH, one effective against *B. mori* and the other effective against *S. cynthia ricini*, he subjected the fractions of early steps of purification to both bioassays with *Bombyx* and *Samia* pupae. The results were remarkable. The PTTHs effective in *B. mori* (PTTH-B) and *S. cynthia ricini* (PTTH-S) were clearly separated in the fractions of fractional precipitation with acetone. Furthermore, the two PTTHs were also clearly separated by gel filtration chromatography of *Bombyx* brain extract. The MW of PTTH-B and PTTH-S was estimated by this chromatography to be approximately 22,000 and 4000, respectively (Ishizaki et al. 1983). Thus, the PTTH purified using the bioassay with *Samia* debrained pupae came to be tentatively called 4K-PTTH. Ishizaki and co-researchers renamed this peptide as bombyxin some years later, because they could not detect its activity to stimulate *Bombyx* PGs in any of their bioassay systems. In fact, however, it was demonstrated in 2015 that bombyxin (4K-PTTH) does stimulate the PGs in other bioassays using a long-term incubation system of *Bombyx* PGs (Gu et al. 2015). A PTTH-like activity of insulin-like peptides (ILPs) has also been demonstrated in *Drosophila melanogaster* (Mirth and Riddiford 2007). Recently, bombyxin (4K-PTTH) and PTTH (22K-PTTH) have been shown to stimulate the PGs in different mechanisms (Gu et al. 2022). Nevertheless, 4K-PTTH was not the PTTH that Ishizaki wanted to purify when he started his study more than 20 years ago. Thus, he and his joint study members immediately started purifying 22K-PTTH, in parallel with their research on 4K-PTTH.

1.1.6 Purification and Structure Determination of 22K-PTTH

Upon starting the purification of 22K-PTTH (simply PTTH hereafter), Ishizaki first established a large-scale culture system for a specific strain (J122 × C115) of the silkworm in his lab. Meanwhile, in the Suzuki lab, Hiroshi Kataoka, a graduate student in the lab, began research on PTTH purification under the supervision of Suzuki and Nagasawa, because Nagasawa was busy isolating 4K-PTTH and determining its structure. The research began with procuring a huge amount of *Bombyx* heads again. The purification process progressed relatively fast, because they had already acquired a wealth of knowledge and skills to purify a specific polypeptide from *Bombyx* heads. However, it took 5 years for Kataoka to establish a procedure isolating PTTH. This hormone was isolated through 16 steps of purification including defatting, salt extraction, fractional precipitations, and 5 steps each of conventional column chromatography and HPLC. Using the polypeptide isolated from 500,000 *Bombyx* heads, he succeeded in determining the N-terminal 13-aa sequence of PTTH (Kataoka et al. 1987). Following the elucidation of the partial aa sequence of PTTH, two lines of approaches were taken to determine the entire sequence of the hormone. One was to obtain by purification the amount of PTTH necessary for complete sequencing with a gas-phase protein sequencer. The other was to clone a cDNA encoding a protein that contains the identified partial sequence of PTTH, as gene cloning technology was becoming widespread at that time. The former work was performed by Kataoka in the Suzuki lab, while the latter was performed by members of Ishizaki lab. Using 3,000,000 *Bombyx* heads, Kataoka determined the N-terminal 104-aa sequence of the hormone and revealed that this hormone had a homodimer structure (Kataoka et al. 1991). About the same time, the cDNA of interest was cloned by screening the expression cDNA library prepared from *Bombyx* brains with a mouse monoclonal antibody raised against the partial sequence of PTTH (Kawakami et al. 1990). The cDNA encoded a 224-aa protein including a PTTH monomer sequence consisting of 109 aa residues at its C-terminal region, where the N-terminal 104 residues were identical to those determined by Kataoka. The cDNA encoding the PTTH monomer was expressed in *Escherichia coli*, and the expression product was air-oxidated for dimerization by disulfide bridge formation to generate a recombinant PTTH (rPTTH). The rPTTH showed almost the same biologic activity as native PTTH in *Bombyx* debrained pupal assay, confirming that the identified protein is *Bombyx* PTTH. This result also showed that a sugar chain present in native PTTH is not essential for its biologic activity. Using rPTTH, locations of three disulfide bridges within the PTTH monomer and one bridge between the two monomers were determined (Ishibashi et al. 1992). Furthermore, a comparison of the primary structure of *Bombyx* PTTH with other known polypeptides and a computer simulation of its 3D structure revealed that this hormone is a member of the cystine knot family of growth factors (Noguchi et al. 1995). In this way, the structure of *Bombyx* PTTH was finally determined. More than 30 years had passed since Ishizaki started purifying *Bombyx* PTTH on his own (Table 1.1).

Table 1.1 History of PTTH purification

Year	Event	Technology
1960		
	PTTH extraction with aqueous solvent	
	Start of PTTH purification by Ishizaki alone	Conventional chromatography
	Achievement of 8000-fold purification	
1970		
	Abandonment of further purification	
	Start of collaboration with chemists	
	Mass procurement of *Bombyx* heads	
	Preparation & characterization of "crude PTTH"	
	Preparation of "highly purified PTTH"	
1980		
	Discovery of two types of PTTH	
	Start of 22K-PTTH purification	HPLC
	Isolation & partial sequencing of 4K-PTTH	
		Gas-phase protein sequencer
	Determination of complete sequence of 4K-PTTH	
	Isolation & partial sequencing of 22K-PTTH	
		cDNA cloning
1990	Determination of complete sequence of 22K-PTTH	Monoclonal antibody

1.2 Purification and Structure Determination of Other Peptide Hormones in the Twentieth Century

The structures of *Bombyx* PTTH and bombyxin were thus determined, and around the same time (at the end of the twentieth century), the structures of other insect peptide hormones were also revealed. The main peptide hormones discovered during the twentieth century are shown in Table 1.2. More than a dozen of hormones with various functions had been found before 1980. However, the structures of these hormones were determined only after 1980, except for proctolin and adipokinetic hormone (AKH), whose structures were determined in 1975 and 1976, respectively.

Table 1.2 Discovery and structure determination of insect peptide hormones in the twentieth century

Hormone	Discovery[a]	Source	Identification[b]
Brain hormone (PTTH)	Kopeć (1922)	Moth, brain	Kataoka et al. (1987)
Allatostatin (A type)	Scharrer (1952)	Cockroach brain	Woodhead et al. (1989)
Diapause hormone	Hasegawa (1957)	Moth SG	Imai et al. (1991)
Hyperglycemic hormone	Steele (1961)	Cockroach CC	Scarborough et al. (1984)
Diuretic hormone (CRH-like)	Maddrell (1962)	Bug mesothoracic ganglia	Kataoka et al. (1989b)
Bursicon	Fraenkel and Catherine (1965)	Fly brain	Luo et al. (2005)
Ovary ecdysteroidogenic hormone	Lea (1967)	Mosquito brain	Brown et al. (1998)
Proctolin	Brown (1967)	Cockroach gut	Starrat and Brown (1975)
Prothoracicostatic peptide	Carlisle and Ellis (1968)	Locust brain	Hua et al. (1999)
Adipokinetic hormone	Mayer and Candy (1969)	Locust CC	Stone et al. (1976)
Eclosion hormone	Truman and Riddiford (1970)	Moth brain	Nagasawa et al. (1985)
Hypoglycemic hormone (ILP)	Norman (1975)	Fly brain	Nagasawa et al. (1984b)
MRCH	Ogura (1975)	Moth SG	Matsumoto et al. (1985)
PBAN	Raina and Klum (1984)	Moth SG	Raina et al. (1989)
Allatotropin	Granger et al. (1984)	Moth head	Kataoka et al. (1989a)
Myokinin (leucokinin)	Holman et al. (1986a)	Cockroach head	Holman et al. (1986a)
Myotropin (pyrokinin)	Holman et al. (1986b)	Cockroach brain	Holman et al. (1986b)
Myosuppressin	Holman et al. (1986c)	Cockroach brain	Holman et al. (1986c)
Neuroparsin	Girardie et al. (1987)	Locust CC (brain)	Girardie et al. (1989)
Corazonin	Veenstra (1989)	Cockroach brain	Veenstra (1989)
Growth-blocking peptide	Hayakawa (1990)	Moth hemolymph	Hayakawa (1990)
Allatostatin (C type)	Kramer et al. (1991)	Moth brain	Kramer et al. (1991)
Ovary maturating parsin	Couillaud et al. (1989)	Locust CC (brain)	Girardie et al. (1991)

(continued)

Table 1.2 (continued)

Hormone	Discovery[a]	Source	Identification[b]
Ion transport peptide	Audsley and Phillips (1990)	Locust CC and ventral ganglion	Audsley et al. (1992)
Allatostatin (B type)	Lorenz et al. (1995)	Cricket brain	Lorenz et al. (1995)
Ecdysis-triggering hormone	Žitňan et al. (1996)	Moth epitracheal glands	Žitňan et al. (1996)

SG subesophageal ganglion, *CC* corpora cardiaca
[a]Discovery: the literature that first described that hormone
[b]Identification: the literature that first described the partial or complete sequence of that hormone. The species in which the structure of a hormone was determined is not necessarily the same species from which that hormone was first discovered

Proctolin is the first insect neuropeptide to be sequenced. This peptide was first found as a "gut factor" to stimulate the contraction of the visceral muscle and considered to be a type of neurotransmitter (Brown 1967), although proctolin is now believed to regulate a variety of physiological processes as a neurohormone or neuromodulator. Proctolin is a small peptide with five aa residues (Starrat and Brown 1975). Although huge efforts of researchers resulted in a successful sequencing of this peptide, its small size was one of the reasons why its structure was determined relatively early. The AKH was also a relatively small peptide with ten aa residues (Stone et al. 1976). This hormone was found as a corpus cardiacum (CC)-derived factor that promotes lipid mobilization to meet an energy demand upon flight of locusts. One of the reasons for the early success of structure determination of this hormone was its high abundance in the CC. In the 1970s, peptide purification was performed through conventional liquid chromatography, and the sequence of isolated peptide was analyzed by manual Edman degradation. The resolution of peptide separation by conventional chromatography was not very good, and sequencing by Edman degradation required amounts of purified material on the order of 10 μg. Therefore, it was tough to determine the structure of many other peptide hormones in this era, despite the researcher's tremendous efforts.

However, the situation changed dramatically at the end of the 1970s by the emergence of HPLC. Furthermore, the emergence of an automatic gas-phase protein sequencer in the early 1980s boosted the structure determination of peptide hormones. This new technique made it possible to sequence polypeptides of interest using small amounts of purified material, 100- or 1000-fold smaller than the amounts required for traditional Edman degradation sequencing. As shown in Table 1.2, many of the peptide hormones discovered before 1980 had their structure determined during the 1980s. Besides, peptide hormones discovered after 1980 were chemically identified within a short period of time after their discovery.

Yet, some large peptide hormones needed another technical innovation, gene cloning, for complete sequencing. For example, the sequence of PTTH monomer was biochemically determined from the N-terminus to the 104th aa residue, but the remaining 5 residues were determined by using gene cloning technology (see Sect. 1.1.6). In this way, almost all peptide hormones discovered in the twentieth century were completely sequenced by the end of the century. Peptide hormones in non-insect invertebrates were also identified one after another in the late twentieth century. Many of the insect peptide hormones identified after 2000 are the homologs of peptide hormones originally identified in non-insect invertebrates or vertebrates. Among them are tachykinin, inotocin, neuropeptide F, calcitonin-like diuretic hormone, orcokinin, pigment-dispersing factor, and FMRFamide. To be precise, some may be better called neuropeptides rather than peptide hormones (see Chap. 4). However, in this chapter, we will refer to neuropeptides as peptide hormones and provide an overview of the recent research on them.

1.3 Research in the Twentieth Century After Chemical Identification of Peptide Hormones

The identification of hormones is the starting point for new developments in the physiological studies of hormones. Such studies include the identification of hormone-producing cells and hormone secretory pathways, the investigation of secretion dynamics, the research on their actions and action mechanisms including identification of receptors and signaling pathways of the hormones, as well as the exploration of the regulatory mechanisms of hormone secretion. For these studies, the production of chemically synthesized or recombinant hormones as well as specific antibodies to each hormone is essential. Indeed, soon after the identification of each hormone, a synthetic hormone has been produced. Synthesis of the identified peptide is also important for confirming that the identified peptide is definitely the hormone of interest, because the sequenced peptide could be an impurity in the purified hormone sample. If the synthetic peptide shows an expected activity in a bioassay, that proves the success of hormone purification. Synthetic or recombinant hormones are used not only as standard hormones to investigate their actions but also as immunogens to produce antibodies to the hormones. The antibodies produced are used to localize hormone-producing cells and hormone secretory pathways by immunohistochemistry and to develop immunoassays for determining hormone titers in the hemolymph or tissues. Immunohistochemistry was performed on tissue sections in early studies, but it is now more commonly performed on whole-mount specimens. Since the 1990s, with the proliferation of molecular biology, in situ hybridization has also become commonly used to identify hormone-producing cells (see Sect. 1.4.2.1). Once neurosecretory cells that produce the hormone of interest have been identified, axonal projections can be traced in more detail by injecting fluorescent dyes such as Lucifer yellow. More recent progress in the anatomy of neurosecretory pathway will be described later (see Sect. 1.4.4).

Examining the changes in hormone titers in hemolymph is essential to understand the real context of hormonal control of biological processes. Hormone titers are determined by radioimmunoassay, enzyme-linked immunosorbent assay, or more sensitive immunoassays, such as time-resolved fluoroimmunoassay. However, measuring insect peptide hormone titers in the hemolymph is generally not easy. Indeed, there are not many reports examining hemolymph titers of insect peptide hormones. One reason may be the presence of substances in the hemolymph that interfere with the immunoassay. Alternatively, some studies have used mass spectrometry to measure hormone levels in hemolymph, which will be discussed in a later section (see Sect. 1.4.3). In the twentieth century, research into hormone effects and action mechanisms was often carried out using in vitro culture systems of specific tissues. For example, in vitro PG culture systems have been used to examine the dose response and time course of the response of PGs to PTTH and to explore the second messenger and downstream signaling components involved in the response of PG cells (Bollenbacher and Granger 1985; Smith and Rybczynski 2012). Although in vitro studies of hormone actions are less popular nowadays, such approaches remain important to compensate for the weaknesses of in vivo studies.

Another line of research after determining the structure of peptide hormones was the identification of hormone genes and analysis of their expression. In other words, it is the molecular biology of hormones. Gene cloning and complementary DNA (cDNA) cloning were performed based on information on the aa sequence of hormones identified, and the nucleic acid sequences of hormone genes and mRNAs as well as the aa sequences of hormone precursors were revealed. The tissues or cells in which hormone genes are expressed and the timing of their expression were also investigated. The abovementioned in situ hybridization is a technique for identifying cells that express a specific gene. These molecular biological techniques will be outlined in Sect. 1.4.

In this way, molecular biology began to have a significant impact on insect endocrinology in the 1990s, and this trend rapidly accelerated with the completion of the *Drosophila* Genome Project in 2000. Research on insect peptide hormones has progressed from the era of biochemistry, through the era of molecular biology, and into the era of genome, omics, and molecular genetics. The second half of this chapter provides an overview of the technological advances in the research on insect peptide hormones in the post-genomic era. These advances in research technology have greatly improved our understanding of the insect endocrine system, and the results are discussed in detail in other chapters of this book.

1.4 Innovation of Research Techniques in the Post-genomic Era

In the twenty-first century, the completion of whole genome sequencing (WGS) has dramatically changed the approach of molecular biology. In the "post-genomic era," researchers are increasingly focused on understanding the in vivo functions of genes and the peptides/proteins they encode. These efforts, coupled with advances in biochemistry and microscopy techniques, have revealed the mysteries of life at cellular and molecular resolution. These changes were no exception in insect endocrinology. In this section, we provide an overview of new biological techniques that have accelerated research into insect peptide hormones.

1.4.1 Bioinformatics

In the early 2000s, a major paradigm shift occurred in insect research. The first insect WGS was reported in *D. melanogaster* in 2000 (Adams et al. 2000; Myers et al. 2000). This was followed by the WGS report for the malaria mosquito *Anopheles gambiae* in 2002 (Holt et al. 2002). Subsequently, WGS of other insects was reported by consortiums of the western honey bee *Apis mellifera* (Honeybee Genome Sequencing Consortium 2006), the red flour beetle *Tribolium castaneum* (Tribolium Genome Sequencing Consortium 2008), and the pea aphid *Acyrthosiphon pisum* (International Aphid Genomics Consortium 2010). Meanwhile, WGS in *B. mori* was reported in 2008 by combining two independent sequence data reported by Japanese and Chinese groups (Mita et al. 2004; Xia et al. 2004; The International Silkworm Genome Consortium 2008). To date, WGS of more than 600 insect species representing 20 orders has been reported by 2021 (Hotaling et al. 2021). Since peptide hormone genes are encoded in the genome, WGS allows the annotation of gene sequences and loci using aa sequences of hormones (Vanden Broeck 2001; Fig. 1.3). Peptide hormone homologs share homology in aa sequences, particularly in the sequence of the mature peptide or motifs within the peptides required for activity (Liu et al. 2006). Therefore, once a gene is identified in an insect, WGS is used to find orthologous genes in other insects by searching for similar nucleotide sequences. For example, 35 peptide hormone genes in *A. gambiae* and more than 41 genes in *T. castaneum* were annotated by homology-based searches using the genome of *D. melanogaster* (Riehle et al. 2002; Li et al. 2008). These comparative studies have also contributed to identifying functional motif sequences within peptides. As described above, conserved sequences contain motifs essential for hormone activity, such as ligand-receptor interactions (Liu et al. 2006). For example, some neuropeptides share conserved aa sequences on their C-termini, such as RYamide and RWamide (Williams 2020; see Chap. 2). In addition to WGS, expression sequence tag libraries have been reported in many insect species. Expression sequence tag libraries contain transcripts of prepropeptide hormones, thus allowing the identification of peptide hormone genes in non-model insects whose genomes have not been fully determined.

The WGS has also contributed to the identification of peptide hormone receptors. Most peptide hormone receptors belong to the large superfamily of G-protein-coupled receptors (GPCRs) (Hauser et al. 2006). Based on sequence similarity, 44 GPCRs have been classified as potential peptide receptors in *D. melanogaster* (Hewes and Taghert 2001). Many peptide hormone receptors have been identified from these "orphan" receptors. For example, receptors of PRXamide peptides,

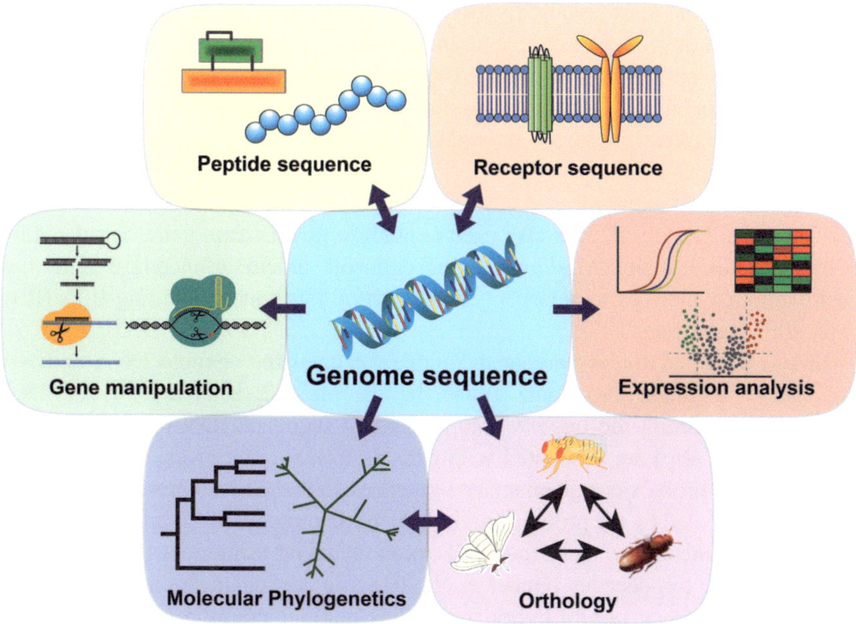

Fig. 1.3 Innovation of biological analyses in the post-genomic era. Genomic sequences are utilized for multiple genetic approaches, such as gene expression analysis, gene manipulation, molecular phylogenetics, and orthology. They also provide aa sequences of peptides and receptors. Conversely, aa sequences are used for gene annotation in genomes. Images of a genome, genetic manipulation techniques, a fruit fly, and a beetle were obtained from TogoTV (© 2016 DBCLS TogoTV, CC-BY-4.0 https://creativecommons.org/licenses/by/4.0/deed.ja). (The image of silk moth was kindly provided by Yusuke Kohmura)

crustacean cardioactive peptide, corazonin (CRZ), and AKH have been identified from orphan receptors, and ligand-receptor interactions were confirmed by electrophysiological assays (Park et al. 2002).

1.4.2 Molecular Biology

1.4.2.1 Gene Expression Analysis

Reverse transcription polymerase chain reaction (RT-PCR) became a fundamental biological technique for detecting target gene transcripts (Bustin 2000; Valasek and Repa 2005). This technique uses cDNA as a template and amplifies the target sequence by PCR using specific primers. Since peptide hormones and their receptors are translated from mRNA, RT-PCR detects the expression of target genes in samples. The RT-PCR is also used to quantify target gene expression by real-time PCR (Bustin 2000), which quantifies the amount of PCR product at each PCR step.

SYBR Green is the most commonly used fluorescent dye in real-time PCR because it binds to the minor groove of double-stranded DNA (Valasek and Repa 2005). These analyses have contributed to many important discoveries, such as identifying hormone-producing sites (high expression of hormonal genes) or hormone target tissues (high expression of receptor genes). For example, many ILPs have been found to be expressed not only in the brain but also in peripheral tissues such as the fat body, based on expression analysis (Okamoto et al. 2009a, b; Aslam et al. 2011; see Chap. 7). The RT-PCR is also used to analyze downstream genes regulated by hormones. For instance, *vitellogenin*, the egg yolk precursor-encoding gene, has been well characterized as a target gene of several hormones including ILPs (Roy et al. 2007; see Chap. 10).

In situ hybridization is another technique for detecting specific DNA or RNA sequences (Lehmann and Tautz 1994). In this method, labeled DNA or RNA fragments containing the target's complementary sequence (DNA probe or RNA probe, respectively) are incubated with cells or tissues. The probe binds to RNA containing the probe's complementary sequences so that cells expressing a specific gene can be labeled by the probe. Although the basic protocol of in situ hybridization has remained unchanged since the 1980s, probe labeling techniques have advanced drastically over the decades (Huber et al. 2018). In particular, the recent development of fluorescence in situ hybridization has made it possible to label multiple mRNAs simultaneously using multiple probes labeled with different fluorophores. For example, multiple RNA probes have demonstrated that genes for γ-aminobutyric acid (GABA) receptor subunits and CRZ are co-expressed in a neurosecretory cell in the pupal brain in *B. mori* (Tsuchiya et al. 2021). This result suggests that the secretion of CRZ is regulated by GABAergic neurons (see Chap. 8). An atlas of neurosecretory hormone- and peptide receptor-expressing cells in *B. mori* determined by in situ hybridization in combination with immunohistochemical analysis using hormone-targeting antibodies (see Sect. 1.3) has been published (Roller et al. 2017; http://bombase.org/; see Chap. 5).

Another approach to determining gene expression is to detect all transcripts in a sample. Transcriptome analysis has developed drastically in recent decades with the development of next-generation sequencers (see Chap. 3). The two main technologies for transcriptome, microarrays and RNA-sequencing (RNA-seq), were developed in the 1990s and 2000s, respectively (Nelson 2001; McGettigan 2013; Lowe et al. 2017). DNA microarrays dominated in the 2000s. Microarrays use nucleic acid probes covalently attached to glass slides. The probes hybridize with fluorescently labeled cDNA to quantify the transcripts containing the target sequence. The amount of DNA containing the target sequence can be analyzed by quantifying the fluorescent intensity of labeled cDNA interacting with each probe (Nelson 2001). RNA-seq became the favored technology in the late 2010s. In this technique, RNA is fragmented before reverse transcription. The sequence of each fragment is analyzed by next-generation sequencing. The fragment sequences are aligned to a reference genome to analyze the expression of specific genes (Lowe et al. 2017). For example, transcriptome analysis of *T. castaneum* has demonstrated that insect parathyroid hormone signaling regulates the expression of cuticle proteins (Xie et al. 2020).

Transcriptome analysis is also applied to de novo assembly to analyze gene expression without a reference genome, although with limited accuracy (Robertson et al. 2010; Liao et al. 2019). Thus, transcriptomics by RNA-seq can be applied to both non-model and model insects.

Recent advances in biotechnology have enabled researchers to quantify transcripts at single-cell resolution (Jovic et al. 2022). This technique, called single-cell RNA sequencing (scRNA-seq), has been applied to many peptide hormone-related studies in *D. melanogaster* and the yellow fever mosquito *Aedes aegypti* (Brunet Avalos et al. 2019; Nässel 2018; Yin et al. 2024). In particular, scRNA-seq of neurosecretory cells provides invaluable information because detecting receptor genes expressed in the cells helps identify the hormones or neurotransmitters that regulate the secretory activity of those cells.

1.4.2.2 RNA Interference

In addition to gain-of-function approaches using synthetic peptides and recombinant proteins, loss-of-function approaches have contributed to elucidating hormone function. RNA interference (RNAi) is a powerful tool to downregulate the expression of target genes. Since the discovery of RNAi in the nematode in 1998 (Fire et al. 1998, Nobel Prize in Physiology or Medicine 2006), RNAi has been found in all eukaryotes (Agrawal et al. 2003; Zhu and Palli 2020). Gene silencing by RNAi was demonstrated in many insect species several years after the discovery of RNAi (Palli 2023). Double-stranded RNA (dsRNA) is often used to induce systemic RNAi in insects. Several techniques have been tested to deliver dsRNA into target cells, including microinjection, feeding, and soaking (Zhu and Palli 2020). The efficiency of RNAi varies among insects due to differences in digestion, cellular uptake, intracellular transport, or processing of dsRNA. For example, RNAi is highly efficient in coleopteran insects, but relatively less efficient in other insects such as lepidopteran insects (Tomoyasu et al. 2008; Palli 2023). The in vivo functions of peptide hormones have been well analyzed by systemic RNAi in *T. castaneum*. For example, knockdown of allatotropin-1 (AT-1), AT-2, allatostatin B (AST-B), and AST-C caused abnormal adult development and reduced egg deposition in *T. castaneum* (Abdel-latief and Hoffmann 2014). In contrast, tissue-specific knockdown has been widely used in *D. melanogaster* using binary expression systems to determine tissue-specific functions of peptide hormones/neuropeptides and their receptors (see Sect. 1.4.2.4).

1.4.2.3 Targeted Gene Disruption

Gene knockdown by RNAi reduces but does not completely eliminate the expression of a target gene. Gene knockout is performed to completely disrupt a targeted gene by manipulating the gene-coding locus in the genome. The *Drosophila* Gene Disruption Project, which began in 1991, has produced many gene knockouts in

D. melanogaster (Bellen et al. 2011). In addition to random mutagenesis with mutagen (ethyl methanesulfonate or X-ray) and random insertion of P elements (Spradling et al. 1999; St Johnston 2002), targeted mutagenesis by homologous recombination was also performed in the 2000s (Gong and Golic 2003; Xie and Golic 2004). By 2011, about two-thirds of all annotated protein-coding gene knockouts were obtained (Bellen et al. 2011). These genetic tools have been used to investigate the in vivo functions of peptide hormones (Clynen et al. 2010).

Although gene knockout approach had been limited to certain insect species such as *D. melanogaster* and *B. mori*, recent innovations in genome editing technology have made it possible to manipulate the genome of any species. Three genome editing technologies have been reported so far: zinc finger nuclease, transcription activator-like effector nuclease (TALEN), and clustered regularly interspaced short palindromic repeat (CRISPR, Nobel Prize in Chemistry 2020) (Khalil 2020). The TALEN and CRISPR have been applied in various insects to knock out target genes or knock in exogenous genes at the target locus (Daimon et al. 2014; Reid and O'Brochta 2016). Both TALEN and CRISPR consist of a nucleotide-targeting component (protein (TALE) in TALEN or RNA (guide RNA) in CRISPR) and an endonuclease that induces double-strand breaks (DSBs) in genomic DNA (Khalil 2020). The target DNA site is selected by TALE or guide RNA, which then recruits endonuclease to induce a DSB on the target site. The DSB is repaired by non-homologous end-joining (NHEJ) or homologous direct repair (HDR). However, the correctly repaired DNA is redigested by nucleases again. As a result, mutations due to nucleotide insertion or deletion occur in the target gene region in the genome. This is because such mutations cannot be redigested by nucleases. Exogenous sequences can be inserted into the target region by HDR. For example, the insertion of molecular tags into hormone gene-encoding regions is performed by HDR. Tagged proteins enable researchers to quantify hormone titers in hemolymph without producing antibodies targeting the hormone. By using this technique, one study demonstrated that immune response inhibits the production of a certain type of ILP in *D. melanogaster* (Suzawa et al. 2019; Chap. 7). Recently, several new techniques have been reported to deliver CRISPR components to target cells. For example, a combination of CRISPR and binary expression systems is used to induce mutagenesis in specific cell types in *D. melanogaster* (Xue et al. 2014; see Sect. 1.4.2.4). In addition, mutagenesis in new non-model insects has been reported year by year (Zhang et al. 2021; Shirai et al. 2022). These efforts are likely to contribute to the understanding of the diversity of hormonal functions.

1.4.2.4 Transgene Expression Systems

Technologies for manipulating gene expression have developed rapidly after the 1990s. In particular, binary expression systems, such as the Gal4-upstream activating sequence (UAS) system, are commonly used in studies of *D. melanogaster* as well as in *B. mori* and mosquitoes (Brand and Perrimon 1993; Imamura et al. 2003; Kokoza and Raikhel 2011; Lynd and Lycett 2012). In this system, two components

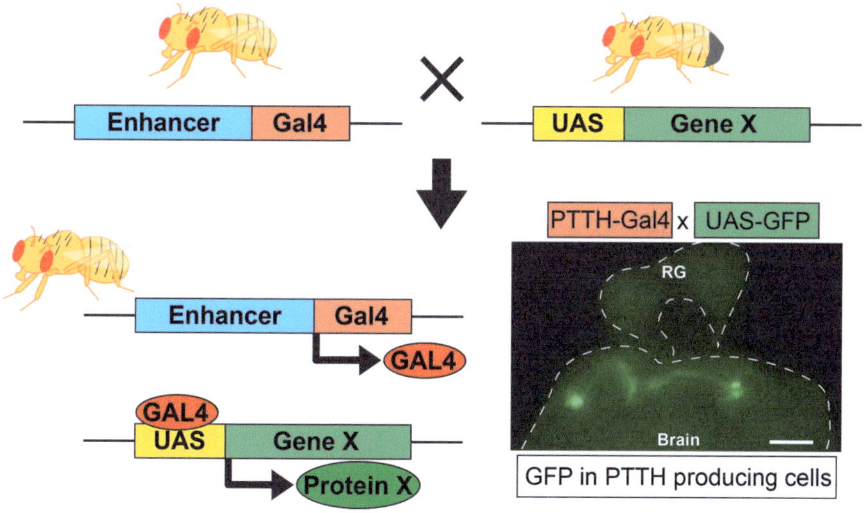

Fig. 1.4 The Gal4-UAS system used for gene manipulation. Gal4 and UAS lines are kept separately. The progeny that have acquired both Gal4 and UAS expresses genes downstream of UAS under the control of Gal4 expression. For example, *PTTH-Gal4*-driven *GFP* expression induces green fluorescence in PTTH-producing cells in the larval brain. The brain-ring gland (RG) complex was dissected from the wandering larva of *D. melanogaster*. The *PTTH-Gal4* line was gifted by Dr. Michael B. O'Connor. Scale bar: 50 μm. (Illustrations of fruit flies were obtained from TogoTV (© 2016 DBCLS TogoTV, CC-BY-4.0 https://creativecommons.org/licenses/by/4.0/deed.ja))

are maintained in separate parental lines. The first line obtains the transcription factor Gal4 under the control of a promoter, while the second line obtains the UAS that controls the expression of a target gene (Fig. 1.4). When a UAS line is crossed with a Gal4 driver line, the progeny acquires both components. Thus, the UAS-targeted gene is expressed with a transcriptional pattern that reflects Gal4 expression in the progeny (Busson and Pret 2007). This system allows researchers to manipulate gene expression in target tissues that express Gal4. For example, when a Gal4 line containing a promoter sequence of *PTTH* is crossed with a UAS-green fluorescent protein (GFP) line, the progeny acquires GFP expression in PTTH-producing cells (McBrayer et al. 2007). Similarly, by using UAS-targeted gene-coding or UAS-targeted gene RNAi (dsRNA or short hairpin RNA coding sequence) lines, the progeny acquires an increased or decreased expression of the target gene in the target tissue. Moreover, recent studies have performed to label the entire secreted proteins from the target cell by overexpressing a modified biotin ligase driven by a cell-specific Gal4 (Droujinine et al. 2021). The labeled proteins are detected by ELISA or mass spectrometry. In this system, secretome is used in some peptide hormone research. For example, the secretome revealed that neuropeptide F (NPF) is secreted from a specific type of enteroendocrine cell in the midgut (Chen et al. 2024). Other binary expression systems, the LexA-LexAop system and the QF-QUAS system, have also been used in *D. melanogaster* studies (del Valle Rodríguez et al. 2011). These alternative systems allow for the simultaneous manipulation of multiple genes in different tissues. Gal80, a suppressor of Gal4, has also been used in many studies to inhibit Gal4 expression at undesirable times or in tissues (Lee and Luo 1999). For example, a temperature-sensitive Gal80, Gal80^{ts}, is inactivated at high temperatures. Thus, when insects are kept at low temperatures, Gal4 transcriptional activity is suppressed, but UAS-targeted genes are expressed at high temperatures (Zeidler et al. 2004). Temporal knockdown of *DILP8* using the Gal4/Gal80^{ts} system revealed that the peptide suppresses systemic larval development during a specific period (Heredia et al. 2021).

1.4.3 Biochemistry

1.4.3.1 Mass Spectrometry

In addition to the advance in HPLC technology (see Sect. 1.2), dramatic advances in mass spectrometry (MS) technology have facilitated the study of peptide hormones (Mehlis and Kertscher 1997; Pitt 2009; Fig. 1.5). Mass spectrometers determine the

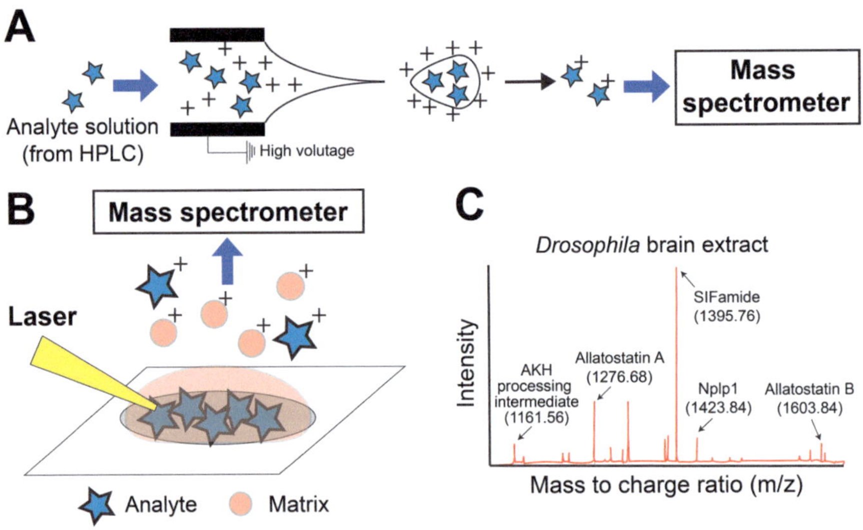

Fig. 1.5 Methods of soft ionization. (**A**) Electrospray ionization, commonly used in LC-MS. Analytes in solution (e.g., HPLC fraction) are injected into a capillary. A high voltage is applied to the tip of the capillary, generating a strong electric field inside the capillary. The strong electric field generates high charged electrospray droplets containing the analytes. The charged droplets evaporate, and the charged analyte is subjected to a mass spectrometer. (**B**) Matrix-assisted laser desorption/ionization. The analyte is mixed with a suitable matrix material and applied to a metal plate. A pulsed laser irradiates the analytes, causing ablation and desorption of the analyte and matrix material. The analyte is ionized while evaporating due to heat generated by absorption of laser photon energy. (**C**) Example of mass spectrum obtained by MALDI-MS. The graph was visualized based on the analysis of brain extract in *D. melanogaster* (Yew et al. 2009)

mass-to-charge ratio of target molecules. Conventional ionization techniques are not suitable for detecting large molecules such as peptides, because they destroy the structure of large molecules. Two ionization techniques reported in the 1980s, electrospray ionization and matrix-assisted laser desorption/ionization (MALDI), made it possible to ionize peptides without destroying their structures (Tanaka et al. 1988; Fenn et al. 1989; Nobel Prize in Chemistry 2002). These "soft ionization" techniques were rapidly adopted by new MS techniques such as MS coupled with liquid chromatography (LC-MS) and MALDI-MS.

After its implementation in the 1990s, LC-MS was quickly adopted to study insect peptide hormones. For example, diuretic hormone 37 in *Tenebrio molitor* and diuretic hormone 41 in *M. sexta*, along with their metabolites, were detected using LC-MS (Furuya et al. 1995; Li et al. 1997). The LC-MS identifies each molecule by determining its mass-to-charge ratio in a sample. Furthermore, the high sensitivity of LC-MS allows researchers to analyze molecules in smaller biological samples than HPLC. Therefore, LC-MS has been applied not only to detect target peptides in samples but also to comprehensively determine proteins and metabolites (known as proteomics and metabolomics, respectively) (Lange et al. 2008; Griffiths and Wang 2009). These "omics" studies have contributed to understanding the functions of peptide hormones. For example, metabolome analysis showed that midgut-derived NPF regulates systemic metabolism of carbohydrates and lipids (Yoshinari et al. 2021; see Chap. 9).

The MALDI-MS is well suited for comprehensive peptide analysis (peptidomics). The relatively gentle ionization of MALDI allows large molecules to be analyzed without significant fragmentation. Moreover, intact cells as well as crude peptide extracts can be applied to MALDI-MS (Albrethsen 2011). Neuropeptides accumulate in secretory granules of neuronal cells and are therefore efficiently detected by irradiating the neuronal tissue with a pulse laser. This technique, called direct MALDI-MS, has been used since the early 2000s to determine neuropeptides in several species (Predel 2001; Predel et al. 2004; Wegener et al. 2006; see Chap. 3). To assign peptides from their mass-to-charge ratio, the aa sequence of the peptide must be known in advance. With the completion of genome projects, it became possible to deduce aa sequences of mature peptides and prohormones (Nässel 2002; Baggerman et al. 2005; Wegener and Gorbashov et al. 2008).

1.4.3.2 Conformational Analysis

The tertiary structures of peptide hormones and their receptors reveal how the ligands interact with the receptor. For example, some research has been conducted to develop species-specific insecticides based on molecular structures of peptide hormones (Jackson et al. 2019; Birgül Iyison et al. 2021; Abdulganiyyu et al. 2020). Nuclear magnetic resonance spectroscopy is a widely used technique to determine the tertiary structure of peptides (Zubrzycki and Gäde 1994; Abdulganiyyu et al. 2020). In contrast, the tertiary structures of insect peptide hormone receptors are currently unavailable due to the difficulty in determining the 3D structure of membrane receptors (Mugumbate et al. 2011; Birgül Iyison et al. 2021). Instead, 3D structures of insect peptide hormone receptors predicted by computational models based on the structure of human GPCRs have been used to analyze ligand-receptor interactions of insect hormones, such as the eclosion hormone in *M. sexta* (Hull et al. 2009; Mugumbate et al. 2011). The technology to determine the 3D structure of proteins has developed rapidly over the past decade, particularly with the advent of cryo-electron microscopy (Herzik Jr et al. 2019; Castells-Graells et al. 2023, Nobel Prize in Chemistry 2017). These breakthroughs are expected to contribute to the determination of the tertiary structure of insect peptide hormone receptors. Prediction algorithms and software, such as AlphaFold2 (Akdel et al. 2022), have also been developed in the past decade (Nobel Prize in Chemistry 2024).

1.4.4 Imaging

1.4.4.1 Innovation of Microscopy

High-resolution microscopes have been developed since the end of the twentieth century. For example, confocal laser scanning microscopy is one of the most widespread fluorescence microscopy techniques in biology (Elliott 2020). Confocal microscopes detect targeted fluorescence of a selected focal object by using a pinhole to eliminate out-of-focus light (Elliott 2020). This system allows the detection of targeted fluorescence in thick tissues. In addition, optical sectioning is also provided, allowing 3D reconstruction of samples from high-resolution image stacks. Fluorescence microscopy has been improved with the development of fluorescent dyes and fluorescent proteins (Lichtman and Conchello 2005; Sanderson et al. 2014). Alexa Fluor family is widely used as a fluorescent label. Target proteins can be labeled by immunostaining using fluorescent dye-conjugated antibodies. These dyes are also used to label cellular components such as mitochondria (Watson et al. 2005). Fluorescent proteins such as GFP have brought new innovations to biology (Chudakov et al. 2010; Nobel Prize in Chemistry 2008). The advantage of fluorescent proteins is that they allow visualization of target proteins or cellular structures in living cells or tissues. Proteins can be fused with fluorescent proteins, allowing visualization of the dynamics of the target protein. For example, a study analyzing the AKH receptor fused with GFP revealed that the AKH receptor is rapidly internalized from the cell membrane upon agonist stimulation (Zhu et al. 2009). Nowadays, many types of fluorescent dyes and proteins are available, covering almost the entire visible spectrum (Chudakov et al. 2010).

More powerful and higher-resolution microscopes are also being developed. Super-resolution microscopes that broke the diffraction limit in image resolution were reported in the first decade in the 2000s (Klar et al. 2000; Betzig et al. 2006; Nobel Prize in Chemistry 2014). Super-resolution microscopes have been used to visualize tiny structures such as neuropeptide cargos in neurosecretory cells (Park et al. 2014; Juarez-Carreño et al. 2021).

Electron microscopy has also drastically developed in the twenty-first century (de Boer et al. 2015). The latest electron microscopes are able to visualize single atoms (Urban et al. 2023). This is used for cryo-electron microscopy to determine the molecular structure of proteins as described above. Furthermore, a recent study using electron microscopy has archived the complete connectome of the central nervous system of *D. melanogaster* (Winding et al. 2023; available online, https://catmaid.virtualflybrain.org/). Most neuropeptides are produced in the brain. Because hormone production and secretion are controlled by other neurons through synaptic modulation, how neurons connect to each other is key information to elucidate the regulation of hormone production and secretion. Studies have already been conducted that focus on the network of neurosecretory cells (Hückesfeld et al. 2021). This study revealed factors that control the activity of input neurons and neurosecretory cells. Connectomics in *D. melanogaster* is expected to bring dramatic advances in research into the regulation of neuropeptides.

1.4.4.2 Calcium Reporter Assay

Calcium ions (Ca^{2+}) are universal second messengers that regulate cell signaling in many cell types, including neurosecretory cells and hormone target cells (Grienberger and Konnerth 2012). In neurons, action potentials trigger rapid changes in cytoplasmic Ca^{2+} concentrations, stimulating the exocytosis of neurotransmitters and neuropeptides (see Chap. 4). Calcium imaging is a microscopy technique to visualize and quantify intracellular Ca^{2+} status using calcium indicators. Although calcium imaging techniques were already used in the twentieth century (Berridge 1998), recent developments in calcium indicators and imaging instruments have enabled sensitive and rapid measurement of Ca^{2+} status in single cells and their microcompartments (Dana et al. 2019). Two types of calcium indicators are used for calcium imaging: chemical indicators and genetically encoded calcium indicators. Chemical indicators, such as fura-2, are injected directly into cells. When the indicator binds to Ca^{2+}, a conformational change occurs in the molecule, which changes the fluorescence emitted. Because of their high sensitivity, chemical indicators have been used for quantitative measurement of Ca^{2+} within target cells (Grienberger and Konnerth 2012). Genetically encoded calcium indicators have been used in culture cells and selected species such as *D. melanogaster*. Calcium-dependent luminescent proteins (e.g., aequorin) or fluorescent proteins (e.g., GCaMP and CaMPARI) are induced into target cells by transfection of gene-coding vectors (culture cells) or by using the Gal4-UAS system (*D. melanogaster*). The calcium status is then visualized microscopically and quantified with an image sensor (Lin and Schnitzer 2016). Calcium imaging has been utilized in various insect peptide hormone studies. For example, calcium dynamics in ecdysis-triggering hormone receptor-expressing neurons was analyzed in *D. melanogaster* using GCaMP (Kim et al. 2006; Fig. 1.6). Furthermore, transcriptional genetically encoded calcium indicators, such as calcium-dependent nuclear import of LexA (CaLexA) and transcriptional reporter of intracellular calcium, have been developed

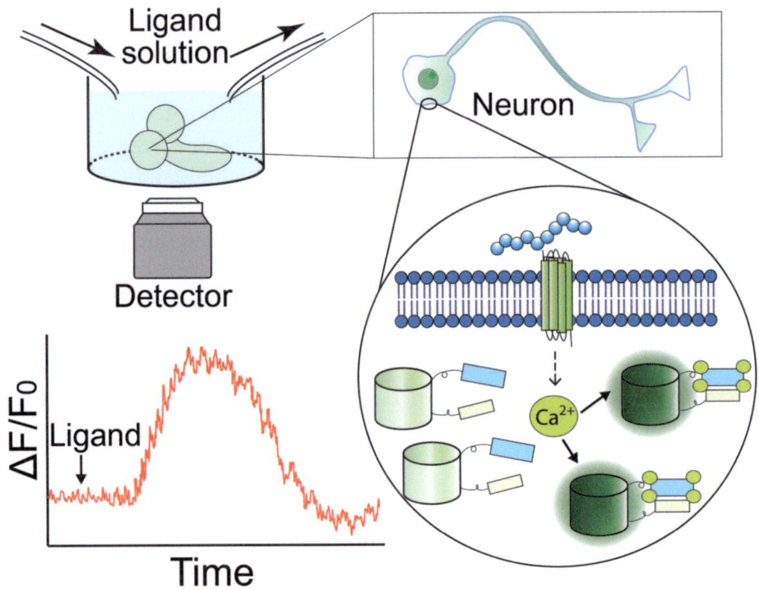

Fig. 1.6 Calcium reporter assay using a genetically encoded calcium indicator. In an in vitro calcium reporter assay, a sample (in this case the central nervous system) is immersed in culture medium and fixed on a Petri dish. Calcium-dependent proteins (e.g., GCaMP) are expressed in the target neuron. Transient application of a ligand increases intracellular calcium through its receptor. High intracellular calcium concentrations cause a fluorescent indicator to exhibit fluorescent properties. Fluorescence activity is typically presented as a relative fluorescence variation ($\Delta F/F0$), where F0 indicates the baseline fluorescence and ΔF indicates deviations from the baseline

in *D. melanogaster* (Bansal and Lin 2023). These systems convert intracellular calcium concentration into transcriptional activity of reporter genes. Thus, this system does not require specific calcium detection tools. CaLexA was utilized to detect nutrient-dependent calcium influx in multiple peptide hormone-producing cells in the midgut in *D. melanogaster* (Park et al. 2016).

1.5 Future Perspectives

With increasing numbers of molecular genetic studies on insect intercellular signaling molecules in *D. melanogaster* and other model insects, various neuropeptides have been shown to be involved in the regulation of insect development, physiology, and behavior. However, in many cases, it is unclear how these neuropeptides are involved in the regulation mechanism of each biological process. For example, even if the disruption of a specific neuropeptide gene results in significant changes in specific insect behavior, it is not known whether this neuropeptide's effect is direct or indirect. Furthermore, even if the direct effect of the peptide to the behavior has been demonstrated, it is still unclear if the peptide acts as a hormone, neuromodulator, or neurotransmitter. Thus, further studies are required for understanding the role of that neuropeptide in the regulation of that behavior. Such studies should include the identification of the neurons that secrete the neuropeptide, the neurons or other types of cells that express its receptor, and the neural circuit responsible for the behavior. Functional analysis of hormones and neuropeptides also requires investigating secretion dynamics of the signal molecules in order to understand the actual state of their action. Highly sensitive immunoassays or calcium reporter assays (see Sect. 1.4.4.2) may be applicable for this purpose. As discussed earlier (see Sect. 1.4.4.1), one new approach to further develop insect neuroendocrinology may be to exploit connectome, a synaptic-resolution connectivity map of the whole central nervous system, because most neuropeptides are released in specific neural circuits. In *D. melanogaster*, connectome studies are now progressing rapidly (Winding et al. 2023; Dorkenwald et al. 2024). We hope that insect endocrinology, which began with the search and identification of insect peptide hormones, will further develop through the development of these new research methods.

References

Abdel-latief M, Hoffmann KH (2014) Functional activity of allatotropin and allatostatin in the pupal stage of a holometablous insect, *Tribolium castaneum* (Coleoptera, Tenebrionidae). Peptides 53:172–184. https://doi.org/10.1016/j.peptides.2013.10.007

Abdulganiyyu IA, Kaczmarek K, Zabrocki, Nachman RJ, Marchal E, Schellens S et al (2020) Conformational analysis of a cyclic AKH neuropeptide analog that elicits selective activity on locust versus honeybee receptor. Insect Biochem Mol Biol 125:103362. https://doi.org/10.1016/j.ibmb.2020.103362

Adams D, Celeniker SE, Holt RA, Evans CA, Gocayne JD, Amanatides PG et al (2000) The genome sequence of *Drosophila melanogaster*. Science 287(5461):2185–2195. https://doi.org/10.1126/science.287.5461.2185

Agrawal N, Dasaradhi PV, Mohmmed A, Malhotra P, Bhatnagar RK, Mukherjee SK (2003) RNA interference: biology, mechanism, and applications. Microbiol Mol Biol Rev 67(4):657–685. https://doi.org/10.1128/MMBR.67.4.657-685.2003

Akdel M, Pires DEV, Pardo EP (2022) A structural biology community assessment of AlphaFold2 applications. Nat Struct Mol Biol 29(11):1056–1067. https://doi.org/10.1038/s41594-022-00849-w

Albrethsen J (2011) The first decade of MALDI protein profiling: a lesson in translational biomarker research. J Proteome 74(6):765–773. https://doi.org/10.1016/j.jprot.2011.02.027

Aslam AF, Kiya T, Mita K, Iwami M (2011) Identification of novel bombyxin genes from the genome of the silkmoth *Bombyx mori* and analysis of their expression. Zool Sci 28(8):609–616. https://doi.org/10.2108/zsj.28.609

Audsley N, Phillips JE (1990) Stimulation of ileal salt transport in neuroendocrine system of the desert locust. Gen Comp Endocrinol 80:127–137. https://doi.org/10.1016/0016-6480(90)90156-G

Audsley N, Mcintosh C, Phillips JE (1992) Isolation of a neuropeptide from locust corpus cardiacum which influence ileal transport. J Exp Biol 173:261–274. https://doi.org/10.1242/jeb.173.1.261

Baggerman G, Boonen K, Verleyen P, De Loof A, Schoofs L (2005) Peptidomic analysis of the larval *Drosophila melanogaster* central nervous system by two-dimensional capillary liquid chromatography quadrupole time-of-flight mass spectrometry. J Mass Spectrom 40(2):250–260. https://doi.org/10.1002/jms.744

Bansal S, Lin S (2023) Transcriptional genetically encoded calcium indicators in *Drosophila*. Cold Spring Harb Protoc 2023(1):8–18. https://doi.org/10.1101/pdb.top107797

Bargmann W, Scharrer E (1951) The site of origin of the hormones of the posterior pituitary. Am Scientist 39:255–259

Bellen HJ, Levis RW, He Y, Carlson JW, Evans-Holm M, Bae E et al (2011) The *Drosophila* gene disruption project: progress using transposons with distinctive site specificities. Genetics 188(3):731–743. https://doi.org/10.1534/genetics.111.126995

Berridge MJ (1998) Neuronal calcium signaling. Neuron 21(1):13–26. https://doi.org/10.1016/s0896-6273(00)80510-3

Betzig E, Patterson GH, Sougrat R, Lindwasser OW, Olenych S, Bonifacino JS et al (2006) Imaging intracellular fluorescent proteins at nanometer resolution. Science 313(5793):1642–1645. https://doi.org/10.1126/science.1127344

Birgül Iyison N, Shahraki A, Kahveci K, Düzgün MB, Gün G (2021) Are insect GPCRs ideal next-generation pesticides: opportunities and challenges. FEBS J 288(8):2727–2745. https://doi.org/10.1111/febs.15708

Bollenbacher WE, Granger NA (1985) Endocrinology of prothoracicotropic hormone. In: Kerkut GA, Gilbert LI (eds) Comprehensive insect physiology biochemistry and pharmacology, vol 7 endocrinology. Pergamon Press, New York

Brand AH, Perrimon N (1993) Targeted gene expression as a means of altering cell fates and generating dominant phenotypes. Development 118(2):401–415. https://doi.org/10.1242/dev.118.2.401

Brown BE (1967) Neuromuscular transmitter substance in insect visceral muscle. Science 155:595–597. https://doi.org/10.1126/science.155.3762.595

Brown MR, Graf R, Swiderek KM (1998) Identification of a steroidogenic neurohormone in female mosquitoes. J Biol Chem 273:3967–3971. https://doi.org/10.1074/jbc.273.7.3967

Brunet Avalos C, Maier GL, Bruggmann R, Sprecher SG (2019) Single cell transcriptome atlas of the *Drosophila* larval brain. eLife 8:e50354. https://doi.org/10.7554/eLife.50354

Busson D, Pret AM (2007) GAL4/UAS targeted gene expression for studying *drosophila* hedgehog signaling. Methods Mol Biol 397:161–201. https://doi.org/10.1007/978-1-59745-516-9_13

Bustin SA (2000) Absolute quantification of mRNA using real-time reverse transcription polymerase chain reaction assays. J Mol Endocrinol 25(2):169–193. https://doi.org/10.1677/jme.0.0250169

Carlisle DB, Ellis PE (1968) Hormonal inhibition of the prothoracic gland by the brain in locusts. Nature 220:706–707. https://doi.org/10.1038/220706a0

Castells-Graells R, Meador K, Arbing MA (2023) Cryo-EM structure determination of small therapeutic protein targets at 3 Å-resolution using a rigid imaging scaffold. Proc Natl Acad Sci USA 120(37):e2305494120. https://doi.org/10.1073/pnas.2305494120

Chen J, Nouzová M, Noriega FG, Tatar M (2024) Gut-to-brain regulation of *drosophila* aging through neuropeptide F, insulin, and juvenile hormone. Proc Natl Acad Sci USA 121(43):e2411987121. https://doi.org/10.1073/pnas.2411987121

Chudakov DM, Matz MV, Lukyanov S, Lukyanov KA (2010) Fluorescent proteins and their applications in imaging living cells and tissues. Physiol Rev 90(3):1103–1163. https://doi.org/10.1152/physrev.00038.2009

Clynen E, Reumer A, Baggerman G, Mertens I, Schoofs L (2010) Neuropeptide biology in *Drosophila*. Adv Exp Med Biol 692:192–210. https://doi.org/10.1007/978-1-4419-6902-6_10

Couillaud F, Girardie A, Girardie J (1989) Identification of gonadotropic and antigonadotropic factors from the nervous part of the corpora cardiaca in the African locust. Invert Reprod Dev 16:17–22. https://doi.org/10.1080/07924259.1989.9672056

Daimon T, Kiuchi T, Takasu Y (2014) Recent progress in genome engineering techniques in the silkworm, *Bombyx mori*. Dev Growth Differ 56(1):14–25. https://doi.org/10.1111/dgd.12096

Dana H, Sun Y, Mohar B, Hulse BK, Kerlin AM, Hasseman JP et al (2019) High-performance calcium sensors for imaging activity in neuronal populations and microcompartments. Nat Methods 16(7):649–657. https://doi.org/10.1038/s41592-019-0435-6

de Boer P, Hoogenboom JP, Giepmans BN (2015) Correlated light and electron microscopy: ultrastructure lights up! Nat Methods 12(6):503–513. https://doi.org/10.1038/nmeth.3400

del Valle Rodríguez A, Didiano D, Desplan C (2011) Power tools for gene expression and clonal analysis in *Drosophila*. Nat Methods 9(1):47–55. https://doi.org/10.1038/nmeth.1800

Dorkenwald S, Matsliah A, Sterling AR, Schlegel P, Yu S-C, MacKellar CE et al (2024) Neuronal wiring diagram of an adult brain. Nature 634:124–138. https://doi.org/10.1038/s41586-024-07558-y

Droujinine IA, Meyer AS, Wang D, Udeshi ND, Hu Y, Rocco D et al (2021) Proteomics of protein trafficking by in vivo tissue-specific labeling. Nat Commun 12(1):2382. https://doi.org/10.1038/s41467-021-22599-x

Elliott AD (2020) Confocal microscopy: principles and modern practices. Curr Protoc Cytom 92(1): e68. https://doi.org/10.1002/cpcy.68

Fenn JB, Mann M, Meng CK, Wong SF, Whitehouse CM (1989) Electrospray ionization for mass spectrometry of large biomolecules. Science 246(4926):64–71. https://doi.org/10.1126/science.2675315

Fire A, Xu S, Montgomery MK, Kostas SA, Driver SE, Mello CC (1998) Potent and specific genetic interference by double-stranded RNA in *Caenorhabditis elegans*. Nature 391(6669): 806–811. https://doi.org/10.1038/35888

Fraenkel G, Catherine H (1965) Bursicon, a hormone which mediates tanning of the cuticle in the adult fly and other insects. J Insect Physiol 11:513–556. https://doi.org/10.1016/0022-1910(65)90137-X

Fukuda S (1940) Hormonal control of molting and pupation in the silkworm. Proc Imp Acad 16: 417–420. https://doi.org/10.2183/pjab1912.16.417

Fukuda S (1944) The hormonal mechanism of larval molting and metamorphosis in the silkworm. J Fac Sci Tokyo Univ Sect IV 6:477–532

Furuya K, Schegg KM, Wang H, Schooley DA (1995) Isolation and identification of a diuretic hormone from the mealworm *Tenebrio molitor*. Proc Natl Acad Sci USA 92(26):12323–12327. https://doi.org/10.1073/pnas.92.26.12323

Girardie J, Boureme D, Girardie A (1987) Production sites of the tree neurosecretory proteins characterized in the corpora cardiaca of the migratory locust. Insect Biochem 17:29–36. https://doi.org/10.1016/0020-1790(85)90048-4

Girardie J, Girardie A, Huet J-C, Pernollet J-C (1989) Amino acid sequence of locust neuroparsins. FEBS Lett 245:4–8. https://doi.org/10.1016/0014-5793(89)80179-6

Girardie J, Richard O, Huet JC, Nespoulous C, Van Dorsselaer A, Pernollet JC (1991) Physical characterization and sequence identification of the ovary maturating parsin: a new neurohormone purified from the nervous corpora cardiaca of the African locust (*Locusta migratoria migratorioides*). Eur J Biochem 202:1121–1126. https://doi.org/10.1111/j.1432-1033.1991.tb16479.x

Gong WJ, Golic KG (2003) Ends-out, or replacement, gene targeting in *Drosophila*. Proc Natl Acad Sci USA 100(5):2556–2561. https://doi.org/10.1073/pnas.0535280100

Granger NA, Mitchell LJ, Janzen WP, Bollenbacher WE (1984) Activation of *Manduca sexta* corpora allata in vitro by a cerebral neuropeptide. Mol Cell Endocrinol 37:349–358. https://doi.org/10.1016/0303-7207(84)90105-9

Grienberger C, Konnerth A (2012) Imaging calcium in neurons. Neuron 73(5):862–885. https://doi.org/10.1016/j.neuron.2012.02.011

Griffiths WJ, Wang Y (2009) Mass spectrometry: from proteomics to metabolomics and lipidomics. Chem Soc Rev 38(7):1882–1896. https://doi.org/10.1039/b618553n

Gu S-H, Chen C-H, Hsieh Y-C, Lin P-L, Young S-C (2015) Modulatory effects of bombyxin on ecdysteroidogenesis in *Bombyx mori* prothoracic glands. J Insect Physiol 72:61–69. https://doi.org/10.1016/j.jinsphys.2014.11.007

Gu S-H, Chang C-H, Lin P-L (2022) Bombyxin-stimulated ecdysteroidogenesis in relation to sugar transporter/trehalase expressions in *Bombyx* prothoracic glands. Insect Biochem Mol Biol 151: 103864. https://doi.org/10.1016/j.ibmb.2022.103864

Hasegawa K (1957) The diapause hormone of the silkworm, *Bombyx mori*. Nature 179:1300–1301. https://doi.org/10.1038/1791300b0

Hauser F, Williamson M, Cazzamali G, Grimmelikhuijzen CJP (2006) Identifying neuropeptide and protein hormone receptors in *Drosophila melanogaster* by exploiting genomic data. Brief Funct Genomic Proteomic 4(4):321–330. https://doi.org/10.1093/bfgp/eli003

Hayakawa Y (1990) Juvenile hormone esterase activity repressive factor in the plasma of parasitized insect larvae. J Biol Chem 265:10813–10816. https://doi.org/10.1016/S0021-9258(19)38517-5

Heredia F, Volonté Y, Pereirinha J, Fernandez-Acosta M, Casimiro AP, Belém CG et al (2021) The steroid-hormone ecdysone coordinates parallel pupariation neuromotor and morphogenetic subprograms via epidermis-to-neuron Dilp8-Lgr3 signal induction. Nat Commun 12(1):3328. https://doi.org/10.1038/s41467-021-23218-5

Herzik MA Jr, Wu M, Lander GC (2019) High-resolution structure determination of sub-100 kDa complexes using conventional cryo-EM. Nat Commun 10(1):1032. https://doi.org/10.1038/s41467-019-08991-8

Hewes RS, Taghert PH (2001) Neuropeptides and neuropeptide receptors in the *Drosophila melanogaster* genome. Genome Res 11(6):1126–1142. https://doi.org/10.1101/gr.169901

Holman GM, Cook BJ, Nachman RJ (1986a) Isolation, primary structure, and synthesis of two neuropeptides from *Leucophaea maderae*: members of a new family of cephalomyotropins. Comp Biochem Physiol 84C:205–211. https://doi.org/10.1016/0742-8413(86)90084-8

Holman GM, Cook BJ, Nachman RJ (1986b) Primary structure and synthesis of a blocked myotropic neuropeptide isolated from the cockroach, *Leucophaea maderae*. Comp Biochem Physiol 85C:219–224. https://doi.org/10.1016/0742-8413(86)90077-0

Holman GM, Cook BJ, Nachman RJ (1986c) Isolation, primary structure and synthesis of leucomyosuppressin, an insect neuropeptide that inhibits spontaneous contractions of the cockroach hindgut. Comp Biochem Physiol 85C:329–333. https://doi.org/10.1016/0742-8413(86)90202-1

Holt RA, Subramanian GM, Halpern A, Sutton GG, Charlab R, Nusskern DR et al (2002) The genome sequence of the malaria mosquito *Anopheles gambiae*. Science 298(5591):129–149. https://doi.org/10.1126/science.1076181

Honeybee Genome Sequencing Consortium (2006) Insights into social insects from the genome of the honeybee *Apis mellifera*. Nature 443(7114):931–949. https://doi.org/10.1038/nature05260

Hotaling S, Sproul JS, Heckenhauer J, Powell A, Larracuente AM, Pauls SU et al (2021) Long reads are revolutionizing 20 years of insect genome sequencing. Genome Biol Evol 13(8):evab138. https://doi.org/10.1093/gbe/evab138

Hua YJ, Tanaka Y, Nakamura K, Sakakibara M, Nagata S, Kataoka H (1999) Identification of a prothoracicostatic peptide in the larval brain of the silkworm, *Bombyx mori*. J Biol Chem 274: 31169–31173. https://doi.org/10.1074/jbc.274.44.31169

Huber D, Voith von Voithenberg L, Kaigala GV (2018) Fluorescence *in situ* hybridization (FISH): history, limitations and what to expect from micro-scale FISH? Micro Nano Eng 1:15–24. https://doi.org/10.1016/j.mne.2018.10.006

Hückesfeld S, Schlegel P, Miroschnikow A, Schoofs A, Zinke I, Haubrich AN et al (2021) Unveiling the sensory and interneuronal pathways of the neuroendocrine connectome in *Drosophila*. eLife 10:e65745. https://doi.org/10.7554/eLife.65745

Hull JJ, Copley KS, Schegg KM, Quilici DR, Schooley DA, Welch WH (2009) *De novo* molecular modeling and biophysical characterization of *Manduca sexta* eclosion hormone. Biochemistry 48(38):9047–9060. https://doi.org/10.1021/bi901078y

Ichikawa M, Isihzaki H (1961) Brain hormone of the silkworm, *Bombyx mori*. Nature 191:933–934. https://doi.org/10.1038/191933b0
Ichikawa M, Isihzaki H (1963) Protein nature of the insect brain hormone. Nature 198:308–309. https://doi.org/10.1038/198308a0
Imai K, Konno T, Nakazawa Y, Koiya T, Isobe M, Koga K et al (1991) Isolation and structure of diapause hormone of the silkworm, *Bombyx mori*. Proc Japan Acad 67(Ser B):98–101. https://doi.org/10.2183/pjab.67.98
Imamura M, Nakai J, Inoue S, Quan GX, Kanda T, Tamura T (2003) Targeted gene expression using the GAL4/UAS system in the silkworm *Bombyx mori*. Genet 165(3):1329–1340. https://doi.org/10.1093/genetics/165.3.1329
International Aphid Genomics Consortium (2010) Genome sequence of the pea aphid *Acyrthosiphon pisum*. PLoS Biol 8(2):e1000313. https://doi.org/10.1371/journal.pbio.1000313
International Silkworm Genome Consortium (2008) The genome of a lepidopteran model insect, the silkworm *Bombyx mori*. Insect Biochem Mol Biol 38(12):1036–1045. https://doi.org/10.1016/j.ibmb.2008.11.004
Ishibashi J, Kataoka H, Isogai A, Kawakami A, Saegusa H, Yagi Y et al (1992) Assignment of disulfide bond location in prothoracicotropic hormone of the silkworm, *Bombyx mori*: a homodimeric peptide. Biochemist 33:5912–5919. https://doi.org/10.1021/bi00185a031
Ishizaki H, Ichikawa M (1967) Purification of the brain hormone of the silkworm *Bombyx mori*. Bio Bull 133:355–368. https://doi.org/10.2307/1539831
Ishizaki H, Suzuki A, Nagasawa H, Tamura S (1977) Enzymatic and chemical inactivation of partially purified prothoracicotropic hormone of the silkworm, *Bombyx mori*. J Insect Physiol 23:1219–1222. https://doi.org/10.1016/0022-1910(77)90062-2
Ishizaki H, Mizoguchi A, Fujishita M, Suzuki A, Moriya I, O'oka H et al (1983) Species specificity of the insect prothoracicotropic hormone (PTTH): the presence of *Bombyx*- and *Samia*-specific PTTHs in the brain of Bombyx mori. Dev Growth Differ 25:593–600. https://doi.org/10.1111/j.1440-169X.1983.00593.x
Jackson GE, Pavadai E, Gäde G, Andersen NH (2019) The adipokinetic hormones and their cognate receptor from the desert locust, *Schistocerca gregaria*: solution structure of endogenous peptides and models of their binding to the receptor. PeerJ 7:e7514. https://doi.org/10.7717/peerj.7514
Jovic D, Liang X, Zeng H, Lin L, Xu F, Luo Y (2022) Single-cell RNA sequencing technologies and applications: a brief overview. Clin Transl Med 12(3):e694. https://doi.org/10.1002/ctm2.694
Juarez-Carreño S, Vallejo DM, Carranza-Valencia J, Palomino-Schätzlein M, Ramon-Cañellas P, Santoro R et al (2021) Body-fat sensor triggers ribosome maturation in the steroidogenic gland to initiate sexual maturation in *Drosophila*. Cell Rep 37(2):109830. https://doi.org/10.1016/j.celrep.2021.109830
Kataoka H, Nagasawa H, Isogai A, Tamura S, Mizoguchi A, Fujiwara Y et al (1987) Isolation and partial characterization of a prothoracicotropic hormone of the silkworm, *Bombyx mori*. Agric Biol Chem 51:1067–1076. https://doi.org/10.1080/00021369.1987.10868149
Kataoka H, Toschi A, Li JP, Carney RL, Schooley DA, Kramer SJ (1989a) Identification of allatotropin from adult *Manduca sexta*. Science 243:1481–1483. https://doi.org/10.1126/science.243.4897.1481
Kataoka H, Troetschler RG, Li JP, Kramer SJ, Carney RL, Schooley DA (1989b) Isolation and identification of a diuretic hormone from the tobacco hornworm, *Manduca sexta*. Proc Natl Acad Sci USA 86:2976–2980. https://doi.org/10.1073/pnas.86.8.2976
Kataoka H, Nagasawa H, Isogai A, Ishizaki H, Suzuki A (1991) Prothoracicotropic hormone of the silkworm, *Bombyx mori*: amino acid sequence and dimeric structure. Agric Biol Chem 55:73–86. https://doi.org/10.1271/bbb1961.55.73
Kawakami A, Kataoka H, Oka T, Mizoguchi A, Kimura-Kawakami M, Adachi T et al (1990) Molecular cloning of the *Bombyx mori* prothoracicotropic hormone. Science 247:1333–1335. https://doi.org/10.1126/science.2315701

Khalil AM (2020) The genome editing revolution: review. J Genet Eng Biotechnol 18(1):68. https://doi.org/10.1186/s43141-020-00078-y

Kim YJ, Zitnan D, Galizia CG, Cho KH, Adams ME (2006) A command chemical triggers an innate behavior by sequential activation of multiple peptidergic ensembles. Curr Biol 16(14): 1395–1407. https://doi.org/10.1016/j.cub.2006.06.027

Klar TA, Jakobs S, Dyba M, Egner A, Hell SW (2000) Fluorescence microscopy with diffraction resolution barrier broken by stimulated emission. Proc Natl Acad Sci USA 97(15):8206–8210. https://doi.org/10.1073/pnas.97.15.8206

Kobayashi M (1955) Relationship between the brain hormone and the imaginal differentiation of silkworm, *Bombyx mori*. J Sericul Sci Japan 24:389–392. https://doi.org/10.11416/kontyushigen1930.24.389

Kobayashi M, Kirimura J (1958) The 'brain hormone' in the silkworm, *Bombyx mori* L. Nature 181: 1217. https://doi.org/10.1038/1811217a0

Kokoza VA, Raikhel AS (2011) Targeted gene expression in the transgenic *Aedes aegypti* using the binary Gal4-UAS system. Insect Biochem Mol Biol 41(8):637–644. https://doi.org/10.1016/j.ibmb.2011.04.004

Kopeć S (1917) Experiments on metamorphosis of insects. Bull Int Acad Sci Cracovie (B):57–60

Kopeć S (1922) Studies on the necessity of the brain for the inception of insect metamorphosis. Bio Bull 42:323–342. https://doi.org/10.2307/1536759

Kramer SJ, Toschi A, Miller CA, Kataoka H, Quistad GB, Li JP et al (1991) Identification of an allatostatin from the tobacco hornworm *Manduca sexta*. Proc Natl Acad Sci USA 88:9458–9462. https://doi.org/10.1073/pnas.88.21.9458

Lange E, Tautenhahn R, Neumann S, Gröpl C (2008) Critical assessment of alignment procedures for LC-MS proteomics and metabolomics measurements. BMC Bioinf 9(375). https://doi.org/10.1186/1471-2105-9-375

Lea AO (1967) The medial neurosecretory cells and egg maturation in mosquitoes. J Insect Physiol 9:793–803. https://doi.org/10.1016/0022-1910(67)90082-0

Lee T, Luo L (1999) Mosaic analysis with a repressible cell marker for studies of gene function in neuronal morphogenesis. Neuron 22(3):451–461. https://doi.org/10.1016/s0896-6273(00)80701-1

Lehmann R, Tautz D (1994) *In situ* hybridization to RNA. Methods Cell Biol 44:575–598. https://doi.org/10.1016/s0091-679x(08)60933-4

Li H, Wang H, Schegg KM, Schooley DA (1997) Metabolism of an insect diuretic hormone by Malpighian tubules studied by liquid chromatography coupled with electrospray ionization mass spectrometry. Proc Natl Acad Sci USA 94(25):13463–13468. https://doi.org/10.1073/pnas.94.25.13463

Li B, Predel R, Neupert S, Hauser F, Tanaka Y, Cazzamali G et al (2008) Genomics, transcriptomics, and peptidomics of neuropeptides and protein hormones in the red flour beetle *Tribolium castaneum*. Genome Res 18(1):113–122. https://doi.org/10.1101/gr.6714008

Liao X, Li M, Zou Y, Wu FX, Pan Y, Wang J (2019) Current challenges and solutions of *de novo* assembly. qBio 7(2):90–109. https://doi.org/10.1007/s40484-019-0166-9

Lichtman JW, Conchello JA (2005) Fluorescence microscopy. Nat Methods 2(12):910–919. https://doi.org/10.1038/nmeth817

Lin MZ, Schnitzer MJ (2016) Genetically encoded indicators of neuronal activity. Nat Neurosci 19(9):1142–1153. https://doi.org/10.1038/nn.4359

Liu F, Baggerman G, Schoofs L, Wets G (2006) Uncovering conserved patterns in bioactive peptides in Metazoa. Peptides 27(12):3137–3153. https://doi.org/10.1016/j.peptides.2006.08.021

Lorenz MW, Kellner R, Hoffmann KH (1995) A family of neuropeptides that inhibit juvenile hormone biosynthesis in the cricket, *Gryllus bimaculatus*. J Biol Chem 270:21103–21108. https://doi.org/10.1074/jbc.270.36.21103

Lowe R, Shirley N, Bleackley M, Dolan S, Shafee T (2017) Transcriptomics technologies. PLoS Comput Biol 13(5):e1005457. https://doi.org/10.1371/journal.pcbi.1005457

Luo CW, Dewey EM, Sudo S, Ewer J, Hsu SY, Honegger HW et al (2005) Bursicon, the insect cuticle-hardening hormone, is a heterodimeric cystine knot protein that activates G protein-coupled receptor LGR2. Proc Natl Acad Sci USA 102:2820–2825. https://doi.org/10.1073/pnas.0409916102

Lynd A, Lycett GJ (2012) Development of the bi-partite Gal4-UAS system in the African malaria mosquito, *Anopheles gambiae*. PLoS One 7(2):e31552. https://doi.org/10.1371/journal.pone.0031552

Maddrell SHP (1962) A diuretic hormone in *Rhodnius prolixus* Stål. Nature 194:605–606. https://doi.org/10.1038/194605b0

Matsumoto S, Isogai A, Suzuki A (1985) N-terminal amino acid sequence of an insect neurohormone, melanization and reddish coloration hormone (MRCH): heterogeneity and sequence homology with human insulin-like growth factor II. FEBS Lett 189:115–118. https://doi.org/10.1016/0014-5793(85)80853-X

Mayer RJ, Candy RJ (1969) Control of haemolymph lipid concentration during locust flight: an adipokinetic hormone from the corpora cardiaca. J Insect Physiol 15:611–620

McBrayer Z, Ono H, Shimell M, Parvy JP, Beckstead RB, Warren JT et al (2007) Prothoracicotropic hormone regulates developmental timing and body size in *Drosophila*. Dev Cell 13(6):857–871. https://doi.org/10.1016/j.devcel.2007.11.003

McGettigan PA (2013) Transcriptomics in the RNA-seq era. Curr Opin Chem Biol 17(1):4–11. https://doi.org/10.1016/j.cbpa.2012.12.008

Mehlis B, Kertscher U (1997) Liquid chromatography/mass spectrometry of peptides of biological samples. Anal Chim Acta 352(1–3):71–83. https://doi.org/10.1016/S0003-2670(97)00169-4

Mirth CK, Riddiford LM (2007) Size assessment and growth control: how adult size is determined in insects. BioEssays 29:344–355. https://doi.org/10.1002/bies.20552

Mita K, Kasahara M, Sasaki S, Nagayasu Y, Yamada T, Kanamori H et al (2004) The genome sequence of silkworm, *Bombyx mori*. DNA Res 11(1):27–35. https://doi.org/10.1093/dnares/11.1.27

Mugumbate G, Jackson GE, van der Spoel D (2011) Open conformation of adipokinetic hormone receptor from the malaria mosquito facilitates hormone binding. Peptides 32(3):553–559. https://doi.org/10.1016/j.peptides.2010.08.017

Myers EW, Sutton GG, Delcher AL, Dew IM, Fasulo DP, Flanigan MJ et al (2000) A whole-genome assembly of *Drosophila*. Science 287:2196–2204. https://doi.org/10.1126/science.287.5461.2196

Nagasawa H, Isogai A, Suzuki A, Tamura S, Ishizaki H (1979) Purification and properties of the prothoracicotropic hormone of the silkworm, *Bombyx mori*. Develop Growth Differ 21:29–38. https://doi.org/10.1111/j.1440-169X.1979.00029.x

Nagasawa H, Kataoka H, Hori Y, Isogai A, Tamura S, Suzuki A et al (1984a) Isolation and some characterization of the prothoracicotropic hormone from *Bombyx mori*. Gen Comp Endocrinol 53:143–152. https://doi.org/10.1016/0016-6480(84)90234-X

Nagasawa H, Kataoka H, Isogai A, Tamra S, Suzuki A, Ishizaki H et al (1984b) Amino-terminal amino acid sequence of the silkworm prothoracicotropic hormone: homology with insulin. Science 266:1344–1345. https://doi.org/10.1126/science.226.4680.1344

Nagasawa H, Kamito T, Takahashi S, Isogai A, Fugo H, Suzuki A (1985) Eclosion hormone of the silkworm *Bombyx mori*: purification and determination of the N-terminal amino acid sequence. Insect Biochem 15:573–578. https://doi.org/10.1016/0020-1790(85)90117-9

Nagasawa H, Kataoka H, Isogai A, Tamura S, Suzuki A, Mizoguchi A et al (1986) Amino acid sequence of a prothoracicotropic hormone of the silkworm *Bombyx mori*. Proc Natl Acad Sci USA 83:5840–5843. https://doi.org/10.1073/pnas.83.16.5840

Nagasawa H, Maruyama K, Sato B, Hietter H, Kataoka H, Isogai A et al (1988) Structure and synthesis of bombyxin from the silkworm *Bombyx mori*. In: Shiba T, Sakakibara S (eds) Peptide Chemistry 1987, Protein Res Found, Osaka. 123–126

Nässel DR (2002) Neuropeptides in the nervous system of *Drosophila* and other insects: multiple roles as neuromodulators and neurohormones. Prog Neurobiol 68(1):1–84. https://doi.org/10.1016/s0301-0082(02)00057-6

Nässel DR (2018) Substrates for neuronal cotransmission with neuropeptides and small molecule neurotransmitters in *Drosophila*. Front Cell Neurosci 12:83. https://doi.org/10.3389/fncel.2018.00083

Nelson NJ (2001) Microarrays have arrived: gene expression tool matures. J Natl Cancer Inst 93(7):492–494. https://doi.org/10.1093/jnci/93.7.492

Noguchi T, Adachi-Yamada T, Katagiri T, Kawakami A, Iwami M, Ishibashi J et al (1995) Insect prothoracicotropic hormone: a new member of the vertebrate growth factor superfamily. FEBS Lett 376:251–256. https://doi.org/10.1016/0014-5793(95)01296-8

Norman TC (1975) Neurosecretory cells in insect brain and production of hypoglycaemic hormone. Nature 254:259–261. https://doi.org/10.1038/254259a0

Ogura N (1975) Hormonal control of larval coloration in the armyworm, *Leucania separata*. J Insect Physiol 21:559–576. https://doi.org/10.1016/0022-1910(75)90160-2

Okamoto N, Yamanaka N, Satake H, Saegusa H, Kataoka H, Mizoguchi A (2009a) An ecdysteroid-inducible insulin-like growth factor-like peptide regulates adult development of the silkmoth *Bombyx mori*. FEBS J 276(5):1221–1232. https://doi.org/10.1111/j.1742-4658.2008.06859.x

Okamoto N, Yamanaka N, Yagi Y, Nishida Y, Kataoka H, O'Connor MB et al (2009b) A fat body-derived IGF-like peptide regulates postfeeding growth in *Drosophila*. Dev Cell 17(6):885–891. https://doi.org/10.1016/j.devcel.2009.10.008

Palli SR (2023) RNAi turns 25: contributions and challenges in insect science. Front Insect Sci 3:1209478. https://doi.org/10.3389/finsc.2023.1209478

Park Y, Kim YJ, Adams ME (2002) Identification of G protein-coupled receptors for *Drosophila* PRXamide peptides, CCAP, corazonin, and AKH supports a theory of ligand-receptor coevolution. Proc Natl Acad Sci USA 99(17):11423–11428. https://doi.org/10.1073/pnas.162276199

Park D, Li P, Dani A, Taghert PH (2014) Peptidergic cell-specific synaptotagmins in *Drosophila*: localization to dense-core granules and regulation by the bHLH protein DIMMED. J Neurosci 34(39):13195–13207. https://doi.org/10.1523/JNEUROSCI.2075-14.2014

Park JH, Chen J, Jang S, Ahn TJ, Kang K, Choi MS et al (2016) A subset of enteroendocrine cells is activated by amino acids in the *Drosophila* midgut. FEBS Lett 590(4):493–500. https://doi.org/10.1002/1873-3468.12073

Pitt JJ (2009) Principles and applications of liquid chromatography-mass spectrometry in clinical biochemistry. Clin Biochem Rev 30(1):19–34

Predel R (2001) Peptidergic neurohemal system of an insect: mass spectrometric morphology. J Comp Neurol 436(3):363–375. https://doi.org/10.1002/cne.1073

Predel R, Wegener C, Russell WK, Tichy SE, Russell DH, Nachman RJ (2004) Peptidomics of CNS-associated neurohemal systems of adult *Drosophila melanogaster*: a mass spectrometric survey of peptides from individual flies. J Comp Neurol 474(3):379–392. https://doi.org/10.1002/cne.20145

Raina AK, Klum JA (1984) Brain factor control of sex pheromone production in the female corn earworm moth. Science 225:531–533. https://doi.org/10.1126/science.225.4661.531

Raina AK, Jaffe H, Kempe TG, Keim P, Blacher RW, Fales HM et al (1989) Identification of a neuropeptide hormone that regulate sex pheromone production in female moths. Science 244:796–798. https://doi.org/10.1126/science.244.4906.796

Reid W, O'Brochta DA (2016) Applications of genome editing in insects. Curr Opin Insect Sci 13:43–54. https://doi.org/10.1016/j.cois.2015.11.001

Riehle MA, Garczynski SF, Crim JW, Hill CA, Brown MR (2002) Neuropeptides and peptide hormones in *Anopheles gambiae*. Science 298(5591):172–175. https://doi.org/10.1126/science.1076827

Robertson G, Schein J, Chiu R, Corbett R, Field M, Jackman SD et al (2010) *De novo* assembly and analysis of RNA-seq data. Nat Methods 7(11):909–912. https://doi.org/10.1038/nmeth.1517

Roller L, Beke G, Stano M, Daubnerova I, Mizoguchi A, Satake H et al (2017) BomBase – web atlas of the neuroendocrine system in *Bombyx mori*. Online database resource. http://bombase.org/

Roy SG, Hansen IA, Raikhel AS (2007) Effect of insulin and 20-hydroxyecdysone in the fat body of the yellow fever mosquito, *Aedes aegypti*. Insect Biochem Mol Biol 37(12):1317–1326. https://doi.org/10.1016/j.ibmb.2007.08.004

Sanderson MJ, Smith I, Parker I, Bootman MD (2014) Fluorescence microscopy. Cold Spring Harb Protoc 2014(10):pdb.top071795. https://doi.org/10.1101/pdb.top071795

Scarborough RM, Jamieson GC, Kalish F, Kramer SJ, Mcenroe GA, Miller CA et al (1984) Isolation and primary structure of two peptides with cardioacceleratory and hyperglycemic activity from the corpora cardiaca of *Periplaneta americana*. Proc Natl Acad Sci USA 81:5575–5579. https://doi.org/10.1073/pnas.81.17.5575

Scharrer B (1952) Neurosecretion XI. The effects of nerve section on the intercerebralis-cardiacum-allatum system of the insect *Leucophaea maderae*. Biol Bull 102:261–272. https://doi.org/10.2307/1538375

Scharrer E, Scharrer B (1939) Secretory cells within the hypothalamus. In: the hypothalamus, Assoc. Res. Nerv. Ment. Dis. V. 20. Hafner Publ. Co., New York, pp 170–194

Shirai Y, Piulachs MD, Belles X, Daimon T (2022) DIPA-CRISPR is a simple and accessible method for insect gene editing. Cell Rep Methods 2(5):100215. https://doi.org/10.1016/j.crmeth.2022.100215

Smith W, Rybczynski R (2012) Prothoracicotropic hormone. In: Gilbert LI (ed) Insect endocrinology. Academic, London. https://doi.org/10.1016/B978-0-12-384749-2.10001-9

Spradling AC, Stern D, Beaton A, Rhem EJ, Laverty T, Mozden N et al (1999) The Berkeley *Drosophila* genome project gene disruption project: single P-element insertions mutating 25% of vital *Drosophila* genes. Genetics 153(1):135–177. https://doi.org/10.1093/genetics/153.1.135

St Johnston D (2002) The art and design of genetic screens: *Drosophila melanogaster*. Nat Rev Genet 3(3):176–188. https://doi.org/10.1038/nrg751

Starrat AN, Brown BE (1975) Structure of the pentapeptide proctolin, a proposed neurotransmitter in insects. Life Sci 17:1253–1256. https://doi.org/10.1016/0024-3205(75)90134-4

Steele JE (1961) Occurrence of a hyperglycaemic factor in the corpus cardiacum of an insect. Nature 192:680–681. https://doi.org/10.1038/192680a0

Stone JV, Mordue W, Batley KE, Morris HR (1976) Structure of locust adipokinetic hormone, a neurohormone that regulates lipid utilization during flight. Nature 263:207–211. https://doi.org/10.1038/263207a0

Suzawa M, Muhammad NM, Joseph BS, Bland ML (2019) The Toll signaling pathway targets the insulin-like peptide Dilp6 to inhibit growth in *Drosophila*. Cell Rep 28(6):1439–1446.e5. https://doi.org/10.1016/j.celrep.2019.07.015

Suzuki A, Isogai A, Horii T, Ishizaki H, Tamura S (1975) A simple procedure for partial purification of silkworm brain hormone. Agric Bio Chem 39:2157–2162. https://doi.org/10.1080/00021369.1975.10861921

Tanaka K, Waki H, Ido Y, Akita S, Yoshida Y, Yoshida T et al (1988) Protein and polymer analyses up to m/z 100 000 by laser ionization time-of-flight mass spectrometry. Rapid Commun Mass Spectrom 2(8):151–153. https://doi.org/10.1002/rcm.1290020802

Tomoyasu Y, Miller SC, Tomita S, Schoppmeier M, Grossmann D, Bucher G (2008) Exploring systemic RNA interference in insects: a genome-wide survey for RNAi genes in *Tribolium*. Genome Biol 9(1):R10. https://doi.org/10.1186/gb-2008-9-1-r10

Tribolium Genome Sequencing Consortium (2008) The genome of the model beetle and pest *Tribolium castaneum*. Nature 452(7190) 949–955. https://doi.org/10.1038/nature06784

Truman JW, Riddiford LM (1970) Neuroendocrine control of ecdysis in silkmoths. Science 167:1624–1626. https://doi.org/10.1126/science.167.3925.1624

Tsuchiya R, Kaneshima A, Kobayashi M, Yamazaki M, Takasu Y, Sezutsu H et al (2021) Maternal GABAergic and GnRH/corazonin pathway modulates egg diapause phenotype of the silkworm

Bombyx mori. Proc Natl Acad Sci USA 118(1):e2020028118. https://doi.org/10.1073/pnas.2020028118
Urban KW, Barthel J, Houben L, Jia CL, Jin L, Lentzen M et al (2023) Progress in atomic-resolution aberration corrected conventional transmission electron microscopy (CTEM). Prog Mater Sci 133:101037. https://doi.org/10.1016/j.pmatsci.2022.101037
Valasek MA, Repa JJ (2005) The power of real-time PCR. Adv Physio Educ 29(3):151–159. https://doi.org/10.1152/advan.00019.2005
Vanden Broeck J (2001) Neuropeptides and their precursors in the fruitfly, *Drosophila melanogaster*. Peptides 22(2):241–254. https://doi.org/10.1016/s0196-9781(00)00376-4
Veenstra JA (1989) Isolation and structure of corazonin, a cardioactive peptide from the American cockroach. FEBS Lett 250:231–234. https://doi.org/10.1016/0014-5793(89)80727-6
Watson P, Jones AT, Stephens DJ (2005) Intracellular trafficking pathways and drug delivery: fluorescence imaging of living and fixed cells. Adv Drug Deliv Rev 57(1):43–61. https://doi.org/10.1016/j.addr.2004.05.003
Wegener C, Gorbashov A (2008) Molecular evolution of neuropeptides in the genus *Drosophila*. Genome Biol 9:R131. https://doi.org/10.1186/gb-2008-9-8-r131
Wegener C, Reinl T, Jänsch L, Predel R (2006) Direct mass spectrometric peptide profiling and fragmentation of larval peptide hormone release sites in *Drosophila melanogaster* reveals tagma-specific peptide expression and differential processing. J Neurochem 96(5):1362–1374. https://doi.org/10.1111/j.1471-4159.2005.03634.x
Wigglesworth VB (1940) The determination of characters at metamorphosis in *Rhodnius prolixus* (Hemiptera). J Exp Biol 17:201–223. https://doi.org/10.1242/jeb.17.2.201
Williams CM (1946) Physiology of insect diapause: the role of the brain in the production and termination of pupal dormancy in the giant silkworm, *Platysamia cecropia*. Biol Bull 90:234–243. https://doi.org/10.2307/1538121
Williams CM (1952) Physiology of insect diapause. IV. The brain and prothoracic glands as an endocrine system in the cecropia silkworm. Biol Bull 103:120–138. https://doi.org/10.2307/1538411
Williams EA (2020) Function and distribution of the Wamide neuropeptide superfamily in metazoans. Front Endocrinol 11:344. https://doi.org/10.3389/fendo.2020.00344
Winding M, Pedigo BD, Barnes CL, Patsolic HG, Park Y, Kazimiers T et al (2023) The connectome of an insect brain. Science 379(6636):eadd9330. https://doi.org/10.1126/science.add9330
Woodhead AP, Stay B, Seidel SL, Khan MA, Tobe SS (1989) Primary structure of four allatostatins: neuropeptide inhibitors of juvenile hormone synthesis. Proc Natl Acad Sci USA 86:5997–6001. https://doi.org/10.1073/pnas.86.15.5997
Xia Q, Zhou Z, Lu C, Cheng D, Dai F, Li B et al (2004) A draft sequence for the genome of the domesticated silkworm (*Bombyx mori*). Science 306(5703):1937–1940. https://doi.org/10.1126/science.1102210
Xie HB, Golic KG (2004) Gene deletions by ends-in targeting in *Drosophila melanogaster*. Genetics 168(3):1477–1489. https://doi.org/10.1534/genetics.104.030882
Xie J, Sang M, Xiaowen, Song, Zhang S, Kim D, Veenstra JA et al (2020) A new neuropeptide insect parathyroid hormone iPTH in the red flour beetle *Tribolium castaneum*. PLoS Genet 16(5):e1008772. https://doi.org/10.1371/journal.pgen.1008772
Xue Z, Wu M, Wen K, Ren M, Long L, Zhang X et al (2014) CRISPR/Cas9 mediates efficient conditional mutagenesis in *Drosophila*. G3 Genes Genomes Genetics 4(11):2167–2173. https://doi.org/10.1534/g3.114.014159
Yew JY, Wang Y, Barteneva N, Dikler S, Kutz-Naber KK, Li L et al (2009) Analysis of neuropeptide expression and localization in adult *Drosophila melanogaster* central nervous system by affinity cell-capture mass spectrometry. J Proteome Res 8(3):1271–1284. https://doi.org/10.1021/pr800601x
Yin C, Morita T, Parrish JZ (2024) A cell atlas of the larval Aedes aegypti ventral nerve cord. Neural Dev 19:2. https://doi.org/10.1186/s13064-023-00178-8

Yoshinari Y, Kosakamoto H, Kamiyama T, Hoshino R, Matsuoka R, Kondo S et al (2021) The sugar-responsive enteroendocrine neuropeptide F regulates lipid metabolism through glucagon-like and insulin-like hormones in *Drosophila melanogaster*. Nat Commun 12(1):4818. https://doi.org/10.1038/s41467-021-25146-w
Zeidler MP, Tan C, Bellaiche Y, Cherry S, Häder S, Gayko U et al (2004) Temperature-sensitive control of protein activity by conditionally splicing inteins. Nat Biotechnol 22(7):871–876. https://doi.org/10.1038/nbt979
Zhang T, Wen T, Yue Y, Yan Q, Du E, Fan S et al (2021) Egg tanning improves the efficiency of CRISPR/Cas9-mediated mutant locust production by enhancing defense ability after microinjection. J Integr Agric 20(10):2716–2726. https://doi.org/10.1016/S2095-3119(21)63736-X
Zhu KY, Palli SR (2020) Mechanisms, applications, and challenges of insect RNA interference. Annu Rev Entomol 65:293–311. https://doi.org/10.1146/annurev-ento-011019-025224
Zhu C, Huang H, Hua R, Li G, Yang D, Luo J et al (2009) Molecular and functional characterization of adipokinetic hormone receptor and its peptide ligands in *Bombyx mori*. FEBS Lett 583(9): 1463–1468. https://doi.org/10.1016/j.febslet.2009.03.060
Žitňan D, Kingan TG, Hermesman JL, Adams ME (1996) Identification of ecdysis-triggering hormone from an epitracheal endocrine system. Science 271:88–91. https://doi.org/10.1126/science.271.5245.88
Zubrzycki IZ, Gäde G (1994) Conformational study on an insect neuropeptide of the AKH/RPCH-family by combined 1H NMR spectroscopy and molecular mechanics. Biochem Biophys Res Commun 198(1):228–235. https://doi.org/10.1006/bbrc.1994.1032

Chapter 2
Diversity of Insect Peptide Hormones/Neuropeptides

Akira Mizoguchi

Abstract A large variety of peptide hormones and neuropeptides (PHs/NPs) have so far been discovered, purified, and characterized in many kinds of insects. In addition, numerous genes and transcripts encoding novel secretory signaling peptides have been found in insect genomes. Some of them are homologs of vertebrate PHs/NPs, and some of others are unique to insects. This chapter summarizes the structural features and known actions of each PH/NP, with special attention to their phylogenetic relationships with vertebrate and non-insect invertebrate PHs/NPs and their evolutionary origins.

Keywords Peptide hormones · Neuropeptides · Receptor · Signaling system · Evolutionary origin · Metazoa · Bilateria · Protostomes · Arthropod · Insect

2.1 Introduction

As seen in the previous chapter, more than 30 peptide hormones (PHs) or neuropeptides (NPs) have been discovered to date based on their specific biological activities. In addition, recent genome/transcriptome sequence analyses in various insect species have identified many other PHs/NPs that are homologous to those found in non-insect invertebrates as well as vertebrates. Although comparative genome/transcriptome analyses disclosed the presence of many PH/NP homologs across metazoans, it is generally not easy to determine the relatedness of the peptides of interest only from their sequences, especially in the case of small peptides, because peptide sequences vary considerably between species. Therefore, the sequences of the precursor proteins for PHs/NPs are often used to analyze the evolutionary relationship of two or more peptides. On the other hand, many receptors for insect PHs/NPs have also been identified since G protein-coupled receptors (GPCRs) encoded by the *Drosophila* genome were comprehensively

A. Mizoguchi (✉)
Nagoya, Aichi, Japan
e-mail: mizoamizo@icloud.com

A. Mizoguchi, N. Okamoto (eds.), *Insect Peptide Hormones and Neuropeptides*, Entomology Monographs, https://doi.org/10.1007/978-981-95-7510-7_2

“deorphanized” in 2001 (Hewes and Taghert 2001). The term “deorphanize” refers to determining the ligand for an “orphan” receptor, that is, a protein that is predicted to be a receptor, but whose ligand has not yet been identified. Most of the receptors for PHs/NPs are GPCRs, and the structure of their ligand-binding domain determines the specificity of their binding to PHs/NPs. Interestingly, it was found that, in many cases, homologous ligands use homologous GPCRs as their receptors. In other words, the set of ligands and their receptors has been well-conserved throughout evolution, although there are exceptions. Therefore, if certain GPCRs in evolutionarily distant species are homologous to each other, we would expect that the ligands of the GPCRs would also be related to each other, even if the sequences of the ligands are not very similar. Thus, information about the receptors is useful to analyze the relationship of PHs/NPs in different species.

In 2013, based on the above ideas, two different research groups comprehensively investigated the relationships between known PHs/NPs in vertebrates and invertebrates using integrated bioinformatic approaches. Jékely investigated evolutionary relationships between PHs/NPs and PH/NP receptors (GPCRs) throughout Metazoa by using similarity-based clustering methods (Jékely 2013), while Mirabeau and Joly used hidden Markov model (HMM)-based programs and phylogenetic reconstructions for the same purpose (Mirabeau and Joly 2013). Both groups identified nearly 30 PH/NP signaling systems (ligand-receptor sets) shared by protostomes and deuterostomes (Elphick et al. 2018). Including these core signaling systems, known PHs/NPs of vertebrates and invertebrates have been grouped into more than 70 PH/NP families (Jékely 2013). Of those, nearly 40 families have been found in Arthropoda (Jékely 2013). In Bilateria, there are several other hormone families whose receptors are not GPCRs but other types of receptors, including receptor tyrosine kinases (RTKs).

Here, we summarize the animal classification and phylogeny to help readers understand the evolutionary origins of each PH/NP. The so called animals are all Metazoa. Metazoa is composed of Bilateria, Cnidaria, Placozoa, Ctenophora, and Porifera (Fig. 2.1). Metazoa excluding Porifera is sometimes called Eumetazoa. Bilateria contains two major clades, deuterostomes and protostomes. Deuterostomes consist of three phyla: phylum Chordata, which includes Vertebrata (such as humans), Urochordata (sea squirts), and Cephalochordata (amphioxi), phylum Hemichordata (acorn worms), and phylum Echinodermata (starfish, sea urchins). Hemichordata and Echinodermata are collectively called Ambulacraria. On the other hand, protostomes are composed of more than 20 phyla and grouped into 2 clades, Lophotrochozoa and Ecdysozoa. The former contains phylum Mollusca (octopuses, snails) and phylum Annelida (earthworms, leeches), while the latter contains phylum Arthropoda (crabs, moths) and phylum Nematoda (nematodes, roundworms). Among non-bilaterians, Cnidaria (jellyfish, sea anemones) and Ctenophora (comb jellies) have a nervous system, whereas Placozoa (flatworms) and Porifera (sponges) do not. The phylogenetic relationships between these phyla are summarized in Fig. 2.1, following Simion et al. (2017).

In this chapter, insect PHs/NPs are overviewed with special focus on their relationship with those of vertebrates and non-insect invertebrates. The existence of closely related hormones in different animal groups suggests that the origin of

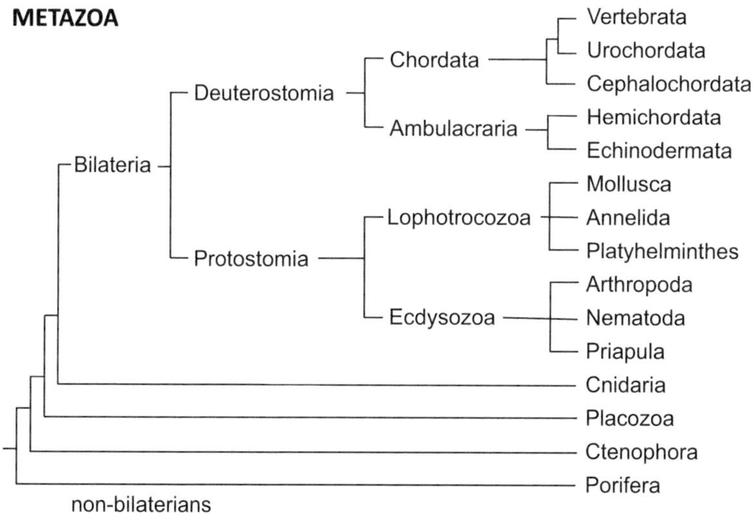

Fig. 2.1 Phylogenetic tree showing phylogenetic relationships of representative phyla of Metazoa. Note that the branch lengths in the tree are arbitrary

these hormones predates a divergence of those animal groups during evolution. For example, if homologous PHs exist in both insects and vertebrates, then we can speculate that a common ancestor peptide of those hormones had already been present before the divergence of protostomes and deuterostomes. On the other hand, hormones that exist in insects but not in other invertebrates and vertebrates are thought to have come to be used as hormones during insect evolution. In most cases, receptors for individual PHs/NPs are not mentioned. All receptors are GPCRs unless otherwise specified.

2.2 Hormone Families Common to Eumetazoans

Genome/transcriptome sequence analyses have disclosed that non-bilaterians (basal metazoans) except poriferans also have some bilaterian hormones including insulin-like peptide (ILP), glycoprotein hormone (GPH), and bursicon and/or their potential receptors. Therefore, the origin of these hormones is thought to be at an early stage in metazoan evolution, before the emergence of bilaterians.

The genes for ILPs and their receptors, a type of RTKs, have been found in some species of Cnidaria (Anctil 2009; Metchell et al. 2021) and Placozoa (Nikitin 2015). The GPH and bursicon are both members of cystine-knot family of proteins, exerting their effects by activating leucine-rich repeat-containing GPCRs (LGRs). Relaxin, a member of insulin superfamily peptides, also uses LGR as its receptor. The type of LGR for these three hormones differs: GPH, bursicon, and relaxin bind to type A, type B, and type C LGR, respectively. All of the poriferan, ctenophoran, placozoan, and cnidarian species examined have genes for LGRs, some of which can be classified as one of the three types of LGRs. Especially in the sea anemone (Cnidaria), all three types of LGRs were identified (Hauser et al. 2022; Roch and Sherwood 2014). On the other hand, genes for cystine-knot hormones with mixed features of GPH, bursicon, and BMP antagonists were identified in the sponge (Porifera) and comb jelly (Ctenophora) genomes (Roch and Sherwood 2014). In contrast, in Cnidaria, which diverged later in metazoan evolution, genes for peptides with mixed features of GPH and bursicon and for several BMP antagonists were identified. Thus, it is assumed that the respective ancestral molecules of cystine-knot hormone and LGR duplicated and diverged throughout the evolution of basal metazoans, resulting in the emergence of distinct signaling systems for GPH and bursicon in the common ancestor of bilaterians (Roch and Sherwood 2014).

As mentioned above, cnidarians have type C LGR, which is a receptor for relaxin. Hence, at least one of the ILPs of cnidarians may have utilized LGR rather than RTK as a receptor.

Insects, like other bilaterians, have the abovementioned three PHs. The characteristics of these hormones in insects and other animals are summarized below.

2.2.1 Insulin-Related Peptides

In vertebrates, insulin, insulin-like growth factors (IGFs), relaxins, and insulin-like peptides (INSLs) constitute an insulin superfamily of peptides. Insulin, relaxins, and INSLs have two-chain structures consisting of A- and B-chains, whereas IGFs are single-chain peptides. Insulin and relaxin are generated by processing of their precursors, where the signal peptide, B-chain, connecting peptide (C-peptide), and A-chain are arranged in this order. Insulin and IGF exert their effects through insulin receptor and IGF receptor, respectively. Both receptors are members of RTKs and resemble each other. In contrast, receptors for relaxins and INSLs are LGRs.

The first insulin family peptide identified in insects is bombyxin, which was purified from *Bombyx mori* heads as a hormone that stimulates the prothoracic glands (PGs) of the heterologous moth *Samia cynthia ricini* to secrete ecdysone (see Sect. 1.1.4). By purification, several molecular species of bombyxin were identified. Their amino acid (aa) sequences were similar to each other and to vertebrate insulin as well. Further structural analyses of bombyxin-II and bombyxin-IV revealed that these polypeptides consisted of A- and B-chains connected by two disulfide bridges, like insulin in vertebrates (see Sect. 1.1.4). Immunohistochemistry and in situ hybridization demonstrated that bombyxin is produced by four pairs of large medial neurosecretory cells (NSCs) of the brain and transferred to and released from the corpora allata (CA) in *B. mori* (Mizoguchi et al. 1987). Thus, in the early stages of research on insect ILPs, it was believed that all insect ILPs are produced by the brain and have a two-chain structure.

In 2009, however, a new type of ILP with a single-chain structure was purified from the hemolymph of *B. mori* pupae. This peptide is produced by the fat body and is present at very high concentrations in the pupal hemolymph. This peptide was called *Bombyx* IGF-like peptide (BIGFLP), because it resembled vertebrate IGFs in its structure, production tissue, and hemolymph titers (Okamoto et al. 2009). However, BIGFLP and bombyxin are considered to bind to the same insulin-like receptor (InR), because the *B. mori* genome has only one copy of the InR gene. Therefore, BIGFLP is not an ortholog of IGFs, but rather a counterpart of proinsulin (insulin precursor). In fact, the BIGFLP gene is also expressed in the medial NSCs in the brain. It is presumed that the BIGFLP secreted from the brain NSCs has a two-chain structure, because the C-domain of BIGFLP is flanked by dibasic cleavage sites at both ends.

Furthermore, an extensive search for genes encoding bombyxin peptides revealed that there are more than 30 bombyxin genes in the *B. mori* genome and some of them are expressed in various tissues, including the fat body, gonads, and midgut, in addition to the brain (Aslam et al. 2011). On the other hand, in the *D. melanogaster* genome, a total of eight genes coding for *Drosophila* ILPs (DILPs) have been identified. Interestingly, six of them, DILP1–6, are thought to use InR as their receptor, whereas DILP7 and DILP8 do not. The receptor for DILP8 has been identified to be *Drosophila* LGR3 (DLGR3), while that for DILP7 is estimated to be DLGR4 (see Chap. 7).

All other insect genomes examined also contain multiple copies (usually fewer than ten) of the ILP genes. In some cases, some of them are called IGF-like or relaxin-like peptides due to their distinctive structural and/or functional characteristics. However, this does not indicate that they are orthologous to vertebrate IGF or relaxin, because their aa sequences are much more closely related to those of other ILPs in the same species than to vertebrate IGF or relaxin.

Recently, it has been proposed that insect (or arthropod) ILPs are classified into four types depending on sequence similarity, presence of a C-terminal extension of the A-chain with conserved motif(s), tissue specificity of expression, or type of receptor (Veenstra 2020). The first type is classical ILPs expressed predominantly in the medial NSCs of the brain and secreted as neurohormones. Many of the well-studied ILPs are of this type. The second type has a long C-terminal extension of the A-domain (chain) with one or two conserved aa sequences or motifs, in which two additional cysteine residues are included. The third type consists of orthologs of DILP8 that utilize GPCRs, more specifically DLGR3, as their receptors. The ILPs of this type are expressed predominantly in the gonads and are therefore called gonadulin in crustaceans, although DILP8 is expressed in imaginal discs as well (Vallejo et al. 2015). The fourth type is orthologs of DILP7, characterized by their notably conserved aa sequences throughout arthropods. DILP7 has been suggested to use LGR as its receptor, like relaxin, and the ILPs of this type are primarily expressed in the gonads. For these reasons, this type of ILP has been suggested to be the arthropod equivalent of relaxin. These four types of ILPs cannot be distinguished in the phylogenetic tree due to high sequence variability, but they are separated in the sequence similarity tree, although the boundary of the first two types may be less precise (Veenstra 2020). The functions of classical ILPs have been extensively studied, but little is known about the functions of other types of ILPs. Various functions of classical ILPs will be discussed in the second part of this book (see Chaps. 7, 8, 9, and 10).

2.2.2 *Bursicon*

Bursicon was discovered as a hormone that triggers sclerotization and melanization (tanning) after metamorphic ecdysis in blowflies (Fraenkel and Hsiao 1965). Its structure was determined in *D. melanogaster* and other insects in 2005 (Luo et al. 2005; Mendive et al. 2005; for a review, see Honegger et al. 2008). This protein hormone has a molecular weight of approximately 30,000 and is a heterodimer of cystine-knot glycoprotein subunits. The subunits are called burs-α and burs-β or burs and pburs (after partner of bursicon). Homologs of the bursicon gene have been found not only in insects but also in a wide range of bilaterians, including deuterostomes, but are absent in vertebrates. Bursicon binds to a receptor named LGR2, a member of type B LGRs. LGR2 is also known as rickets, as the *Drosophila* gene *rickets* encodes this protein. Bursicon also forms a homodimer. Burs or pburs homodimer has been shown to have pleiotropic actions, including regulation of cell proliferation, energy homeostasis, immune response, and reproduction (Li et al. 2021). The action of bursicon in tanning is discussed in detail in Chap. 12.

2.2.3 *Glycoprotein Hormone*

In vertebrates, GPHs include follicle-stimulating hormone (FSH), luteinizing hormone (LH), thyroid-stimulating hormone (TSH), chorionic gonadotropin (CG), and thyrostimulin (TS). These hormones are heterodimers of cystine-knot-containing glycoprotein subunits. Each of FSH, LH, TSH, and CG consists of a common subunit (GPα1) and a hormone-specific subunit (GPβ1–4). These four hormones are thought to be paralogs generated by gene duplications of the TS gene. The TS consists of GPα2 and GPβ5. Homologs of TS are widespread in the genomes of most metazoans and are called GPHs (Roch and Sherwood 2014) or GPA2/GPB5 (Al-Dailami et al. 2022). Insect GPH binds to a receptor named LGR1, which is a homolog of vertebrate GPH receptors. Studies of GPH function in insects are still limited, as this hormone was discovered through the search for TS homologs in insect genomes. However, the previous studies suggest that GPH plays a role in regulating development, salt and water balance, and reproduction (Al-Dailami et al. 2022).

2.3 Hormone Families Common to Bilaterians

Recent advances in comparative genome studies have revealed the existence of a number of peptidergic signaling systems common to both protostomes and deuterostomes. These signaling systems are thought to have already been present in the common ancestor of bilaterians. Although a number of potential NPs and GPCRs have been found in non-bilaterians, homologs of most bilaterian PHs/NPs have not been identified. Therefore, most of these peptidergic signaling systems are thought to have evolved after bilaterians diverged from other metazoans. Insect PHs/NPs common to bilaterians are summarized below, with special attention to their structural features and relationships with vertebrate hormones (if present). The heading of each subsection indicates the PHs/NPs of insects and other invertebrates and vertebrates, if any, belonging to each hormone family or superfamily.

2.3.1 *Adipokinetic Hormone/Corazonin/ACP/RPCH/GnRH*

Adipokinetic hormone (AKH), corazonin (CRZ), and AKH/CRZ-related peptide (ACP) are different insect hormones produced by separate NSCs. Gonadotropin-releasing hormone (GnRH) is a vertebrate hormone released from the hypothalamus to stimulate the secretion of gonadotropic hormones (FSH and LH) from the pituitary. Although the sequences of these hormones are not very similar to each other (Table 2.1), their receptors are closely related (Jékely 2013; Mirabeau and Joly 2013). Therefore, these hormones are thought to have a common ancestral origin, dating back to the split of Protostomia and Deuterostomia approximately 700 million years ago. Based on the comparative analysis of the sequences of the receptors and their ligands, it has been proposed that in the protostome lineage, the ancestral GnRH-like receptor gene duplicated as did its GnRH-like ligand gene, and subsequent diversification led to the development of AKH, CRZ, and ACP signaling systems (Hauser and Grimmelikhuijzen 2014). Red pigment-concentrating hormone (RPCH) of crustaceans is also a member of this superfamily of peptides (Marco et al. 2024).

The AKH mobilizes lipids and carbohydrates from their storage site, the fat body, into the hemolymph to meet energy demands, such as during long flights. The effects of this hormone will be discussed in detail in Chap. 9. The CRZ was isolated as a cardioactive peptide from the corpora cardiaca (CC) of the cockroach *Periplaneta americana* and sequenced (Veenstra 1989). This hormone has been shown to exhibit multiple functions, including the control of cuticle melanization and shift to the gregarious phase in locusts, inhibition of ecdysis sequence in moths, modulation of stress responses, and male reproduction in flies. These functions of CRZ will be discussed in relevant chapters of this book. The ACP has been reported to play a role in the regulation of lipid metabolism in the fat body of the locust *Locusta migratoria* (Hou et al. 2024).

Table 2.1 Amino acid sequences of GnRH superfamily peptides

Locusta migratoria AKH-I	**pE**LNFTPN**W**GT-NH_2
Locusta migratoria [His[7]]CRZ	**pE**TFQYSHG**W**TN-NH_2
Locusta migratoria ACP	**pE**VTFSRD**W**SP-NH_2
Pandalus borealis RPCH	**pE**LNFSTK**W**-NH_2
Mus musculus GnRH	**pE**HWSYGLRPG-NH_2

Amino acid residues characteristic of peptides in this superfamily are shown in bold

2.3.2 Allatostatin-A/Buccalin/Galanin

Allatostatin A (AST-A) was first isolated in 1989 as four similar peptides that inhibit juvenile hormone (JH) biosynthesis in the CA of the cockroach *Diploptera punctata* (Woodhead et al. 1989). Subsequent study revealed that the four peptides, along with nine other similar peptides, are derived from a single precursor protein (Bendena et al. 1999). The AST-A peptides have also been found in many insect species and other invertebrates. In any species, many AST-A peptides are produced from a single precursor protein. The AST-A consists of approximately ten aa residues and has a conserved C-terminal structure (FGL-NH_2) (Table 2.2). Buccalin, isolated from the buccal ganglia of the mollusk *Aplysia californica*, and a vertebrate hormone galanin are homologs of this NP. The CA-inhibiting activity of AST-A is restricted to specific insect groups such as cockroaches and crickets. The AST-A exhibits muscle contraction inhibitory activity on the gut and heart in many insect species. In contrast, in specific insects, it exerts a variety of effects, including inhibition of vitellogenin release from the fat body, promotion of carbohydrate-metabolizing enzyme activity in the midgut, and stimulation of AKH secretion from the CA (Gäde 2002). The function of this peptide will be described in detail in relevant chapters.

2.3.3 Allatostatin C/Somatostatin

Allatostatin C (AST-C) was isolated in 1991 from the brain of the tobacco hornworm *Manduca sexta* as a hormone that inhibits the biosynthesis of JH by the CA (Kramer et al. 1991; Stay and Tobe 2007). Although homologous peptides have been found in many other insect species, they can be grouped into two types. Members of the first type are 15-aa peptides with a blocked N-terminus (the N-terminus is pyroglutamic acid) and a non-amidated C-terminus (Table 2.2), whereas those of the second type are 14-aa peptides with an unblocked N-terminus and an amidated C-terminus (Audsley et al. 2008). Both types of AST-C contain two cysteine residues, which form a cyclic structure. The sequences of each type of AST-C are highly conserved among species and are represented as pEXRXRQCYFNPISCF and SYWKQCAFNAVSCF-NH_2, respectively. The AST-C has two paralogs, AST-CC and AST-CCC, which are thought to have arisen through gene duplication. The C-terminus of AST-CC is not amidated, while that of AST-CCC is amidated (Veenstra 2016; see Sect. 3.3.1 for details). The vertebrate hormone somatostatin is a homolog of AST-C. Although AST-C inhibits CA activity in lepidopteran insects, the more general function of this peptide is to inhibit muscle contraction in the intestine and heart.

Table 2.2 Primary structure of a representative peptide of each insect PH/NP family

PH/NP	Species	Structure/cysteine arrangement
AKH	*L. migratoria*	(AKH-I)** **pELNF**TPN**W**GT-**NH$_2$**
ADF	*T. molitor*	VVNTPGHAVSYHVY
AST-A	*D. punctata*	(AST-1) L**YDFGL-NH$_2$**
AST-C	*M. sexta*	pEVRFR**QCYFNP**I**SCF**
AT	*M. sexta*	**GFKN**VEMMT**ARGF-NH$_2$**
Bursicon α-subunit β-subunit	*D. melanogaster*	(140 aa) (123 aa)
CAP2b	*M. sexta*	pE**LY**A**FPRV-NH$_2$**
CCHa	*D. melanogaster*	(CCHa-1) S**C**LEYGHS**C**WGA**H-NH$_2$**
Corazonin	*P. americana*	**pETF**Q**YS**R**GWTN-NH$_2$**
CCAP	*M. sexta*	**PFCNAFTGC-NH$_2$**
CRF-DH	*M. sexta*	RM**PSLSI**DL**PM**S**VLRQ**KL**SLE**KERKVHALRAAA**NRNFLNDI-NH$_2$**
CT-DH	*D. punctata*	**GLDLGL**S**RGFSG**S**QAAKHLMGLAAANYAGGP-NH$_2$**
DH	*B. mori*	T**DMKD**ESDR**GAHSER**GA**LWFGPRL**-**NH$_2$**
ETH	*M. sexta*	**SN**EAISPF**D**QGMMGYVI**K**TN**K**NI**PR**M-**NH$_2$**
EH	*M. sexta*	-------------C---C--C------------C---C---------C------------- (62 aa)
FMRFa	*D. melanogaster*	(FMRFa-2) DPKQ**DFMRF-NH$_2$**
GPH α-subunit β-subunit	*D. melanogaster*	(141 aa) (169 aa)
ILP	*B. mori*	(Bombyxin-II) A-chain: **GI**VD**ECC**LRP**CS**VDV**L**LS**YC** B-chain: pEQPQAVHT**YCG**RH**L**AR**TL**ADL**C**WEAGVD
ITP	*S. gregaria*	------C---------------C-C------------C---C--------C-------------------NH$_2$ (72 aa)
Inotocin	*L. migratoria*	**CLITNCPRG-NH$_2$**
Leucokinin	*L. migratoria*	(Leucokinin-1) DPA**F**N**SWG-NH$_2$**
MIP	*L. migratoria*	(MIP-1) **AW**Q**DL**NA**GW-NH$_2$**
MS	*L. maderae*	pE**D**V**DH**V**FLRF-NH$_2$**

(continued)

Table 2.2 (continued)

PH/NP	Species	Structure/cysteine arrangement
Neuroparsin/OEH	*L. migratoria*	(Neuroparsin B) -C----C------C------------C-------C---------C-----C-C--C--C------C---------- C (78 aa)
NPF	*D. melanogaster*	SNSRP**P**RKNDVNTM**A**DAYKF**L**QD**LD**T**YY**GDRA**R**V**RF-NH$_2$**
Orcokinin	*D. melanogaster*	(OK-A) **NFDEID**KASASFSILNQLV
PBAN	*H. zea*	**LS**D**DMPATPADQEMY**RQ**DPE**QID**SRT**K**YFSPRL-NH$_2$**
Pyrokinin	*L. maderae*	pETS**FT****PRL-NH$_2$**
PDF	*R. microptera*	**NSE**I**INSLLGLPK**LL**NDA-NH$_2$**
Proctolin	*P. americana*	**RYLPT**
PTTH	*B. mori*	(109 aa/monomer)
SIFa	*N. bullata*	A**YRKPPFNGSIF-NH$_2$**
sNPF	*D. melanogaster*	(sNPF-1) AQ**RSPSLRLRF-NH$_2$**
SP	*D. melanogaster*	WEWPWNRKP*TKFP*IP*SP*NP*RDKWCRLNLGPAWGGRC
Sulfakinin	*L. maderae*	(SK-1) EQ**F**E**DY*GHMRF-NH$_2$**
Tachykinin	*L. migratoria*	(TK-1) GPSG**F**Y**G**V**R-NH$_2$**

Amino acid residues that are well conserved in each family of peptides in insects are shown in bold
P* in the SP sequence denotes a hydroxyproline residue
Y* in the sulfakinin sequence denotes a sulfated tyrosine residue
**PH/NP with number indicates that it is one of the multiple homologous peptides derived from a single precursor, except for bombyxin-II and CCHa-1, which are one of the homologous peptides encoded by different genes

2.3.4 Allatotropin/Orexin

Allatotropin (AT) was first identified in *M. sexta* as a hormone that promotes JH synthesis by the CA (Kataoka et al. 1989a; Stay and Tobe 2007). Subsequently, homologous peptides have been found in many other insect species and other invertebrates. *M. sexta* AT is a 13-aa peptide with an amidated C-terminus (Table 2.2). Several N- and C-terminal aa residues are highly conserved among insect ATs. The vertebrate hormone orexin is a homolog of AT, although the sequences of both hormones are not very similar. The CA-stimulating effect of AT is limited to a specific group of insects, including moths, honey bees, and blowflies. In *B. mori*, the action of AT on the CA has been reported to be mediated by short NPF (Yamanaka et al. 2008; see Chap. 6). Other actions of this hormone include the regulation of ion transport, promotion of heart rate and muscle contraction, and regulation of the secretion of digestive enzymes in the midgut (Elekonich and Horodyski 2003).

2.3.5 CCHamide/EP/NMB

CCHamide (CCHa) was first discovered as a parturition hormone in the tsetse fly *Glossina morsitans* (Zdarek et al. 2000). The CCHa consists of 13 aa residues, with cysteines at positions 2 and 9 forming a cyclic structure, and the C-terminus is amidated (Table 2.2). The CCHa was named as such because of the presence of two cysteine residues and a C-terminal histidine amide. Insects typically have two CCHa peptides (CCHa-1 and CCHa-2) encoded by different genes, and each peptide binds to a different receptor. The two CCHa genes are suggested to be generated through duplication in a close ancestor of insects, because other arthropods have only one CCHa gene. The CCHa-1 and CCHa-2 each bind to highly homologous but different receptors. The CCHa has been suggested to be involved in various physiological functions and related behaviors, including feeding, diuresis, reproduction, metabolism, and memory (Yamagata et al. 2020). Excitatory peptide (EP, also called GGNG peptide) in annelids is a homolog of CCHa, although it has a different C-terminal structure (Thiel et al. 2019). Its receptor is also homologous to the CCHa receptor. The CCHa/EP signaling system is sometimes grouped together with the neuromedin B (NMB) signaling system in vertebrates based on the orthology of receptors (Jékely 2013).

2.3.6 CRF-Like Diuretic Hormone/ELH/CRH/Urotensin-1/Urocortin

Insects have several types of diuretic hormones (DHs). One of them was first isolated from the adult heads of *M. sexta* (Kataoka et al. 1989b). This DH is called corticotropin-releasing factor (CRF)-like DH (CRF-DH or CRF/DH) because of its sequence similarity to vertebrate CRF, which is now called CRH. Homologous peptides have since been identified in many other insects and invertebrates. This type of DH is sometimes referred to as DH44 due to the number of aa residues that make up this peptide in a given species. In fact, however, the number varies considerably among insect species. The CRF-DH and CRH, along with the vertebrate hormones urotensin-1 and urocortin and egg-laying hormone (ELH) in mollusks, are thought to be members of the same family of peptides, based on the structural features of their precursors and close association of their receptors. The CRF-DH not only modulates diuresis by controlling fluid secretion from the Malpighian tubules (Coast 1996; see Sect. 9.3.2) but also controls drinking behavior (Audsley et al. 1997; see Sect. 11.3.3).

2.3.7 Crustacean Cardioactive Peptide/Neuropeptide S

Crustacean cardioactive peptide (CCAP) was first isolated from the shore crab *Carcinus maenas* as an NP that increases heart rate (Stangier et al. 1987), and the same peptide was subsequently found in many other crustaceans and insects. The CCAP is the most conserved arthropod NP and has an identical aa sequence, PFCNAFTGC-NH_2. The two cysteine residues are linked by a disulfide bridge, forming a cyclic peptide. The precursor peptide contains a single copy of CCAP. The CCAP is sometimes regarded as a peptide in the same family as neuropeptide S in vertebrates and NG peptides such as NGFFFamide in sea urchins and SFRNGVamide in cephalochordates, although their structure is quite different. This is because their receptors are homologous and the precursors of these peptides are thought to have evolved from a common ancestral peptide (Semmens et al. 2015). In addition to cardioacceleratory activity, CCAP has multiple functions, including accelerating the frequency and amplitude of oviduct contractions in *L. migratoria* (Donini et al. 2001) and regulating the release of digestive enzymes in *P. americana* (Sakai et al. 2006). The CCAP also plays a role in regulating ecdysis behavior in insects and crustaceans (see Sect. 11.1).

2.3.8 CT-Like Diuretic Hormone/Calcitonin

Two types of DHs were isolated from the cockroach *Diploptera punctata* in 2000 (Furuya et al. 2000). One of them was CRF-DH; the other was not. This new type of DH is called calcitonin-like DH (CT-DH), because of its relatedness to the vertebrate hormone calcitonin (CT), although their sequences are not very similar. The receptors of CT-DH and CT are closely related. This type of DH is widely found in invertebrates and is sometimes referred to as DH31 based on the number of aa residues that make up this peptide. Interestingly, this number is the same in most insect species, unlike CRF-DH (DH44).

Two types of CT-related peptides exist in invertebrates. One has two cysteine residues in the N-terminus of the peptide, like CT (CT-type), whereas the other does not. The CT-DH (DH31) is the latter type. Among protostomes, some have both types of CT-related peptides (e.g., arthropods and some annelids), while others have only one type. In contrast, deuterostomes have only CT-type peptides. Therefore, it is believed that the CT gene was duplicated in the common ancestor of protostomes, and one of them was subsequently lost in some animal groups during evolution. The role of insect CT-DH in regulating water balance is discussed in detail in Chap. 9. The CT-DH has also been reported to inhibit foregut and hindgut contractions (Lajeunesse et al. 2010) and control circadian rhythms (see Sect. 13.4.1 for details).

2.3.9 Eclosion Hormone

Eclosion hormone (EH) was first discovered as a hormone that induces pupa-adult ecdysis (eclosion) of lepidopteran insects and was named as such. However, this hormone was later shown to also induce larva-larva and larva-pupa ecdysis (Truman 2005). Homologs of EH have been identified in many other insect groups and crustaceans. The EH consists of approximately 60 aa residues, and its sequence is well conserved, with 6 cysteine residues at conserved positions. Although the existence of EH was previously thought to be limited to arthropods, recent studies have revealed that EH-type peptides are also present in other protostomes, deuterostomes, and even cnidarians. Furthermore, the gene for the EH receptor, guanylyl cyclase, has been identified in a wide variety of protostomes and deuterostomes, although not found in cnidarians, suggesting that the origin of this ligand-receptor pair dates back to the cnidarian-bilaterian split (de Oliveira et al. 2019). The roles of EH in regulating ecdysis are discussed in detail in Chap. 11.

2.3.10 Inotocin/Vasopressin/Oxytocin

Vasopressin (VP) and oxytocin (OT) are mammalian neurosecretory hormones released from the posterior lobe of the pituitary gland. The structure of both hormones resembles each other and is related to the fish hormone vasotocin. It is thought that the genes for VP and OT have arisen through duplication of the vasotocin gene. Homologs of VP/OT have been found in a wide range of bilaterians. Among them are inotocin in insects, annetocin in earthworms, and conopressin in snails. All these peptides consist of nine aa residues, with cysteines at positions 1 and 6 forming a cyclic structure, and the C-terminus is amidated (Table 2.2). Their precursor proteins have a common neurophysin domain on the C-terminal side of the hormone domain. This domain is thought to have a chaperone-like function to put the hormones on the secretory pathway. The first identified VP/OT-like peptide in insects was a locust DH (Lom-DH) isolated from the subesophageal and thoracic ganglia of *L. migratoria* (Proux et al. 1987). Its homologs have since been identified in many other insect species and collectively called inotocin or arginine-vasopressin-like DH. However, the inotocin genes are absent from some insect orders, including Trichoptera, Lepidoptera, Siphonaptera, Mecoptera, and Diptera, suggesting that the inotocin signaling system was probably lost in their common ancestor ~280 million years ago (Liutkeviciute et al. 2020). Inotocin plays a role in regulating water balance in locusts and beetles and in regulating metabolism, activity, and feeding behavior in ants (Muratspahić et al. 2020).

2.3.11 Neuropeptide F/Neuropeptide Y

Neuropeptide F (NPF) was first isolated from the tapeworm *Moniezia expansa* as a peptide immunoreactive with the antibody against the vertebrate hormone pancreatic polypeptide (PP) (Maule et al. 1991). Sequencing revealed a 39-aa peptide with a similar structure to neuropeptide Y (NPY), which, together with PP and peptide YY, constitutes a peptide hormone family in vertebrates. This tapeworm peptide was named NPF because the C-terminal tyrosine (Y) was substituted by phenylalanine (F) (Table 2.2). Subsequently, NPF-type peptides have been identified in other protostomes, including mollusks, annelids, and arthropods. The NPFs are typically composed of 36–40 aa residues, having a conserved C-terminal tetrapeptide, RXRF-NH_2 (Fadda et al. 2019). In insects, NPF has pleiotropic actions such as regulating feeding and stress responses (Nässel and Wegner 2011). Its role in the regulation of feeding and other behaviors is discussed in Chap. 11.

2.3.12 Orcokinin/Pedal Peptide

Orcokinin (OK) was first isolated and sequenced as a potent myotropic peptide of the crayfish *Orconectes limosus* (Stangier et al. 1992) and has since been identified in other crustaceans and various insects. Its name comes from the genus name of the crayfish. In 2012, the presence of an OK-related peptide produced by alternative splicing of the OK gene was reported in the kissing bug *Rhodnius prolixus* (Sterkel et al. 2012). Alternative splicing of the OK gene is common in invertebrates, and the classical OK is now called OK-A and its splice variant OK-B (Jiang et al. 2015). Insect OK-A peptides typically consist of 6–14 aa residues, and their N-terminal sequence (NFDEIDR-) is well conserved (Table 2.2). The C-terminal sequence is variable and generally not amidated. One or a few copies of OK-As are produced from the OK-A precursor protein. On the other hand, the sequence of OK-B is less conserved, and many more paracopies (1–12) of OK-B are produced from its precursor protein (Jiang et al. 2015). Pedal peptide isolated from *Aplysia* pedal ganglia is a homolog of OK-A. Homologs of the OK gene have also been identified in annelids, tardigrades, and echinoderms (Lin et al. 2017). In insects, OK is involved in various biological processes, including circadian rhythms, development, and reproduction (Silva et al. 2021).

2.3.13 PBAN/MRCH/DH/PK/MT/CAPA/NMU

Several PHs/NPs with the conserved C-terminal sequence, -FXPRL-NH_2, have been isolated and sequenced from various kinds of insects and crustaceans. Pheromone biosynthesis-activating neuropeptide (PBAN) was discovered as a hormone that stimulates biosynthesis of sex pheromones in the moth (Raina and Klun 1984) and sequenced (Raina et al. 1989; Kitamura et al. 1989). Melanization and reddish coloration hormone (MRCH) was isolated as a hormone that regulates the body color of moth larvae (Matsumoto et al. 1984) and was found to be identical to PBAN after sequencing. Diapause hormone (DH) was isolated and sequenced as a hormone that induces embryonic diapause in *B. mori* (Yamashita 1996). Please note that the abbreviation for this hormone is the same as that for diuretic hormone. Pyrokinin (PK) and myotropin (MT) were first discovered in the cockroach as NPs that stimulate hindgut contractions and were subsequently found in many other insects (Holman et al. 1986a). All these peptides share the conserved C-terminal sequence (Table 2.2), constituting the FXPRLamide family of peptides. The PBAN/MRCH, DH, and multiple PKs are generated from a common precursor protein (Yamashita 1996). Cardioacceleratory peptide 2b (CAP2b) in *M. sexta* and CAPA in *D. melanogaster* also have similar C-terminal sequences (-FXPRV-NH_2) and are also generated along with similar peptides from their precursor proteins. A similar C-terminal structure (-FRPRN-NH_2) is also found in the vertebrate hormone neuromedin-U, and all these PHs/NPs are thought to be members of the same peptide family. The functions of PBAN, MRCH, DH, and CAPA are discussed in Chaps. 10, 12, 8, and 9, respectively.

2.3.14 *RYamide/Luqin/RWamide*

RYamide, named for its C-terminal motif, is the arthropod homolog of luqin, a myotropic peptide with a C-terminal RFamide motif discovered in *A. californica* (Aloyz and DesGroseillers 1995). The name of luqin comes from the left upper quadrant (LUQ) cells of the abdominal ganglion, where luqin is expressed. RWamide found in echinoderms is also a member of this family of peptides (Jékely 2013). These luqin-type neuropeptides consist of approximately ten aa residues, and their precursors typically contain one or two copies of the peptide. Based on the comparative studies on their precursors and receptors, luqin-type signaling system is thought to have originated in the common ancestor of Bilateria as a paralog of the tachykinin-type signaling system. However, it has been lost in chordates (Yañez-Guerra and Elphick 2020). In insects, RYamide has been reported to have roles in regulating feeding and diuresis (Yañez-Guerra and Elphick 2020).

2.3.15 *Short NPF/PrRH*

Short neuropeptide F (sNPF) was isolated from the Colorado potato beetle *Leptinotarsa decemlineata* as a peptide recognized by antibodies generated against *M. expansa* NPF (Spittaels et al. 1996). The identified peptides (ARGPQLRLRF-NH_2 and APSLRLRF-NH_2) were initially called NPF-related peptides, but were later renamed sNPFs due to their much shorter length than NPF. Subsequently, sNPFs have been identified in many other insects and other invertebrates. Interestingly, a comparative study of NPYs, NPFs, sNPFs, and their precursors has revealed that NPF is closely related to NPY, whereas sNPF is not (Jékely 2013). Furthermore, a comparative study of receptors (GPCRs) revealed that the sNPF receptor is orthologous to the prolactin-releasing hormone (PrRH) receptor, but not to the NPY receptor (Mirabeau and Joly 2013). The PrRH is a member of the vertebrate RFamide superfamily of peptides, like NPY. It has been suggested that PrRH and sNPF signaling systems are orthologous and arose as paralogs of NPY/NPF-type signaling in the common ancestor of bilaterians (Yañez-Guerra et al. 2020). The sNPF has been reported to be involved in the regulation of various biological processes, including feeding, molting, courtship, social interaction, learning, circadian clock, reproduction, and development (Fadda et al. 2019).

2.3.16 *SIFamide/IMFamide/SMYamide/NPFF/GnIH*

SIFamide (originally LFamide) was first isolated from the crude extract of the gray fleshfly *Neobellieria bullata* as a peptide with myostimulatory activity (Janssen et al. 1996). The identified sequence (AYRKPPFNGSLF-NH_2) was later corrected to

AYRKPPFNGSIF-NH_2, and this peptide was named SIFamide after its C-terminal sequence. The sequences of SIFamides are highly conserved among arthropods. Homologous peptides are widespread in protostomes and are all called SIFamides (Verleyen et al. 2004). This peptide is considered to be homologous to the vertebrate neuropeptides NPFF and GnIH, because the SIFamide receptor is an ortholog of the NPFF and GnIH receptors (Bauknecht and Jékely 2015). SIFamide has been shown to play a role in regulating courtship behavior, feeding, and sleep in *D. melanogaster* (Martelli et al. 2017) and in regulating reproduction in the yellow fever mosquito *Aedes aegypti* (Keyes-Scott et al. 2023). Recently, two paralogs of SIFamide, IMFamide and SMYamide, have been discovered in insects and other arthropods (Veenstra 2021), but little is known about their functions.

2.3.17 Sulfakinin/CCK/Gastrin

Sulfakinin (SK) was first isolated from the head extracts of the cockroach *Leucophaea maderae* as an NP with myotropic activity on the isolated hindgut (Nachman et al. 1986). The SK is a homolog of the vertebrate gastrointestinal hormones cholecystokinin (CCK) and gastrin and is composed of 10–17 aa residues (Table 2.2). Most insects have two copies of the SK peptides in their SK precursors. This peptide is characterized by a C-terminal consensus sequence DYGHMRF-NH_2, with the tyrosine residue (Y) sulfated. The name SK is derived from this feature of the peptide. The SK has an RFamide motif at its C-terminus, indicating that this peptide is a member of the RFamide superfamily of peptides, although the C-terminus of CCK/gastrin is GWMDF-NH_2. It has been demonstrated that SK inhibits contractions of the foregut, heart, ejaculatory duct, and oviduct. Furthermore, recent studies in *D. melanogaster* have shown that this peptide is involved in the regulation of satiety and feeding, locomotor activity, as well as aggression and reproductive behaviors (Nässel and Wu 2022).

2.3.18 Tachykinin/Tachykinin-Related Peptide/Eledoisin/Substance P

Tachykinins (TKs) are NPs of approximately ten aa residues characterized by a conserved C-terminal sequence of -FXGXR-NH_2 in protostomes and -FXGKM-NH_2 in deuterostomes. Protostome TKs are sometimes referred to as tachykinin-related peptide (TRP). The first TK family peptide to be discovered was eledoisin, which was isolated from the salivary glands of octopuses. When eledoisin was sequenced (Erspamer and Anastasi 1962), the structure of substance P, a representative TK in mammals, had not yet been determined. In 1990, four kinds of TKs were found from the central nervous system (CNS) of locusts, and to date, more than

350 TK family peptides have been identified from more than 50 insect species (Nässel et al. 2019). The TK is produced in many types of neurons in the CNS and colocalizes with various other NPs or neurotransmitters, acting as a neuromodulator or cotransmitter. It is also produced by the endocrine cells of the intestine. Exhibiting muscle contractile activity, this NP has pleiotropic effects. In *D. melanogaster*, TK has been reported to play roles in early olfactory processing, neuromodulation of circuits controlling locomotion and food search, nociception, aggression, metabolic stress, and hormone release (Nässel et al. 2019).

2.4 Hormone Families Common to Protostomes

Homologs of some insect PHs/NPs are widespread in protostomes but not found in deuterostomes. These PHs/NPs are thought to have evolved in the common ancestor of protostomes after both lineages diverged. Below is the overview of these hormones.

2.4.1 *Ecdysis Triggering Hormone/Pre-ecdysis-Triggering Hormone*

Ecdysis-triggering hormone (ETH) was isolated from the epitracheal gland extract of *M. sexta* and sequenced as a hormone that releases ecdysis behavior (Žitňan et al. 1996). Later, a related peptide, pre-ecdysis triggering hormone (PETH), was identified from the same glands. *Manduca* ETH and PETH are C-terminally amidated peptides consisting of 26 and 11 aa residues, respectively. Both peptides are derived from the same precursor, having similar but not identical C-terminal sequences. The genes encoding ETH and its receptor have been identified in many arthropods and other protostomes, although they are lacking in some lineages (de Oliveira et al. 2019). Notably, ETH is produced by non-neuronal Inka cells attached to the tracheal system (see Sect. 4.1.2). The roles of ETH in controlling ecdysis behavior are described in detail in Chap. 11.

2.4.2 *FMRFamide/FaRP*

FMRFamide is the first of many NPs now known as the RFamide superfamily peptides. Isolated and sequenced in 1977 as a molluscan NP with cardioexcitatory activity, this peptide, as its name suggests, is composed of four aa residues (FMRF) with its C-terminus amidated (Price and Greenberg 1977). Subsequently, peptides of the same length or up to about ten aa residues with a similar C-terminal structure were discovered one after another, and these peptides are called FMRFamide-related peptides (FaRPs). FMRFamides and/or FaRPs are encoded by a single gene, and in most cases, the precursor protein generates more than ten copies of these peptides. The predominant effect of FaRPs is the reduction of spontaneous muscle contractions. In insects, FaRPs have been shown to exhibit diverse actions, such as accelerating heart rate, inhibiting gut motility, and modulating synaptic activity (Orchard et al. 2001a).

2.4.3 Leucokinin/Myokinin/Kinin

Leucokinin was isolated and sequenced in 1986 as a peptide that stimulates hindgut contractions in *L. maderae* (Holman et al. 1986b) and was so named after the insect's genus name. Subsequently, homologous peptides were identified in many insects, crustaceans, and other protostomes and called by different names depending on the animal species, such as locustakinin in *L. migratoria*, aedeskinin in *A. aegypti*, and lymnokinin in the snail *Lymnaea stagnalis*. Because these peptides are all orthologs, they are now collectively referred to as myokinins, invertebrate kinins, or simply kinins. Most of this type of peptides in insects consist of six to eight aa residues and have a FXXWGamide motif in their C-terminus (Table 2.2). The precursor protein generally produces multiple kinin peptides, but the number varies between species, with less than 12 copies in insects, but approximately 40 copies in *A. californica*. In insects, kinin-type peptides are known to exhibit diuretic effects (see Sect. 9.3.5 for details) in addition to myotropic effects. Furthermore, in *D. melanogaster*, this type of peptide is involved in the regulation of ecdysis behavior (see Chap. 11).

2.4.4 Myoinhibitory Peptide/Allatostatin B/PTSP/Wamide

Myoinhibitory peptide (MIP) was discovered in locusts as a peptide that inhibits muscle contractions of the hindgut and oviduct (Schoofs et al. 1991). Allatostatin B (AST-B) was identified in crickets as a peptide that inhibits JH synthesis by the CA (Lorenz et al. 1995). Prothoracicostatic peptide (PTSP) was isolated from silkworms as a peptide that inhibits the PG activity to secrete ecdysone (Hua et al. 1999). Sequencing of these PHs isolated using different bioassays revealed that they were all the same peptide. This type of peptide typically consists of nine aa residues and is characterized by two tryptophan residues (W) located at the second position from the N-terminus and at the C-terminus (Table 2.2). The C-terminus is amidated. Therefore, this peptide is also called W(X_6) Wamide or Wamide. The precursor protein contains many copies of this type of peptide. Homologs of this peptide are also found in other arthropods, mollusks, annelids, and other protostomes.

2.4.5 Myosuppressin/Myomodulin

Myosuppressin (MS) was first isolated as a factor that reduces the frequency of hindgut spontaneous contractions (Holman et al. 1986c). Homologous peptides are widely distributed not only in arthropods but also in other protostomes. The MSs are typically decapeptides and characterized by a C-terminal FLRFamide motif (Table 2.2). The precursor protein contains one copy of the MS peptide. The MS is sometimes considered a member of FaRP, but the MS receptor is distinct from the FMRF receptor. Therefore, MS is thought to constitute a separate peptide family. It has been suggested that in addition to the C-terminal FLRFamide motif, the N-terminal region of MS is required to activate the MS receptor (Schöller et al. 2005). The MS has the effect of suppressing the contraction of smooth muscles in various organs, including the gut, heart, oviduct, Malpighian tubule, and salivary gland (Orchard et al. 2001b). In addition, MS inhibits the release of AKH from the CC of *L. migratoria* (Vullings et al. 1998) and ecdysteroid secretion from the PG of *B. mori* (Yamanaka et al. 2005). It has also been reported that this hormone inhibits feeding activity in some insects (Audsley and Weaver 2009). Myomodulin with the sequence PMSMLRLamide, which was discovered in mollusks, is considered to be a peptide in the same family as MS, as its receptor is an ortholog of the MS receptor (Bauknecht and Jékely 2015).

2.4.6 PDF/PDH

Pigment-dispersing hormone (PDH) was first isolated from the eyestalks of the shrimp *Pandalus borealis* as a light-adapting hormone and sequenced (Fernlund 1976). Later, it was shown that this hormone has the effect of diffusing pigment granules within epithelial chromatophores, and it came to be called PDH. The PDH consists of 18 aa residues and is C-terminally amidated. Crustacean PDHs are classified into α-PDHs and β-PDHs by substitutions of a few aa residues on the C-terminal side of the peptide. However, overall sequences are highly conserved among PDHs and even among homologous peptides found in insects. The homolog of PDH in insects is called pigment-dispersing factor (PDF). This is because this peptide does not have pigment-dispersing activity in insects that lack chromatophores in their integument. The best-known function of PDF is its association with circadian rhythms. This function of PDF is discussed in Chap. 13. The PDF has also been reported to be involved in regulating courtship behavior and sex pheromone synthesis in *D. melanogaster* (Krupp et al. 2013) and ecdysteroid synthesis in *B. mori* (Iga et al. 2014).

2.4.7 *Proctolin*

Proctolin is the first bioactive peptide to be identified in insects. This peptide was isolated from the whole body of the American cockroach *Periplaneta americana* as a factor that stimulates hindgut contractions (Starratt and Brown 1975). Its aa sequence is RYLPT. This sequence is highly conserved among homologous peptides subsequently identified in many other insects and crustaceans, although Lepidoptera and Diptera do not have this type of peptide. The proctolin gene is thought to have been lost during the evolution of these insect groups. In addition to stimulating the contraction of smooth, cardiac, and skeletal muscles, proctolin has been suggested to stimulate the secretion of AKH and JH in locusts (Clark et al. 2006). In the grasshopper *Chorthippus biguttulus*, proctolin plays a modulatory role in a neuronal circuit controlling sexual behavior (Vezenkov and Danalev 2009).

2.5 Hormone Families Common to Arthropods

Some insect PHs/NPs are widely distributed in arthropods but absent from other animal phyla. These PHs/NPs are thought to have evolved in the common ancestor of arthropods after arthropods diverged from other animal groups of protostomes.

2.5.1 *Ion Transport Peptide/CHH*

Ion transport peptide (ITP) was isolated from the locust CC as a hormone that stimulates chloride transport across the ileal epithelium (Audsley et al. 1992) and sequenced by cDNA cloning in the desert locust *Schistocerca gregaria* (Meredith et al. 1996). The ITP gene also produces a longer ITP isoform called ITP-like (ITPL) through alternative splicing. The ITP/ITPL or their homologs have been identified in approximately 20 arthropod species. Most ITPs consist of 72 aa residues and are C-terminally amidated, whereas ITPLs are non-amidated peptides of 79–94 residues. The ITP and ITPL have the same sequence in their N-terminal halves, but differ in sequence and length in their C-terminal halves. The ITP/ITPL shares structural features with crustacean hyperglycemic hormone (CHH). Six cysteine residues that form three intramolecular disulfide bridges are located in corresponding positions (Table 2.2), indicating that they are homologous peptides. In crustaceans, CHH, together with molt-inhibiting hormone (MIH), vitellogenesis-inhibiting hormone (VIH), and mandibular organ-inhibiting hormone (MOIH), constitutes the CHH family of peptides. Although receptors for ITP and ITPL have been identified (Nagai et al. 2014), those four CHH family peptides remain unidentified. The role of ITP/ITPL in regulating ion and water balance is discussed in detail in Chap. 9. The ITP has also been suggested to affect reproduction (Begum et al. 2009).

2.5.2 Neuroparsin/OEH

Neuroparsin was isolated from the CC of *L. migratoria* as a hormone that inhibits vitellogenesis (Girardie et al. 1989), but was subsequently shown to also have antidiuretic and neuritogenic effects (Fournier et al. 1987; Vanhems et al. 1990). The name comes from the fact that this hormone is produced in the pars intercerebralis (dorsomedial protocerebrum; see Sect. 4.2.1). Ovary ecdysteroidogenic hormone (OEH) was isolated from *A. aegypti* as a hormone that stimulates ecdysteroidogenesis in the ovary (Brown et al. 1998). Through stimulation of ecdysone synthesis, this hormone promotes vitellogenin synthesis by the fat body and, as a result, yolk accumulation in the ovary. Sequencing of this hormone revealed that OEH is homologous to neuroparsin. Subsequently, similar peptides have been identified in the genomes of other insects and crustaceans, but *D. melanogaster* and *B. mori* lack this type of hormone. Both hormones are large cysteine-rich peptides consisting of approximately 80 or more aa residues. Although the overall sequences of these polypeptides are not very similar, the spacing pattern of 11 cysteine residues is well conserved (Table 2.2). It has been pointed out that neuroparsin/OEH is similar to vertebrate IGF-binding protein and its insect homologs in the number and arrangement of cysteine residues (Badisco et al. 2007). Interestingly, the OEH receptor is not a GPCR but one of the venus kinase receptors (VKRs), a subfamily of RTKs (Vogel et al. 2015). The VKRs are widely distributed among protostomes and echinoderms, but are absent from chordates. The effects of neuroparsin and OEH on egg maturation are discussed in detail in Chap. 10.

2.6 Hormone Families Unique to Insects

2.6.1 Prothoracicotropic Hormone

Prothoracicotropic hormone (PTTH), as its name suggests, is a hormone that stimulates the PG to secrete ecdysone. Discovery, structure, synthesis, release, and action of PTTH are described in detail in Chaps. 1 and 6. Here, we will only touch on the evolutionary aspect of PTTH. The PTTH is a heterodimer of cystine-knot glycoprotein subunits and is thought to be a paralog of "trunk," an extracellular signaling molecule involved in the regulation of embryonic development in *D. melanogaster* (Rewitz et al. 2009). The PTTH and trunk share the common receptor "torso," which is a member of the RTKs. The PTTH gene is suggested to have evolved through duplication of the trunk gene in the lineage of insects (de Oliveira et al. 2019). In contrast, trunk genes are present not only in a wide range of bilaterians but also in cnidarians, one of non-bilaterians. Furthermore, a trunk-like aa sequence has been found even in ctenophore (de Oliveira et al. 2019). On the other hand, its receptor torso has been identified in ecdysozoans, lophotrochozoans, deuterostomes, and cnidarians, indicating that the origin of the trunk-torso signaling system dates back at least the last common ancestor of Cnidaria and Bilateria (de Oliveira et al. 2019). For these reasons, although PTTH may be an innovation in insects, in a broad sense, it may be considered to be a hormone with a very ancient origin.

2.6.2 Antidiuretic Factor

Antidiuretic factor (ADF) was isolated from the head extract of the beetle *Tenebrio molitor* as a peptide that strongly inhibits fluid secretion by Malpighian tubules (Eigenheer et al. 2002). The ADF consists of 14 aa residues (Table 2.2), and no similar peptide has been identified in insects of other orders. The ADF appears to exert its effect via cGMP as a second messenger.

2.7 PHs/NPs Whose Evolutionary Origin Has Been Poorly Investigated

The evolutionary origins of recently identified insect PHs/NPs are poorly understood, because few comparative genomic studies have been conducted. These PHs/NPs include agatoxin-like peptide (ALP), CNMamide, EFLamide, elevenin, hansolin, IDL-like, insect parathyroid hormone (iPTH), ITG-like, limostatin, natalisin, neuropeptide-like precursor (NPLP), NVP, RFLamide, and trissin. As information on some of them is available in Chap. 3, this section only describes some others whose functions have been analyzed.

2.7.1 Elevenin

Elevenin was first discovered in 1984 in the identified neuron L11 of the *Aplysia* abdominal ganglion and was named L11 neuropeptide (Taussig et al. 1984). It consisted of 20 aa residues, including 2 cysteine residues at positions 6 and 15. Recently, homologous peptides were identified in various protostomes, including mollusks, annelids, nematodes, and arthropods (Fu et al. 2023), and it was renamed elevenin. They all have two cysteine residues with eight residues between them (see Table 2.3), although the total numbers of residues in these peptides vary between species. In insects, elevenin has been suggested to be involved in the regulation of body color, based on the results of RNAi experiments using the brown plant hopper *Nilaparvata lugens* (Uchiyama et al. 2017; see Sect. 12.3.2).

Table 2.3 Primary structures of recently identified insect neuropeptides

Neuropeptide	Species	Primary structure
Elevenin	*S. gregaria*	PQKR**VDC**RMYP**FAPICRG**IMT
Limostatin	*D. melanogaster*	(LST-15)[a] AIVFRPLFVYKQQEI
Natalisin	*D. melanogaster*	(NTL-1) EKLFDGYQFGEDMSKEND**PF**IPP**R-NH$_2$**
NPLP	*D. melanogaster*	(MTYamide peptide) YIGSLARAGGLMTY-NH$_2$ (IPNamide peptide) NVGTLARDFQLPIPN-NH$_2$ (APK peptide) SVAALAAQGLLNAPK

Conserved aa residues are shown in bold
[a]LST-15 is one of the peptides produced from a precursor protein encoded by the LST gene

2.7.2 *Limostatin*

Limostatin (LST) was identified as a peptide that suppresses insulin secretion through an ectopic expression screen of selected starvation-regulated *Drosophila* genes (Alfa et al. 2015). A 15-aa peptide (LST-15) contained in the precursor protein (see Table 2.3) is thought to be responsible for the action of this peptide. The name of this peptide comes from the Greek goddess of starvation, Limos. The LST is colocalized with AKH in CC cells. The LST receptor is suggested to be a GPCR orthologous to the neuromedin-U receptor in vertebrates.

2.7.3 *Natalisin*

Natalisin (NTL) was first discovered in the *D. melanogaster* genome as one of the predicted neuropeptides (Jiang et al. 2013). Similar peptides have since been identified in many other insects and crustaceans, but not in any other phyla other than Arthropoda. The NTLs consist of 6–25 aa residues and are characterized by an FXXXRamide C-terminal motif. In any species, several NTL peptides are generated from precursor proteins. The sequence of its C-terminal motif is similar to that of TRPs (FXGXRamide), but NTL binds to a GPCR that is closely related to but distinct from TRP receptor, indicating that NTLs and TRPs constitute different peptide families. The close relationship between the two ligands and their receptors strongly suggests that the NTL and TRP genes diverged during arthropod evolution through ancestral gene duplication (Jiang et al. 2013). The name of this peptide is derived from the Latin word "natalis," meaning birth, due to its function to promote reproduction (Jiang et al. 2013). The NTL has been reported to be involved in the regulation of mating behavior in several insect species (Gong et al. 2024).

2.7.4 Neuropeptide-Like Precursor

A peptidomic analysis on *D. melanogaster* CNS performed in 2002 identified several unknown predicted NPs consisting of 10–22 aa residues, and subsequent search for the genes encoding these NPs in the *Drosophila* genome revealed the presence of 4 neuropeptide-like precursors (NPLP1–4) (Baggerman eta al. 2002). NPLP1, NPLP2, NPLP3, and NPLP4 contained three, one, two, and one of the newly discovered NPs, respectively. The structures of three peptides (MTYamide peptide, IPNamide peptide, and APK peptide) contained in NPLP1 are shown in Table 2.3. The NPLPs have since been identified in several other insects as well. Although the functions of individual NPs contained in these precursors remain unknown, gene expression analyses have suggested that NPLP1 is involved in the regulation of ecdysis (Baggerman et al. 2002), while NPLP4 is related to diapause (Li et al. 2009).

2.8 Future Perspectives

In this chapter, we have categorized insect PHs/NPs based on the antiquity of their evolutionary origin as estimated by the distribution of PH/NP precursors across animal phyla. However, this categorization is provisional. Finding homologs of a particular PH/NP precursor can be difficult because the sequences of homologous proteins sometimes change significantly during evolution. Additionally, as discussed in Chap. 3, genes for homologous proteins can be lost in certain animal species, groups, or phyla. Therefore, if deeper mining of known genomes or sequencing of unsearched genomes discovers the desired homologs of PH/NP precursors, it is quite possible that some hormones will recategorized in the future. In order to determine the true evolutionary origin of each hormone, it is necessary to pay attention to further advances in comparative genomic research.

Insect PHs/NPs are not limited to those described in this chapter. In fact, many more PHs/NPs have been identified in insect genomes, as shown in Table 3.1. However, most of them are paralogs of known insect PHs/NPs, homologs of PHs/NPs found in non-insect animals, or NPs predicted in transcriptomic and/or neuropeptidomic analysis, and their functions have not yet been analyzed extensively. Functional analysis of these new PHs/NPs, together with further analysis of the functions of known PHs/NPs, will further expand our understanding of the roles of PHs/NPs in the regulation of insect development, homeostasis, reproduction, and behavior.

References

Al-Dailami AN, Orchard I, Lange AB (2022) Investigating the role of glycoprotein hormone GPA2/GPB5 signaling in reproduction in adult female *Rhodnius prolixus*. Front Insect Sci 2:1096089. https://doi.org/10.3389/finsc.2022.1096089

Alfa RW, Park S, Skelly K-R, Poffenberger G, Jain N, Gu X et al (2015) Suppression of insulin production and secretion by a decretin hormone. Cell Metab 21:323–333. https://doi.org/10.1016/j.cmet.2015.01.006

Aloyz RS, DesGroseillers L (1995) Processing of the L5-67 precursor peptide and characterization of LUQIN in the LUQ neurons of *Aplysia californica*. Peptides 16:331–338. https://doi.org/10.1016/0196-9781(94)00140-5

Anctil M (2009) Chemical transmission in the sea anemone *Nematostella vectensis*: a genomic perspective. Comp Biochem Physiol Part D Genomics Proteomics 4:268–289. https://doi.org/10.1016/j.cbd.2009.07.001

Aslam AFM, Kiya T, Mita K, Iwami M (2011) Identification of novel bombyxin genes from the genome of the silkmoth *Bombyx mori* and analysis of their expression. Zool Sci 28:609–616. https://doi.org/10.2108/zsj.28.609

Audsley N, Weaver RJ (2009) Neuropeptides associated with the regulation of feeding in insects. Gen Comp Endocrinol 162:93–104. https://doi.org/10.1016/j.ygcen.2008.08.003

Audsley N, McIntosh C, Phillips JE (1992) Isolation of a neuropeptide from locust corpus cardiacum which influences ileal transport. J Exp Biol 173:261–274. https://doi.org/10.1242/jeb.173.1.261

Audsley N, Goldsworthy GJ, Coast GM (1997) Circulating levels of *Locusta* diuretic hormone: the effects of feeding. Peptides 18:59–65. https://doi.org/10.1016/s0196-9781(96)00234-3

Audsley N, Matthews HJ, Weaver RJ (2008) Allatoregulatory peptides in lepidoptera, structures, distribution and functions. J Insect Physiol 54:969–980. https://doi.org/10.1016/j.jinsphys.2008.01.012

Badisco L, Claeys I, Van Loy T, Van Hiel M, Franssens V, Simonet G et al (2007) Neuroparsins, a family of conserved arthropod neuropeptides. Gen Comp Endocrinol 153:64–71. https://doi.org/10.1016/j.ygcen.2007.03.008

Baggerman G, Cerstiaens A, De Loof A, Schoofs L (2002) Peptidomics of the larval *Drosophila melanogaster* central nervous system. J Biol Chem 277:40368–40374. https://doi.org/10.1074/jbc.M206257200

Bauknecht P, Jékely G (2015) Large-scale combinatorial deorphanization of platynereis neuropeptide GPCRs. Cell Rep 12:684–693. https://doi.org/10.1016/j.celrep.2015.06.052

Begum K, Li B, Beeman RW, Park Y (2009) Functions of ion transport peptide and ion transport peptide-like in the red flour beetle *Tribolium castaneum*. Insect Biochem Mol Biol 39:717–725. https://doi.org/10.1016/j.ibmb.2009.08.005

Bendena WG, Donly BC, Tobe SS (1999) Allatostatins: a growing family of neuropeptides with structural and functional diversity. Ann N Y Acad Sci 891:311–329. https://doi.org/10.1111/j.1749-6632.1999.tb07902.x

Brown MR, Graf R, Swiderek KM (1998) Identification of a steroidogenic neurohormone in female mosquitoes. J Bio Chem 273:3967–3971. https://doi.org/10.1074/jbc.273.7.3967

Clark L, Zhang JR, Tobe S, Lange AB (2006) Proctolin: a possible releasing factor in the corpus cardiacum/corpus allatum of the locust. Peptide 27:559–566. https://doi.org/10.1016/j.peptides.2005.06.028

Coast GM (1996) Neuropeptides implicated in the control of diuresis in insects. Peptides 17:327–336. https://doi.org/10.1016/0196-9781(95)02096-9

de Oliveira AL, Calcino A, Wanninger A (2019) Ancient origins of arthropod molting pathway components. elife 8:e46113. https://doi.org/10.7554/eLife.46113

Donini A, Agricola H, Lange AB (2001) Crustacean cardioactive peptide is a modulator of oviduct contractions in *Locusta migratoria*. J Insect Physiol 47:277–285. https://doi.org/10.1016/s0022-1910(00)00112-8

Eigenheer RA, Nicolson SW, Schegg KM, Hull JJ, Schooley DA (2002) Identification of a potent antidiuretic factor acting on beetle Malpighian tubules. Proc Natl Acad Sci USA 99 z:84–89. https://doi.org/10.1073/pnas.012436199

Elekonich MM, Horodyski FM (2003) Insect allatotropins belong to a family of structurally-related myoacitive peptides present in several invertebrate phyla. Peptides 24:1623–1632. https://doi.org/10.1016/j.peptides.2003.08.011

Elphick MR, Mirabeau O, Larhammar D (2018) Evolution of neuropeptide signaling system. J Exp Biol 221:jeb151092. https://doi.org/10.1242/jeb.151092

Erspamer V, Anastasi A (1962) Structure and pharmacological actions of eledoisin, active endecapeptide of posterior salivary glands of eledone. Experientia 18:58–59. https://doi.org/10.1007/BF02138250

Fadda M, Hasakiogullari I, Temmerman L, Beets I, Zels S, Schoofs L (2019) Regulation of feeding and metabolism by neuropeptide F and short neuropeptide F in invertebrates. Front Endocrinol 10:64. https://doi.org/10.3389/fendo.2019.00064

Fernlund P (1976) Structure of a light-adapting hormone from the shrimp, *Pandalus borealis*. Biochem Biophys Acta 439:17–25. https://doi.org/10.1016/0005-2795(76)90155-0

Fournier B, Herault JP, Proux J (1987) Antidiuretic factor from the nervous corpora cardiaca of the migratory locust: improvement of an existing in vitro bioassay. Gen Comp Endocrinol 68:49–56. https://doi.org/10.1016/0016-6480(87)90059-1

Fraenkel G, Hsiao C (1965) Bursicon, a hormone which mediates tanning of the cuticle in the adult fly and other insects. J Insect Physiol 11:513–556. https://doi.org/10.1016/0022-1910(65)90137-X

Fu P, Mei Y-S, Liu W-J, Chen P, Jin Q-C, Guo S-Q et al (2023) Identification of three elevenin receptors and roles of elevenin disulfide bond and residues in receptor activation in *Aplysia californica*. Sci Rep 13:7662. https://doi.org/10.1038/s41598-023-34596-9

Furuya K, Milchak RJ, Schegg KM, Zhang J, Tobe SS, Coast GM et al (2000) Cockroach diuretic hormones: characterization of a calcitonin-like peptide in insects. Proc Natl Acad Sci USA 97: 6469–6474. https://doi.org/10.1073/pnas.97.12.6469

Gäde G (2002) Allatoregulatory peptides-molecules with multiple function. Invert Rep Dev 41: 127–135. https://doi.org/10.1080/07924259.2002.9652743

Girardie J, Girardie A, Huet JC, Pernollet JC (1989) Amino acid sequence of locust neuroparsins. FEBS Lett 245:4–8. https://doi.org/10.1016/0014-5793(89)80179-6

Gong W, Linghu J-H, Xu H-M, Luo LL, Samgghe G, Liu T-X et al (2024) Neuropeptide natalisin regulates reproductive behaviors in *Spodoptera frugiperda*. Sci Rep 14:15122. https://doi.org/10.1038/s41598-024-66031-y

Hauser F, Grimmelikhuijzen CJP (2014) Evolution of AKH/corazonin/ACP/GnRH receptor superfamily and their ligands in the Protostomia. Gen Comp Endocrinol 209:35–49. https://doi.org/10.1016/j.ygcen.2014.07.009

Hauser F, Koch TL, Grimmelikhuijzen CJP (2022) Review: the evolution of peptidergic signaling in Cnidaria and Placozoa, including a comparison with Bilateria. Front Endocrinol:973862. https://doi.org/10.3389/fendo.2022.973862

Hewes RS, Taghert PH (2001) Neuropeptides and neuropeptide receptors in *Drosophila melanogaster* genome. Genome Res 11:1126–1142. https://doi.org/10.1101/gr.169901

Holman GM, Cook BJ, Nachman RJ (1986a) Primary structure and synthesis of a blocked myotropic neuropeptide isolated from the cockroach, *Leucophaea maderae*. Comp Biochem Physiol C 85:219–224. https://doi.org/10.1016/0742-8413(86)90077-0

Holman GM, Cook BJ, Nachman RJ (1986b) Isolation, primary structure, and synthesis of two neuropeptides from *Leucophaea maderae*: members of a new family of cephalomyotropins. Comp Biochem Physiol 84C:205–211. https://doi.org/10.1016/0742-8413(86)90084-8

Holman GM, Cook BJ, Nachman RJ (1986c) Isolation, primary structure and synthesis of leucomyosuppressin, an insect neuropeptide that inhibits spontaneous contractions of the cockroach hindgut. Comp Biochem Physiol 85C:329–333. https://doi.org/10.1016/0742-8413(86)90202-1

Honegger H-W, Dewey EM, Ewer J (2008) Bursicon, the tanning hormone of insects: recent advances following the discovery of its molecular identity. J Comp Physiol A 194:989–1005. https://doi.org/10.1007/s00359-008-0386-3

Hou L, Guo S, Wang Y, Liu S, Wang X (2024) Neuropeptide ACP is required for fat body lipid metabolism homeostasis in locusts. Insect Sci 31:1453–1465. https://doi.org/10.1111/1744-7917.13321

Hua Y-J, Tanaka Y, Nakamura K, Sakakibara M, Nagata S, Kataoka H (1999) Identification of a prothoracicostatic peptide in the larval brain of the silkworm, *Bombyx mori*. J Biol Chem 274: 31169–31173. https://doi.org/10.1074/jbc.274.44.31169

Iga M, Nakaoka T, Suzuki Y, Kataoka H (2014) Pigment dispersing factor regulates ecdysone biosynthesis via bombyx neuropeptide G protein coupled receptor-B_2 in the prothoracic glands of *Bombyx mori*. PLoS One 9:e103239. https://doi.org/10.1371/journal.pone.0103239

Janssen I, Schoofs L, Spittaels K, Neven H, Vanden Broeck J, Devreese B et al (1996) Isolation of NEB-LFamide, a novel myotropic neuropeptide from the grey fleshfly. Mol Cell Endocrinol 117:157–165. https://doi.org/10.1016/0303-7207(95)03746-2

Jékely G (2013) Global view of the evolution and diversity of metazoan neuropeptide signaling. Proc Natl Acad Sci USA 110:8702–8707. https://doi.org/10.1073/pnas.1221833110

Jiang H, Lkhagva A, Daubnerová I, Chae H-S, Simo L, Jung S-H et al (2013) Natalisin, a tachykinin-like signaling system, regulates sexual activity and fecundity in insects. Proc Natl Acad Sci USA 110:E3526–E3534. https://doi.org/10.1073/pnas.1310676110

Jiang H, Kim HG, Park Y (2015) Alternatively spliced orcokinin isoforms and their functions in *Tribolium castaneum*. Insect Biochem Mol Biol 65:1–9. https://doi.org/10.1016/j.ibmb.2015.07.009

Kataoka H, Toschi A, Li JP, Carney R, Schooley DA, Kramer SJ (1989a) Identification of an allatotropin from adult *Manduca sexta*. Science 243:1481–1483. https://doi.org/10.1126/science.243.4897.1481

Kataoka H, Troetschler RG, Li JP, Kramer SJ, Carney RL, Schooley DA (1989b) Isolation and identification of diuretic hormone from the tobacco hornworm, Manduca sexta. Proc Natl Acad Sci USA 86:2976–2980. https://doi.org/10.1073/pnas.86.8.2976

Keyes-Scott NI, Swade KR, Allen LR, Vogel KJ (2023) RNAi-mediated knockdown of two orphan G protein-coupled receptors reduces fecundity in the yellow fever mosquito *Aedes aegypti*. Front Insect Sci 3:1197945. https://doi.org/10.3389/finsc.2023.1197945

Kitamura A, Nagasawa H, Kataoka H, Inoue T, Matsumoto S, Ando T et al (1989) Amino acid sequence of pheromone-biosynthesis-activating neuropeptide (PBAN) of the silkworm, *Bombyx mori*. Biochem Biophys Res Commun 163:520–526. https://doi.org/10.1016/0006-291x(89)92168-2

Kramer SJ, Toschi A, Miller CA, Kataoka H, Quistad GB, Li JP et al (1991) Identification of an allatostatin from the tobacco hornworm *Manduca sexta*. Proc Natl Acad Sci USA 88:9458–9462. https://doi.org/10.1073/pnas.88.21.9458

Krupp JJ, Billeter J-C, Wong A, Choi C, Nitabach MN, Levine JD (2013) Pigment-dispersing factor modulates pheromone production in clock cells that influence mating in *Drosophila*. Neuron 79: 54–68. https://doi.org/10.1016/j.neuron.2013.05.019

LaJeunesse DR, Johnson B, Presnell JS, Catignas KK, Zapotoczny G (2010) Peristalsis in the junction region of the *Drosophila* larval midgut is modulated by DH31 expressing neuroendocrine cell. BMC Physiol 10:14. https://doi.org/10.1186/1472-6793-10-14

Li A, Rinehart JP, Denlinger DL (2009) Neuropeptide-like precursor 4 is uniquely expressed during pupal diapause in the flesh fly. Peptides 30:518–521. https://doi.org/10.1016/j.peptides.2008.10.008

Li J, Zhu Z, Bi J, Feng Q, Beerntsen BT, Song Q (2021) Neuropeptide bursicon influences reproductive physiology in *Tribolium castaneum*. Front Physiol 12:717437. https://doi.org/10.3389/fphys.2021.717437

Lin M, Egertová M, Zampronio CG, Jones AM, Elphick MR (2017) Pedal peptide/orcokinin-type neuropeptide signaling in a deuterostome: the anatomy and pharmacology of starfish

myorelaxant peptide in *Asterias rubens*. J Comp Neurol 525:3890–3917. https://doi.org/10.1002/cne.24309
Liutkeviciute Z, Koehbach J, Eder T, Gil-Mansilla E, Gruber CW (2020) Global map of oxytocin/vasopressin-like neuropeptide signalling in insects. Sci Rep 6:39177. https://doi.org/10.1038/srep39177
Lorenz MW, Kellner R, Hoffmann KH (1995) A family of neuropeptides that inhibit juvenile hormone biosynthesis in the cricket, *Gryllus bimaculatus*. J Biol Chem 270:21103–21108. https://doi.org/10.1074/jbc.270.36.21103
Luo CW, Dewey EM, Sudo S, Ewer J, Hsu SY, Honegger HW et al (2005) Bursicon, the insect cuticle-hardening hormone, is a heterodimeric cystine knot protein that activate G protein-coupled receptor LGR2. Proc Natl Acad Sci USA 102:2820–2825. https://doi.org/10.1073/pnas.0409916102
Marco HG, Glendinning S, Ventura T, Gäde G (2024) The gonadotropin-releasing hormone (GnRH) superfamily across Pancrustacea/Tetraconata: a role in metabolism? Mol Cell Endocrinol 590:112238. https://doi.org/10.1016/j.mce.2024.112238
Martelli C, Pech U, Kobbenbring S, Pauls D, Bahl B, Sommer MV et al (2017) SIFamide translates hunger signals into appetitive and feeding behavior in *Drosophila*. Cell Rep 20:464–478. https://doi.org/10.1016/j.celrep.2017.06.043
Matsumoto S, Isogai A, Suzuki A (1984) Isolation of the melanization and reddish coloration hormone (MRCH) of the armyworm, *Leucania separata*, from the silkworm, *Bombyx mori*. Agr Biol Chem 48:2401–2403. https://doi.org/10.1271/bbb1961.48.2401
Maule AG, Shaw C, Halton DW, Thim L, Johnston CF, Fairweather I et al (1991) Neuropeptide F: a novel parasitic flatworm regulatory peptide from *Moniezia expansa* (Cestoda: Cyclophyllidea). Parasitology 102:309–316
Mendive FM, Van Loy T, Claeysen S, Poels J, Williamson M, Hauser F et al (2005) *Drosophila* molting neurohormone bursicon in a heterodimer and the natural agonist of the orphan receptor DLGR2. FEBS Lett 579:2171–2176. https://doi.org/10.1016/j.febslet.2005.03.006
Meredith J, Ring M, Macins A, Marschall J, Cheng NN, Theilmann D et al (1996) Locust ion transport peptide (ITP): primary structure, cDNA and expression in a baculovirus system. J Exp Biol 199:1053–1061. https://doi.org/10.1242/jeb.199.5.1053
Metchell ML, Hossain MA, Lin F, Pinheiro-Junior EL, Peigneur S, Wai DCC et al (2021) Identification, synthesis, conformation and activity of an insulin-like peptide from a sea anemone. Biomolecules 11:1765. https://doi.org/10.3390/biom11121785
Mirabeau O, Joly J-S (2013) Molecular evolution of peptidergic signaling systems in bilaterians. Proc Natl Acad Sci USA 110:E2028–E2037. https://doi.org/10.1073/pnas.1219956110
Mizoguchi A, Ishizaki H, Nagasawa H, Kataoka H, Isogai A, Tamura S et al (1987) A monoclonal antibody against a synthetic fragment of bombyxin (4K-prothoracicotropic hormone) from the silkmoth, Bombyx mori: characterization and immunohistochemistry. Mol Cell Endocrinol 51(3):227–235. https://doi.org/10.1016/0303-7207(87)90032-3
Muratspahić E, Monjon E, Duerrauer L, Rogers SM, Cullen DA, Vanden Broeck J et al (2020) Oxytocin/vasopressin-like neuropeptide signaling in insects. Vitam Horm 113:29–53. https://doi.org/10.1016/bs.vh.2019.08.011
Nachman RJ, Holman GM, Haddon WF, Ling N (1986) Leucosulfakinin, a sulfated insect neuropeptide with homology to gastrin and cholecystokinin. Science 234:71–73. https://doi.org/10.1126/science.3749893
Nagai C, Mabashi-Asazuma H, Nagasawa H, Nagata S (2014) Identification and characterization of receptors for ion transport peptide (ITP) and ITP-like (ITPL) in the silkworm *Bombyx mori*. J Boil Chem 289:32166–32177. https://doi.org/10.1074/jbc.M114.590646
Nässel DR, Wegner CA (2011) Comparative review of short and long neuropeptide F signaling in invertebrates: any similarities to vertebrate neuropeptide Y signaling? Peptides 32:1335–1355. https://doi.org/10.1016/j.peptides.2011.03.013

Nässel DR, Wu S-F (2022) Cholecystokinin/sulfakinin peptide signaling: conserved roles at the intersection between feeding, mating and aggression. Cell Mol Life Sci 79:188. https://doi.org/10.1007/s00018-022-04214-4

Nässel DR, Zandawala M, Kawada T, Satake H (2019) Tachykinins: neuropeptides that are ancient, diverse, widespread and functionally pleiotropic. Front Neurosci 13:1262. https://doi.org/10.3389/fnins.2019.01262

Nikitin M (2015) Bioinformatic prediction of *Trichoplax adhaerens* regulatory peptides. Gen Comp Endocrinol 212:145–155. https://doi.org/10.1016/j.ygcen.2014.03.049

Okamoto N, Yamanaka N, Satake H, Saegusa H, Kataoka H, Mizoguchi A (2009) An ecdysteroid-inducible insulin-like growth factor-like peptide regulates adult development of the silkmoth *Bombyx mori*. FEBS J 276:1221–1232. https://doi.org/10.1111/j.1742-4658.2008.06859.x

Orchard I, Lange AB, Bendena WG (2001a) FMRFamide-related peptides: a multifunctional family of structurally related neuropeptides in insects. Adv Insect Physiol 28:267–329. https://doi.org/10.1016/S0065-2806(01)28012-6

Orchard I, Lange AB, Bendena WG (2001b) FMRFamide-related peptides: a multifunctional family of structurally-related neuropeptides in insect. Adv Insect Physiol 28:267–329. https://doi.org/10.1016/S0065-2806(01)28012-6

Price DA, Greenberg MJ (1977) Purification and characterization of a cardioexcitatory neuropeptide from the central ganglia of a bivalve mollusc. Prep Biochem 7:261–281. https://doi.org/10.1080/00327487708061643

Proux JP, Miller CA, Li JP, Carney RL, Girardie A, Delaage M et al (1987) Identification of an arginine vasopressin-like diuretic hormone from *Locusta migratoria*. Biochem Biophys Res Comm 149:180–186. https://doi.org/10.1016/0006-291x(87)91621-4

Raina AK, Klun JA (1984) Brain factor control of sex pheromone production in the corn earworm moth. Science 225:531–533. https://doi.org/10.1126/science.225.4661.531

Raina AK, Jaffe H, Kempe TG, Keim P, Blacher RW, Fales HM et al (1989) Identification of a neuropeptide hormone that regulates sex pheromone production in female moths. Science 244:796–798. https://doi.org/10.1126/science.244.4906.796

Rewitz KF, Yamanaka N, Gilbert LI, O'Connor MB (2009) The insect neuropeptide PTTH activates receptor tyrosine kinase torso to initiate metamorphosis. Science 326:1403–1405. https://doi.org/10.1126/science.1176450

Roch GJ, Sherwood NM (2014) Glycoprotein hormones and their receptors emerged at the origin of metazoans. Genome Bio Evo 6:1466–1479. https://doi.org/10.1093/gbe/evu118

Sakai T, Satake H, Takeda M (2006) Nutrient-induced alpha-amylase and protease activity is regulated by crustacean cardioactive peptide (CCAP) in the cockroach midgut. Peptides 27:2157–2164. https://doi.org/10.1016/j.peptides.2006.04.009

Schöller S, Belmont M, Cazzamali G, Hauser F, Williamson M, Grimmelikhuijzen CJP (2005) Molecular identification of a myosuppressin receptor from the malaria mosquito *Anopheles gambiae*. Biochem Biophys Res Comm 327:29–34. https://doi.org/10.1016/j.bbrc.2004.11.139

Schoofs L, Holman GM, Hayes TK, Nachman RJ, De Loof A (1991) Isolation, identification and synthesis of locustamyoinhibiting peptide (LOM-MIP), a novel biologically active neuropeptide from *Locusta migratoria*. Regul Pept 36:111–119. https://doi.org/10.1016/0167-0115(91)90199-q

Semmens DC, Beets I, Rowe ML, Blowes LM, Oliveri P, Elphick MR (2015) Discovery of sea urchin NGFFFamide receptor unites a bilaterian neuropeptide family. Open Biol 5:150030. https://doi.org/10.1098/rsob.150030

Silva V, Palacios-Muñoz A, Volonté M, Frenkel L, Ewer J, Ons S (2021) Orcokinin neuropeptides regulate reproduction in the fruit fly, *Drosophila melanogaster*. Insect Biochem Mol Biol 139:103676. https://doi.org/10.1016/j.ibmb.2021.103676

Simion P, Phillippe H, Baurain D, Jager M, Richter DJ, Di Franco A et al (2017) A large and consistent phylogenomic dataset supports sponges as the sister group to all other animals. Curr Biol 27:958–967. https://doi.org/10.1016/j.cub.2017.02.031

Spittaels K, Verhaert P, Shaw C, Johnston RN, Devreese B, Van Beeumen J et al (1996) Insect neuropeptide F (NPF)-related peptides: isolation from Colorado potato beetle (*Leptinotarsa decemlineata*) brain. Insect Biochem Mol Biol 26:375–382. https://doi.org/10.1016/0965-1748(95)00104-2

Stangier J, Hilbich C, Beyruether K, Keller R (1987) Unusual crustacean cardioactive peptide (CCAP) from pericardial organs of the shore crab *Carcinus maenas*. Proc Natl Acad Sci USA 84:575–579. https://doi.org/10.1073/pnas.84.2.575

Stangier J, Hilbich C, Burdzik S (1992) Orcokinin: a novel myotropic peptide from the nervous system of the crayfish, *Orconectes limosus*. Peptides 13:859–864. https://doi.org/10.1016/0196-9781(92)90041-z

Starratt AN, Brown BE (1975) Structure of the pentapeptide proctolin, a proposed neurotransmitter in insects. Life Sci 17:1253–1256. https://doi.org/10.1016/0024-3205(75)90134-4

Stay B, Tobe SS (2007) The role of allatostatins in juvenile hormone synthesis in insects and crustaceans. Annu Rev Entomol 52:277–299. https://doi.org/10.1146/annurev.ento.51.110104.151050

Sterkel M, Oliveira PL, Urlaub H, Hernandez-Martinez S, Rivera-Pomar R, Ons S (2012) OKB, a novel family of brain-gut neuropeptides from insects. Insect Biochem Mol Biol 42:466–473. https://doi.org/10.1016/j.ibmb.2012.03.003

Taussig R, Kaldany RR, Scheller RH (1984) A cDNA clone encoding neuropeptides isolated from *Aplysia* neuron L11. Proc Natl Acad Sci USA 81:4988–4992. https://doi.org/10.1073/pnas.81.15.4988

Thiel D, Bauknecht P, Jékely G, Hejnol A (2019) A nemertean excitatory peptide/CCHamide regulates ciliary swimming in the larvae of *Lineus Longissimus*. Front Zool 16:28. https://doi.org/10.1186/s12983-019-0326-9

Truman JW (2005) Hormonal control of insect ecdysis: endocrine cascades for coordinating behavior with physiology. Vitam Horm 73:1–30. https://doi.org/10.1016/S0083-6729(05)73001-6

Uchiyama H, Maehara S, Ohta H, Seki Y, Tanaka Y (2017) Elevenin regulates the body color through a G protein-coupled receptor NiA42 in the brown planthopper *Nilaparvata lugens*. Gen Comp Endocrinol 258:33–38. https://doi.org/10.1016/j.ygcen.2017.07.017

Vallejo DM, Zuarez-Carreño S, Bolivar J, Morante J, Dominguez M (2015) A brain circuit that synchronizes growth and maturation revealed through Dilp8 binding to Lgr3. Science 350: aac6767. https://doi.org/10.1126/science.aac6767

Vanhems E, Delbos M, Girardie J (1990) Insulin and neuroparsin promote neurite outgrowth in cultured locust CNS. Eur J Neurosci 2:776–782. https://doi.org/10.1111/j.1460-9568.1990.tb00468.x

Veenstra JA (1989) Isolation and structure of corazonin, a cardioactive peptide from the American cockroach. FEBS Lett 250:231–234. https://doi.org/10.1016/0014-5793(89)80727-6

Veenstra JA (2016) Allatostatins C, double C and triple C, the result of a local gene triplication in an ancestral arthropod. Gen Com Endocrinol 230–231:153–157. https://doi.org/10.1016/j.ygcen.2016.04.013

Veenstra (2020) Arthropod IGF, relaxin and gonadulin, putative orthologs of *Drosophila* insulin-like peptides 6, 7, 8, likely originated from an ancient gen triplication. PeerJ 8:e 9534. https://doi.org/10.7717/peerj.9534

Veenstra JA (2021) The neuropeptide SMYamide, a SIFamide paralog, is expressed by salivary gland innervating neurons in the American cockroach and likely functions as a hormone. Peptides 136:170466. https://doi.org/10.1016/j.pepties.2020.170466

Verleyen P, huybrechts J, baggerman G, van Lommel A, de Loof A, Schoofs L (2004) SIFamide is a highly conserved neuropeptide: a comparative study in different insects. Biochem Biophys Res Commun 320:334–341. https://doi.org/10.1016/j.bbrc.2004.05.173

Vezenkov SR, Danalev DL (2009) From molecule to sexual behavior: the role of the neuropentapeptide proctolin in acoustic communication in the male grasshopper *Chorthippus biguttulus*. Eur J Pharmachol 619:57–60. https://doi.org/10.1016/j.ejphar.2009.08.010

Vogel KJ, Brown MR, Strand MR (2015) Ovary ecdysteroidogenic hormone requires a receptor tyrosine kinase to activate egg formation in the mosquito *Aedes aegypti*. Proc Natl Acad Sci USA 112:5057–5062. https://doi.org/10.1073/pnas.1501814112

Vullings HG, Ten Voorde SE, Passier PC, Diederen JH, Van Der Horst DJ, Nässel DR (1998) A possible role of SchistoFLRFamide in inhibition of adipokinetic hormone release from locust corpora cardiaca. J Neurocytol 27:901–913. https://doi.org/10.1023/a:1006901123566

Woodhead AP, Stay B, Seidel SL, Khan MA, Tobe SS (1989) Primary structure of four allatostatins: neuropeptide inhibitors of juvenile hormone synthesis. Proc Natl Acad Sci USA 86:5997–6001. https://doi.org/10.1073/pnas.86.15.5997

Yamagata N, Imanishi Y, Wu H, Kondo S, Sano H, Tanimoto H (2020) Nutrient responding PH CCHamide-2 consolidates appetitive memory. Front Behav Neurosci 16:986064. https://doi.org/10.3389/fnbeh.2022.986064

Yamanaka N, Hua YJ, Mizoguchi A, Watanabe K, Niwa R, Tanaka Y et al (2005) Identification of a novel prothoracicostatic hormone and its receptor in the silkworm *Bombyx mori*. J Biol Chem 280:14684–14690. https://doi.org/10.1074/jbc.M500308200

Yamanaka N, Yamamoto S, Žiňan D, Watanabe K, Kawada T, Satake H et al (2008) Neuropeptide receptor transcriptome reveals unidentified neuroendocrine pathways. PLoS One 3:e3048. https://doi.org/10.1371/journal.pone.0003048

Yamashita O (1996) Diapause hormone of the silkworm, *Bombyx mori*: structure, gene expression and function. J Insect Physiol 42:669–679. https://doi.org/10.1016/0022-1910(96)00003-0

Yañez-Guerra LA, Elphick MR (2020) Evolution and comparative physiology of luqin-type neuropeptide signaling. Front Neurosci 14:130. https://doi.org/10.3389/fnins.2020.00130

Yañez-Guerra LA, Zhong X, Moghul I, Butts T, Zampronio CG, Jones AM et al (2020) Echinoderms provide missing link in the evolution of PrRP/sNPF-type neuropeptide signaling. elife 9: e57640. https://doi.org/10.7554/eLife.57640

Zdarek J, Nachman RJ, Denlinger DL (2000) Parturition hormone in the tsetse *Glossina morsitans*: activity in reproductive tissues from other species and response of tsetse to identified neuropeptides and other neuroactive compounds. J Insect Physiol 46:213–219. https://doi.org/10.1016/s0022-1910(99)00114-6

Žitňan D, Kingan TG, Adams ME (1996) Identification of ecdysis triggering hormone from an epitracheal endocrine system. Science 271:88–91. https://doi.org/10.1126/science.271.5245.88

Chapter 3
Comprehensive Analysis of Insect Peptide Hormones/Neuropeptides by Next-Generation Sequencing

Yoshiaki Tanaka

Abstract DNA sequencing technologies have been pivotal in genome analysis. Technological advances over the years, particularly the advent of next-generation sequencing, have drastically increased the scale and speed of genomic and transcriptomic data collection. These developments have enabled in-depth studies of peptide hormones/neuropeptides (PHs/NPs) and receptors across various insect species without whole-genome sequences. Techniques such as RNA sequencing and mass spectrometry have further enhanced our understanding of insect NPs, leading to the discovery of new PHs/NPs and expanding our knowledge of their evolutionary biology and physiology. Insect PHs/NPs show remarkable diversity and evolutionary adaptations across different lineages. The desert locust has many PHs/NPs that regulate complex behaviors and survival strategies. The PH/NP profiles vary among Hemiptera, possibly to adapt to different habitats of each species. Some PHs/NPs that are commonly present in many insects have been lost in holometabolous insects, which may be due to neutral evolution or functional redundancy. Variable and rapid evolutionary changes are observed within some PH/NP families, such as AST-CC, PK/PBAN, and EFLamide. However, many unknown aspects exist regarding the relationship between the loss/gain of PHs/NPs in individual species and their adaptation to diverse habitats.

Keywords Allatostatin C (AST-C) · Apterygote · CAPA/pyrokinin (PK) · EFLamide (EFL) · Hemiptera · Holometabolous insects · MALDI-TOF MS · Next-generation sequencing (NGS) · Orthoptera

Y. Tanaka (✉)
Institute of Agrobiological Sciences, NARO, Tsukuba, Japan
e-mail: tanaka.yoshiaki660@naro.go.jp

A. Mizoguchi, N. Okamoto (eds.), *Insect Peptide Hormones and Neuropeptides*, Entomology Monographs, https://doi.org/10.1007/978-981-95-7510-7_3

3.1 Progress in the Comprehensive Analysis of Insect PHs/NPs by Next-Generation Sequencing

DNA sequencing analysis forms the foundation of many areas of modern biology and has been widely used since the introduction of the Sanger method in the late 1970s (Sanger and Coulson 1975). However, this method requires a known sequence to initiate sequencing and can obtain only a few hundred bases from one analysis. Consequently, the "shotgun method," which involves randomly introducing numerous genomic DNA fragments into vector DNA and then sequencing them, was initially used for whole-genome sequence analysis. The results of whole-genome analysis of the fruit fly *Drosophila melanogaster* using a shotgun approach were published in 2000 (Adams et al. 2000; Myers et al. 2000). The knowledge gained from *D. melanogaster* has been extensively referenced in annotating subsequent insect whole-genome analyses. The insect PHs/NPs and their receptors (mainly G-protein-coupled receptors, GPCRs) were comprehensively analyzed in *D. melanogaster* using the known PHs/NPs and GPCRs as reference sequences (Hewes and Taghert 2001). This was the first report of a comprehensive analysis of PHs/NPs on insects, after which comprehensive analyses were performed on the African malaria mosquito *Anopheles gambiae* (Riehle et al. 2002), the western honey bee *Apis mellifera* (Hummon et al. 2006), the red flour beetle *Tribolium castaneum* (Li et al. 2008; Hauser et al. 2008), the silkworm *Bombyx mori* (Roller et al. 2008), the blood-sucking bug *Rhodnius prolixus* (Ons et al. 2009), the jewel wasp *Nasonia vitripennis* (Hauser et al. 2010), the pea aphid *Acyrthosiphon pisum* (Huybrechts et al. 2010), the termite *Zootermopsis nevadensis* (Veenstra 2014), and the body louse *Pediculus humanus* (Li et al. 2014).

The number of whole-genome analyses increased significantly in the 2010s with the emergence of next-generation sequencing (NGS) technologies in 2007. Short-read sequencing technologies, represented by Illumina's HiSeq system, allow the acquisition of a large amount of short sequence information in one sequence analysis. In contrast, long-read sequencing technologies, such as the PacBio sequencing system (Pacific Biosciences) and the MinION system (Nanopore Technologies), excel at analyzing single long-chain DNA sequences. A draft 6.5 Gb genome sequence of the migratory locust *Locusta migratoria*, the largest animal genome sequenced at that time, was obtained using the Illumina HiSeq 2000 sequencing platform (Wang et al. 2014). Following the release of the *Locusta* genome, a comprehensive list of neuropeptide genes was compiled based on genome information (Veenstra 2014). As of 2020, more than 50 families of PHs/NPs identified from 669 insect species have been registered in NCBI. There has been an approximately 30-fold increase in the number of insect species with registered peptide genes over the past decade.

The NGS-based sequence analysis has made significant contributions, particularly in the analysis of RNA rather than genomic DNA itself. RNA sequencing (RNA-seq) allows us to analyze gene expression on a large scale without needing genome information, which has been a huge step forward for molecular biological research in insect science.

An international consortium aimed at conducting evolutionary discussions through the transcriptome analysis of over 1000 insect species was formed (1KITE: 1K Insect Transcriptome Evolution). One result of this effort is a study on phylogenetic tree construction on the basis of the coding region sequences of 1478 genes (Misof et al. 2014).

The NGS has revolutionized genomics and transcriptomics, allowing for high-throughput identification and characterization of PHs/NPs and their receptors, even in non-model organisms. Through RNA-seq, researchers can now explore the complete set of genes encoding PHs/NPs and their receptors in various insects without a whole-genome sequence. The PHs/NPs and GPCRs have been comprehensively identified using RNA-seq and whole-genome sequencing in the brown plant hopper *Nilaparvata lugens* (Tanaka et al. 2014) and the desert locust *Schistocerca gregaria* (Ragionieri et al. 2022). In the rice stem borer *Chilo suppressalis* (Walker, Xu et al. 2016) and the stick insect *Carausius morosus* (Duan Şahbaz and Birgül Iyison 2019), the genes encoding PHs/NPs and GPCRs have been comprehensively identified without reference genome sequencing. These transcriptome analyses have promoted comprehensive studies of PHs/NPs and GPCRs in many insects. The Database for Insect Neuropeptide Research (https://www.neurostresspep.eu/diner) provides a consolidated resource for PH/NP information across numerous insect species, detailing isoform sequences, physiological functionality, and receptor-binding sites. This database covers 50 well-described PH/NP families from over 400 different insect species, illustrating the diversity and complexity of PH/NP functions in insects (Yeoh et al. 2017).

The emergence of highly sensitive mass spectrometry and the miniaturization of separation techniques have enabled the analysis of the entire peptidome of tissues at the amino acid sequence level from very little material, a process known as peptidomics. This analysis confirmed the presence of mature peptides predicted from the NGS sequence. The matrix-assisted laser desorption/ionization time-of-flight mass spectrometry (MALDI-TOF MS) involves ionizing peptides without fragmentation and measuring their mass-to-charge ratio. Peptide profiling by MALDI-TOF mass spectrometric analysis in adult *D. melanogaster* revealed the occurrence of 32 different peptides. This demonstrated a tagma-specific distribution of PHs/NPs within the central nervous system (CNS)-associated neurohemal system (Predel et al. 2004).

3.2 Diversity and Evolution of Insect PHs/NPs Among Insect Species

The evolution of insect PHs/NPs, as revealed by comprehensive analyses, encompasses a variety of topics that highlight the diversity and complexity of these molecules across different insect lineages. Table 3.1 shows the list of genes for PHs/NPs among representative insects. In a previous paper (Sturm et al. 2016), the genes encoding LQDVamide and agatoxin-like peptide were described as separate

Table 3.1 List of neuropeptide genes in representative insects

		Holometabolous				Hemimetabolous				Apterygote
		Lepidoptera	Hymenoptera	Diptera	Coleoptera	Hemiptera			Orthoptera	Zygentoma
Peptide precursor gene (abbreviation)	Consensus sequence	*B. mori*	*A. mellifera*	*D. melanogaster*	*T. castaneum*	*A. pisum*	*N. lugens*	*R. prolixus*	*S. gregaria*	*T. domestica*
Adipokinetic hormone/corazonin-related peptide (ACP)	pEVTFSRDWxxamide	+	nd	nd	+	nd	+	+	+	+
Adipokinetic hormones (AKH)	pExxFxPxWamide	2	+	+	2	+	+	+	3	+
Antidiuretic factor b (ADF)	YDDGSYKPHIYGF	nd	nd	nd	+	nd	nd	nd	nd	nd
Agatoxin-like peptide/LQDVamide (ALP)[1]	xYIDPxDxDxExxxxxYxxDPxxxQx LxDVamide (LQDV), GGxCDHRxxDCCxxSSCRCNLWxxNCRCQRM GLFQxWamide (ALP)	+	+	nd	+	nd	+	+	+	+
Allatostatin A (AST-A)	-FGLamide	+	+	+	nd	+	+	+	+	+
Allatostatin C (AST-C)	*(KR)QxRxxCYFNPxxCF	+	nd	+	+	nd	nd	nd	2	nd
Allatostatin CC (AST-CC)	*(RR)------RCxFNAVSCF	+	+	+	+	+	+	+	+	+
Allatostatin CCC (AST-CCC)	*(RKR)xYWKQCAFNAVSCFamide	nd	+	nd	nd	+	+	+	+	+
Allatotropin (AT)	-RGFKNxxxxTARGFamide	+	+	nd	+	+	+	+	+	+
Bursicon α subunit (Burs)	See Fig. 1 in Van Loy et al. (2007)	+	+	+	+	+	+	+	+	+
Capability/cardioacceleratory peptide 2b/periviscerokinin (CAPA/PK)	-FPRxamide	2	+	+	+	+	+	+	3	+
CCHamide 1 (CCH1)	xxCLxYGHxCWGxHamide	+	+	+	+	+	+	+	+	+
CCHamide 2 (CCH2)	GCxxxGHxCYGxxamide	+	+	+	+	+	+	+	2	+
CNMamide (CNM)	-CHFKICNMamide	nd	+	+	+	nd	+	+	+	+
Corazonin (CRZ)	pETFQYSxGWTNamide	+	+	+	nd	nd	+	+	+	+
Corticotropin-releasing factor (CRF)-like diuretic hormone (CRF-DH)	See Fig. 5 in Schooley et al. (2012)	3	+	+	3	+	+	+	+	+

Crustacean cardioactive peptide (CCAP)	PFCNAFTGCamide	+	+	+	+	+	+	+	+	+
Calcitonin-like diuretic hormone/DH31 (CT-DH)	See Table 1 in Zandawala (2012)	+	+	+	+	+	+	+	+	+
Ecdysis-triggering hormone (ETH)	-PRxamide	+	+	+	+	+	+	+	+	+
Eclosion hormone (EH)	See Fig. 3 in Tang et al. (2022)	+	+	+	+	2	2	+	3	2
EFLamide (EFL)	-GxEFLamide	nd	nd	nd	nd	nd	nd	+	+	+
Elevenin (ELE)	-CxxxxxxxxCRGxxx	nd	+	nd	+	nd	+	+	+	+
FMRFamide (FMRF)	-FMRFamide	+	+	+	+	+	+	+	+	+
Glycoprotein hormone alpha 2 (GPA2)	See Fig. 1 in Paluzzi et al. (2014)	+	nd	+	+	+	+	+	+	+
Glycoprotein hormone beta 5 (GPB5)	See Fig. 1 in Paluzzi et al. (2014)	+	nd	+	+	+	+	+	+	+
Gonadulin/ILP 8 (ILP8)	See Fig. S2 in Leyria et al. (2022)	nd	+	+	nd	+	nd	+	+	nd
Hansolin	-GxPLRWamide	nd	nd	nd	+	+	+	nd	+	+
IDL-like (IDL)	-xDLxR-	+	+	+	+	+	+	nd	+	+
IMFamide/SMYamide (IMF)	-PxNGIMFamide/-PxNGSMYamide	+	nd	nd	nd	nd	nd	nd	+	nd
Insulin-like peptide (ILP)	See Fig. 2 in Mizoguchi and Okamoto (2013)	>32	2	7	4	10	4	4	+	+
Ion-transport peptide (ITP)	See Fig. 2 in Yu et al. (2016)	+	+	+	+	+	+	+	+	+
Ion-transport peptide-like (ITPL)[2]	See Fig. 2 in Yu et al. (2016)	+	+	+	+	+	+	+	+	+
Inotocin (ITC)	CLITNCPRGamide	nd	nd	nd	+	nd	+	nd	+	+
Insect parathyroid hormone (iPTH)	See Fig. 3.1 in this chapter	nd	+	nd	+	nd	nd	nd	+	nd
ITG-like (ITG)	ITGQGNRxF	+	+	+	+	nd	+	+	2	nd
Kinin (K)	-FxxWGamide	+	+	+	nd	+	+	+	+	+
Limostatin (LST)	AlxFRPLFVYKQQxx	nd	+	+	nd	nd	nd	nd	nd	nd
Myoinhibitory peptide (MIP)	-WxxxxxxWamide	+	nd	+	+	+	+	+	2	+
Myosupressin (MS)	xDVxHxFLRFamide	2	+	+	+	+	+	+	2	+
Natalisin (NTL)	-FxxxRamide	+	nd	+	+	+	+	+	+	+

(continued)

Table 3.1 (continued)

Peptide precursor gene (abbreviation)	Consensus sequence	Holometabolous				Hemimetabolous				Apterygote
		Lepidoptera	Hymenoptera	Diptera	Coleoptera	Hemiptera			Orthoptera	Zygentoma
		B. mori	*A. mellifera*	*D. melanogaster*	*T. castaneum*	*A. pisum*	*N. lugens*	*R. prolixus*	*S. gregaria*	*T. domestica*
Neuroparsin/ovary ecdysteroidogenic hormone (OEH)	See Fig. 1 in Veenstra (2010)	+	+	nd	+	nd	+	+	4	+
Neuropeptide F (NPF)	-AxxxxxLxxLxxxxxxxARPRFamide	3	+	2	+	+	+	+	3	3
Neuropeptide-like precursor 1 (NPLP1)	-*(KR)xxxxLAR-	+	3	4	+	+	+	+	+	+
NVP-like (NVP)	xVPxYQExRx	+	+	nd	+	nd	nd	+	2	nd
Orcokinin (OK)	NxDEIDR- (OKA) -LDxxGGGxL- (OKB)[3]	2	+	+	+	+	2	4	2	2
Partner of bursicon/bursicon β subunit (Pburs)	See Fig. 1 in Van Loy et al. (2007)	+	+	+	+	+	+	+	+	+
Pyrokinin/pheromone biosynthesis-activating neuropeptide (PK/PBAN)	-FxPRLamide	+	+	+	+	+	+	+	+	+
Pigment-dispersing factor (PDF)	NSELINSLLGxPKxxNxxamide	+	+	+	+	nd	+	+	+	+
Proctolin (PRT)	RYLPT	nd	nd	+	+	+	+	+	+	+
Prothoracicotropic hormone (PTTH)	See Fig. 2 in Barberà and Martínez-Torres (2017)	+	nd	+	+	+	+	+	+	nd
RFLamide (RFL)	-RFLamide	nd	nd	+	+	nd	+	nd	+	nd
RYamide (RY)	-RYamide	+	+	+	+	nd	+	+	+	+
Short neuropeptide F (sNPF)	HxxxRKxxRxPxLRLRFamide	3	+	+	+	+	+	+	2	+
SIFamide (SIF)	-RxPPFNGSIFamide	+	+	+	+	+	+	+	+	+
Sulfakinin (SK)	-FDDYGHMRFamide	+	+	+	+	nd	+	+	+	+
Tachykinin-related peptide (TRP)	-FxxGxRamide	+	+	+	+	+	+	+	+	+
Trissin (TR)	-CDxCGxECxxACGTxHFRxCCFNYLR	2	nd	+	+	nd	nd	nd	+	+

(1) In Lepidoptera, the ALP gene encodes both LQDV and ALP in a precursor protein. (2) Alternative splicing variant of ITP. (3) OKA and/or OKB are produced by alternative splicing of a single gene. +, detected; nd, not detected. The number represents the quantity of detected neuropeptide precursors. *(), convertase cleavage site

genes. As LQDVamide is encoded by an agatoxin-like peptide precursor gene in Lepidoptera (Roller et al. 2017), this peptide is termed an agatoxin-like peptide/ LQDVamide (ALP) in this chapter.

3.2.1 Orthoptera

In *S. gregaria* (Acrididae, an infraorder of Orthoptera), 81 precursor genes encoding PHs/NPs typical of insects were found (Ragionieri et al. 2022). The dataset for *S. gregaria* confirms significant conservation of the entire neuropeptidergic gene set typical of insects, except for limostatin (LST, see Sect. 5.2.2) and coleopteran-specific antidiuretic factor (ADF). Most insects have lost one of the allatostatin C genes (*AST-C, AST-CC, AST-CCC*), but *S. gregaria* still has all three genes (Veenstra 2016). All three AST genes are also present in the two-spotted cricket *Gryllus bimaculatus* (Ensifera, suborder of Orthoptera), but CNMamide (CNM), EFLamide (EFL), inotocin (ITC), insect parathyroid hormone (iPTH), and tachykinin are not found in the cricket (Mochizuki et al. 2023).

Acrididae has a larger genome size than other Orthoptera (Hawlitschek et al. 2023), and *S. gregaria* has the largest genome size (approximately 8.55 Gb) among the insects sequenced and assembled to date (Verlinden et al. 2020). Large genomes may support complex regulatory mechanisms crucial for phase changes between solitary and gregarious forms, swarm formation, and long-distance migration. The conservation of the entire PH/NP gene set in Acrididae may be necessary for their critical survival and adaptive strategies, as suggested in the migratory locust (Wang et al. 2014).

3.2.2 Hemiptera

Hemiptera represents the largest and most diverse group of hemimetabolous insects (Stork 2018), including many harmful species to humans, animals, and plants, and a few essential natural enemies, such as predatory bugs. Most hemipterans are sap-sucking insects that feed on plant nutrients, and some are blood-sucking insects. The PHs/NPs of three representative species, *A. pisum* (Sternorrhyncha), *N. lugens* (Stål, Auchenorrhyncha), and *R. prolixus* (Heteroptera), are compared because they have distinct habitats and environmental adaptations. The AST-C, trissin (TR), and the TR receptor appear to be commonly absent in Hemiptera (Table 3.1 and Huang et al. 2024). The TR signaling pathway regulates evening activity offset in *D. melanogaster* (Sekiguchi et al. 2024) and appears to be involved in the regulation of feeding activity in *B. mori* (Roller et al. 2016). The function of the TR in Hemiptera might be redundant or could be substituted by other signals. Three receptors (kininR, SIFR, and NPFR) are commonly duplicated in the three hemipteran insects, suggesting that the duplication of these receptors may be a common evolutionary event for most hemipteran species (Gao et al. 2024). These three receptor types primarily regulate water balance, sexual behavior, and eating behavior (Terhzaz et al. 2007; Al-Anzi et al. 2010; Nässel and Wegener 2011). The significance of gene duplication for the survival of hemipteran insects remains to be explored.

The presence/absence of some PH/NP systems varies among the hemipteran insects. Parathyroid hormone (PTH) is a vertebrate peptide hormone that regulates the serum calcium concentration through its effects on bone, kidney, and intestine. Insect PTH receptors (iPTHRs) have been previously described as counterparts of vertebrate PTHRs; however, their authentic ligands and biological functions in insects were unknown. Recently, the neuropeptide originally described as PXXXamide was identified as the authentic ligand for iPTHR by using a GPCR functional assay in *T. castaneum* (Xie et al. 2020). This peptide is named insect PTH (iPTH), but iPTH has no sequence similarity to vertebrate PTH (Fig. 3.1). The genes encoding both iPTH and iPTHR are present in several hemipteran insects, including the bed bug *Cimex lectularius* (Xie et al. 2020), the brown marmorated stink bug *Halyomorpha halys*, and the reduviid predator *Rhynocoris fuscipes* (Fabricius, Fig. 3.1b). However, genes homologous to *iPTH* have not been identified in species such as the brown plant hopper (*N. lugens*) and the whitefly (*Bemisia tabaci*), where only iPTHR has been found (Tanaka et al. 2014; Xie et al. 2020). In *T. castaneum*, iPTH signaling is likely involved in the regulation of cuticle maturation at the time of adult eclosion and egg production (Xie et al. 2020), but the function of iPTH signaling has not yet been explored in hemimetabolous insects.

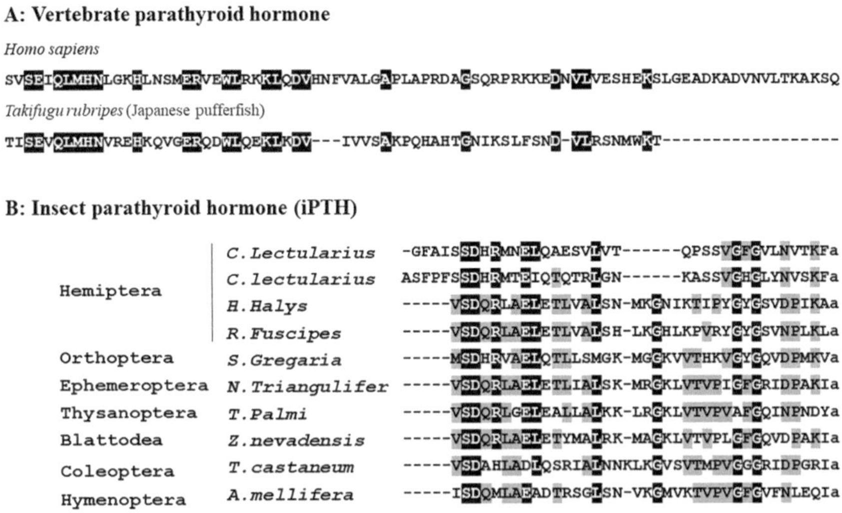

Fig. 3.1 Structure of parathyroid hormones. (**a**) Vertebrate parathyroid hormones. (**b**) Insect parathyroid hormones (iPTH). The letters with black backgrounds are identical (more than 80%), and those with gray backgrounds are similar amino acids, with the 50% majority rule

Compared with other hemipteran insects, aphids seem more prone to losing PH/NP signaling systems. The ALP, corazonin (CRZ), adipokinetic hormone (AKH)/CRZ-related peptide (ACP), ITC, and sulfakinin (SK) systems (both the peptide and receptor) appear to be completely lost in *A. pisum* (Gao et al. 2024). The AKH, CRZ, and ACP are thought to be members of the invertebrate gonadotropin-releasing hormone (GnRH) superfamily (see Sect. 2.3.10). The loss of CRZ and ACP is a common phenomenon in Sternorrhyncha (Gao et al. 2024). The SK bears structural homology to the vertebrate gastrin/cholecystokinin family (see Sect. 2.3.17), inhibits food uptake in several insects, and stimulates the release of digestive enzymes into the gut (Schoofs and Nachman 2006; Downer et al. 2007). The loss of SK signaling may underpin aphids' prolonged phloem-feeding behavior (Gao et al. 2024). Conversely, several GPCRs associated with feeding and water balance, including CCH1R, CT-DHR, kininR, NPFR, SIFR, and type A leucine-rich repeat-containing G protein-coupled receptors (LGRs) (receptors for GPA2/GPB5) in *A. pisum*, have undergone duplication (Gao et al. 2024). The PH/NP signaling systems of aphids may be specialized to adjust to their unique lifestyles, such as ant or endosymbiont–aphid mutualism, sexual and wing polymorphism, and honeydew production.

Despite their completely different feeding habits, the PH/NP signaling systems are very similar but slightly different between *N. lugens* and *R. prolixus*. The EFL, gonadulin, and NVP-like (NVP) are not found in the plant-feeding insect *N. lugens*, whereas hansolin, ITC, and RFLamide (RFL) are not found in the blood-feeding insect *R. prolixus*. Gonadulin and NVP play essential roles in the development of *R. prolixus*. Gonadulin is an ortholog of *Drosophila* insulin-like peptide 8 (DILP8; Veenstra 2020; see Sect. 2.2.1) and inhibits ovulation and oviposition in *R. prolixus* (Leyria et al. 2022). The NVP compounds are speculated to be involved in food intake in *R. prolixus* (Sterkel et al. 2011). LGR3 is a putative receptor for gonadulin, and the ortholog of LGR3 is not found in *N. lugens* (Tanaka et al. 2014). These findings suggest that gonadulin signaling is absent in *N. lugens*. Hansolin and RFL have recently been identified as putative PHs/NPs in *C. morosus* by mass spectrometry (Liessem et al. 2018), and these PHs/NPs are present in Polyneoptera (Bläser and Predel 2020). A BLAST search revealed that RFL appears to be common among insects, including Heteroptera, whereas hansolin is not found in Heteroptera. Hansolin is probably absent in *R. prolixus*. The presence of four splicing variants of the hansolin gene may indicate some function in *N. lugens*. In the Cuban subterranean termite *Prorhinotermes simplex*, hansolin is essential for reproduction and caste determination (Veenstra 2023).

3.2.3 Holometabolous Insects

A characteristic feature of PHs/NPs in holometabolous insects is the lack of some PHs/NPs that are ubiquitous in insects. For example, in the beetle *T. castaneum*, the PHs/NPs present in most insect species, such as AST-A, CRZ, and kinin, are lost in *T. castaneum* (Li et al. 2008).

The loss of many PHs/NPs has also been observed in other holometabolous insects. The AST-C, GPA2/GPB5, MIP, and proctolin (PRT) appear to be commonly lost in Hymenoptera (Jindal et al. 2022; Habenstein et al. 2021; Hauser et al. 2010; Hummon et al. 2006). Furthermore, the presence or absence of specific PH/NP systems varies among Hymenoptera. Prothoracicotropic hormone (PTTH) is a hormone that stimulates the prothoracic glands to secrete ecdysteroid in insects (see Sect. 2.6.1). The PTTH and its receptor torso are present in all available genomes of solitary bees, but both PTTH and torso appear to have been lost in all fully sequenced genomes of social bees (Costa et al. 2022). The SK is present in *A. mellifera* and ants but is absent in most wasps, except the ichneumonid *Diadromus collaris*. Conversely, ITC is absent in bees but present in sawflies, wasps, and ants (Muratspahić et al. 2020). The TR is lacking in *A. mellifera* but is present in wasps and ants. This diversity in the presence or absence of specific PHs/NPs within the same order was also observed in Diptera. Allatotropin (AT), ACP, and neuroparsin/ovary ecdysteroidogenic hormone (OEH) are lost in flies but present in mosquitoes. These peptides are functional in mosquitoes. The AT regulates the immune response (Hernández-Martínez et al. 2017). The OEH is important in mosquito reproduction (Brown et al. 1998). The ACP appears to regulate glycogen and carbohydrates in adult mosquitoes (Afifi et al. 2023). The ACP signaling system is thought to have been abandoned several times during insect evolution. In Diptera, the ACP signaling system is lost in flies, whereas mosquitoes appear to need the ACP signaling system to survive in extreme habitats or complex ecological niches (Hansen et al. 2010). The same hypothesis as for ACP may also apply to AT and neuroparsin/OEH.

The types of genes encoding PHs/NPs are almost identical among Lepidoptera (Luo et al. 2024). The AST-CCC, CNM, and PRT are lost in Lepidoptera. The CNM is present in many insect species except for Lepidoptera and aphids and has diverse functions, such as the promotion of warmth-induced wakefulness (Jin et al. 2021), the regulation of feeding decisions (Peng et al. 2021), and the regulation of oviposition (Grmai et al. 2024). The CNM receptor (CNMR)-like GPCR is present in certain lepidopteran species, including *B. mori* and *C. suppressalis*. However, the CNMR-like GPCR of *B. mori* shows only weak sensitivity to CNM, suggesting that CNMR has evolved as a receptor for another unknown ligand in *B. mori* (Jung et al. 2014). The PRT is considered a potent stimulator of visceral and skeletal muscles, directly affecting insect behavior (Fiandra et al. 2010; Ormerod et al. 2016), but it is absent in both Lepidoptera and Hymenoptera. In the phylogenetic tree, the PRT receptor is located near the CNMR, but only the CNMR-like GPCR is found in Lepidoptera (Xu et al. 2016). These findings suggest that Lepidoptera has lost the PRT signaling system.

Holometabolous insects often present relatively high evolutionary rates, which correlate with relatively high rates of gene loss. This is a stochastic process in which genes can be lost through neutral evolution, meaning that their losses do not necessarily confer a disadvantage to the organism (Wyder et al. 2007). The substantial loss of neuropeptide genes in holometabolous insects is due primarily to increased evolutionary rates leading to neutral gene losses, functional redundancy, adaptive evolution, and specific environmental adaptations. This dynamic evolutionary process ensures that insects retain only those essential genes for survival and ecological success.

3.2.4 *Apterygote Insects*

The apterygote is a wingless insect and includes the most evolutionary primitive orders, Archaeognatha (jumping bristletails) and Zygentoma (silverfish, firebrats). Most insect PHs/NPs that are commonly present among insects are present in these insects, except for AST-C, ITG-like (ITG), and PTTH (Table 3.1 in this chapter and Derst et al. 2016). The PTTH signaling may be absent because the torso is also unidentified in apterygote insects. The amino acid substitutions of the mature peptides are found in their precursors, but the amino acid substitutions are restricted to a few positions within each peptide. Several insect neuropeptide precursors contain multiple bioactive peptides. In apterygote insects, more than ten paracopies of putative bioactive peptides are encoded by the precursor genes for AST-A, MIP, and EFL (Fig. 3.2). Gene duplication occurs in the EH, ITC, NPF, and SIF genes in Zygentoma (Derst et al. 2016). The sequences of neuropeptide precursor genes in apterygote insects show a variety of modifications, such as amino acid substitutions, increasing numbers of gene duplications and splice variants. These modifications appear to remain within the limits of variation hitherto known from insects (Wang et al. 2015). This may reflect their gradual evolutionary adaptation to stable environments. Apterygote insects primarily inhabit stable environments such as soil or surface layers, where dramatic evolution or adaptations may not have been needed.

3.3 Diversity and Evolution of the Neuropeptide Family

3.3.1 *AST-C, AST Double C (AST-CC), and AST Triple C (AST-CCC)*

The AST-C is the arthropod homolog of vertebrate somatostatin (see Sect. 2.3.3). The AST-C is a disulfide bridge-containing neuropeptide that was initially identified from the tobacco hornworm *Manduca sexta* and inhibits the biosynthesis of juvenile hormone (JH) in holometabolous insects, except for Hymenoptera (Kramer et al. 1991; Li et al. 2006; Abdel-Latif and Hoffmann 2010). Later, AST-CC and AST-CCC were identified as paralogs of AST-C. A notable characteristic of AST-C is the presence of a predicted N-terminal pyroglutamate residue. The AST-CC has the most variable structure of the three, and its precursor has an Arg-Arg (RR) convertase cleavage site instead of a Lys-Arg (KR) convertase site on the N-terminus of the predicted peptide (Table 3.1). The AST-CCC is the only one that is consistently amidated at its C-terminus. In several articles (Tanaka et al. 2014; Derst et al. 2016; Christie 2020; Işbilir et al. 2020), the amidated AST-C-like peptides are also termed AST-C, but they should be corrected to AST-CCC. These three peptides possibly arose from local gene triplication early during arthropod evolution (Veenstra 2016). The majority of insect species have AST-CC and AST-CCC (apterygote, hemimetabolous insects, and Hymenoptera) or AST-CC and AST-C (Lepidoptera, Diptera, and Coleoptera). Locusts and the crickets have all three peptides, and the primitive hymenopteran *Athalia rosae*, the sawfly, has all three peptides, although the structure of AST-C in this species is somewhat incomplete and may show signs (a lack of N-terminal pyroglutamate residue) of becoming a pseudogene (Veenstra 2016). Most insect species have only one AST-C receptor. The speculation that two different peptides may act through the same receptor has been confirmed for AST-C and AST-CC in *T. castaneum* (Audsley et al. 2013) and for AST-CC and AST-CCC in *A. mellifera* (Urlacher et al. 2016). Locusts have all three peptides, but only one AST-C receptor has been identified. Three different peptides may act through the same AST-C receptor in locusts.

The different distributions of these peptides may suggest functional differences among insect species. In *B. mori*, *AST-C* and *AST-CC* are expressed in different cells of the brain, abdominal ganglia (AG), and terminal abdominal ganglion (TAG; Roller et al. 2017). In the brain, only AST-C is produced in neurosecretory cells (NSCs) and delivered to the CA to inhibit JH biosynthesis in the CA (Kaneko and Hiruma 2014), whereas *AST-CC* is not expressed in NSCs of the brain. In holometabolous insects, with the exception of Hymenoptera, only AST-C shows allatostatic activity, and AST-A and MIP (AST-B) have no allatostatic effect (Stay and Tobe 2007). The AST-C appears to be a regulator of JH synthesis in three holometabolous insects (Lepidoptera, Diptera, and Coleoptera). In contrast, AST-A and MIP appear to be regulators of JH synthesis in hemimetabolous insects because they show allatostatic activity (Stay and Tobe 2007; Matsumoto et al. 2017), and AST-C is lost in hemimetabolous insects except the locust and cricket (Fig. 3.2). *AST-CC* is expressed in the ecdysis-triggering hormone (ETH) receptor A (ETHR-A)-expressing cells of the AG and initiates ecdysis behavior, but *AST-C* may not be involved in the regulation of ecdysis behavior because *AST-C* is not expressed in the ETHR-expressing cells in *B. mori* (Daubnerová et al. 2021). These data show that differential expressions result in distinct functions between the two peptides. The distinct functions elicited by the different distributions of the three AST-C peptides have also been reported in the American lobster *Homarus americanus* (Dickinson et al. 2018). The AST-CCC ubiquitously occurs among insects except Lepidoptera, Diptera, and Coleoptera (Fig. 3.2), but its function remains unknown in insects. The AST-CCC is detected in the CNS (Boerjan et al. 2010) and in the enteroendocrine cells (EECs) of the midgut of *A. mellifera* (Christie 2020). The EECs in the insect midgut respond to physiological changes in the intestine by releasing multiple peptides to control food intake, gastrointestinal activity, and systemic metabolism (see Sect. 5.3.2). One of the distinct functions of AST-CCC may be related to the regulation of feeding and digestive activity in insects (Fig. 3.3).

Lygus hesperus **(Hemiptera) MT210023.1**

MSCSHSALVLAAIGVVCLAWPHSVMA**IDLSRLYGHVAV**KRNSEACHPYEPFKCPGDGACISIQYLCDGAPDCLDGYDEDSR

LCTAAVYFDKRNRWARKKCSSPGSDCDGKR**TVGTEFL**GKRRLPDYKSKR**TVGTEFL**GKRRYPPI

Thermobia domestica **(Zygentoma)**

MVDMKSRFFHLFILICLSTGVTPENTEESPEIYDAEEKRGLTARELLRFFLMAFHNQPDSGLGMHGIPYRPV

RR**LGSEFL**GKRSSTDDHVDDKR**VMGSELL**GKR**ALGSEFL**GKR**ALGSEFL**GKR**ALGSEFL**GKR**ALGSEFL**GKR

ALGSEFLGKR**ALGSEFL**GKR**ALGSEFL**GKR**ALGSEFL**GKR**SLGSEFL**GKRSDEFDKR**VMGSEFL**GKR**ALGSE**

FLGKR**ALGSEFL**GKRDSESDYLEKK**ALGSEFL**GKR**ALGSEFL**GKR**ALGSEFL**GKRTPEQEENLENHKEVVAE

PWTFEGSPEHNSTQILLLERKKRFA

IDL (enclosed in a square), EFL (bold), Cleavage site (underline),
Signal peptide (marked in gray).

Fig. 3.2 Precursor sequences containing EFLamide peptides of Hemiptera and apterygote insects. (Upper) Lygus hesperus (the western tarnished plant bug, sequence ID: MT210023.1). (Lower) Thermobia domestica (the firebrat; Derst et al. 2016). Gray background signal peptides, boxed text mature IDL-like peptide, bold mature EFLamide

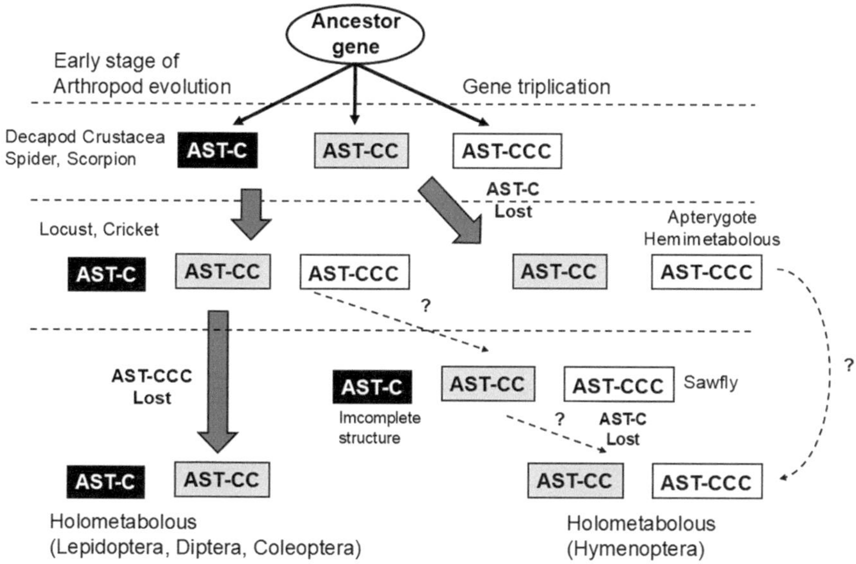

Fig. 3.3 Evolution of AST-C family genes. AST-C allatostatin C, AST-CC allatostatin double C, AST-CCC allatostatin triple C

3.3.2 PK/PBAN Peptide Family

The CAPA/pyrokinin (PK) genes provide an interesting example of rapid evolution, including duplication of a neuropeptide gene encoding different ligands. The ancient CAPA/PK gene in arthropods is thought to encode two types of PH/NP: periviscerokinins (PVKs) and PKs. This basic pattern is typical of various taxa of Myriapoda (millipedes, centipedes). Two separate genes, *CAPA* and *PK*, seem to have "suddenly" appeared in Diplura (two-pronged bristletails, non-insect hexapods) because of gene duplication of the original *CAPA/PK* (Derst et al. 2016). Both novel genes underwent extensive differentiation in insects (Fig. 3.4). The *PK* encodes a tryptopyrokinin (TPK) and PKs in Zygentoma, although Diplura *PK* encodes only PKs. In insects, *PK* is termed *PK/PBAN* (in *Drosophila*, termed *hugin*). Both the *CAPA* and *PK/PBAN* genes typically encode a TPK, and these TPKs are designated diapause hormones DH1 and DH2, respectively (Jurenka 2015). DH2 is termed diapause hormone in *B. mori* (see Sect. 8.2.1). *L. migratoria* appears to have entirely lost coding sequences for DH2, leaving DH1 peptides solely responsible for inducing egg diapause under a short photoperiod in *L. migratoria* (Hao et al. 2019). Loss of DH2 has also occurred independently within Heteroptera and Neodiptera (Redeker et al. 2017; Ahn and Choi 2018; Hao et al. 2019). Large-scale rearrangements in peptide coding sequences coupled with gene duplications have been reported in species of two polyneopteran groups, Dictyoptera (superorders including Blattodea and Mantodea) and Orthoptera (Predel et al. 2001; Redeker et al. 2017; Hao et al. 2019). In *L. migratoria*, five genes encode only TPKs in addition to *CAPA* and *PK/PBAN*.

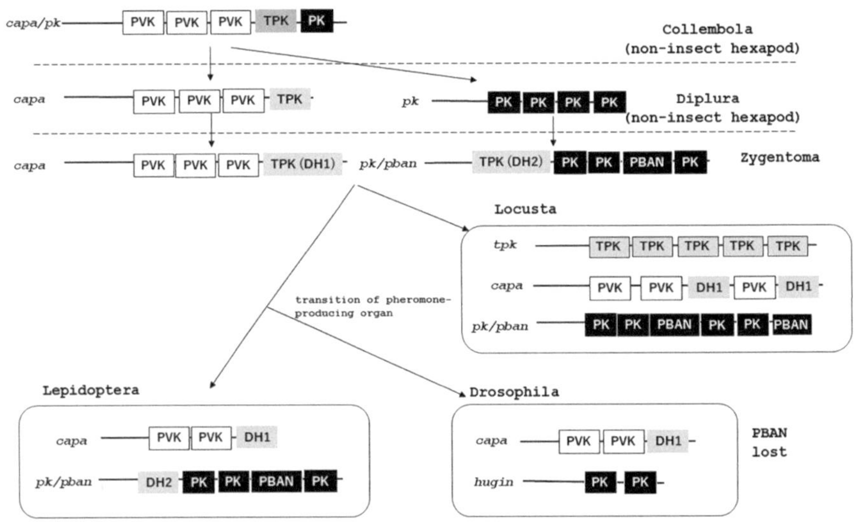

Fig. 3.4 Evolution of CAPA and the pyrokinin peptide gene. PVK periviscerokinin, PK pyrokinin, TPK trypto-PK, DH diapause hormone, PBAN pheromone biosynthesis-activating neuropeptide

PBAN is nearly universally present in insects. In its precursor protein, PBAN resides between the two PRXamide neuropeptides (β-subesophageal ganglion neuropeptide (β-SGNP) and γ-SGNP), but PBAN is lost in the higher Diptera. The absence of PBAN in higher Diptera may be related to the fact that this peptide no longer plays a functional role in these insects. Pheromone biosynthesis in *D. melanogaster* is directly regulated by the release of PDF from circadian rhythm-generating clock neurons in the brain (Krupp et al. 2013). Furthermore, pheromone biosynthesis occurs in large epidermal secretory cells called oenocytes in higher flies, whereas basal Diptera and most other insects produce pheromones in PBAN-responsive abdominal pheromone glands (Blomquist et al. 1987; Spiegel et al. 2016). Perhaps the transition of pheromone-producing organs contributes to the obsolescence of the PBAN peptide in higher Diptera (Farris 2023).

3.3.3 EFLamide (EFL)

The EFL is an ortholog of vertebrate thyrotropin-releasing hormone (Bauknecht and Jékely 2015; Van Sinay et al. 2017). The presence of EFL signaling illustrates the diverse evolution of PH/NP signaling among insects. The EFL precursor proteins of apterygote insects include more than ten paracopies of mature EFL, and the green drake mayfly *Ephemera danica* (Ephemeroptera) also has a similar EFL precursor (Sequence ID: KAF4533180.1). However, in Hemiptera and Orthoptera, the EFL precursor includes only one or two copies of mature EFL. In some Hemiptera and Orthoptera, the EFL-encoding exon is fused downstream of the IDL-like (IDL) gene (*IDL-EFL*). A single gene generates two splicing variants, *IDL* and *IDL-EFL*, in these insects. On the other hand, *IDL* and *EFL* exist as separate genes in chelicerates, crustaceans, basal hexapods, and basal insects, including early winged Ephemeroptera and Zygentoma (Kotwica-Rolinska et al. 2020). In holometabolous insects, both EFL and the EFL receptor (EFLR) are lost, and the precursor protein of the IDL does not include the EFL-like sequence. This finding suggests that the connection of the EFL-encoding exon with *IDL* is an evolutionary novelty found only in Hemiptera and Locusta.

The EFLR was first identified in *L. migratoria* (Veenstra and Šimo 2020). In Hemiptera, the occurrence of EFL and its receptor is diverse. Both EFLR and EFL are found in Heteroptera (*R. prolixus*), Aleyrodoidea, Psylloidea (infraorder of Sternorrhyncha, whitefly, psyllid), and Cicadomorpha (infraorder of Auchenorrhyncha, sharpshooter), whereas only the receptor is identified in Fulgoromorpha (infraorder of Auchenorrhyncha, *N. lugens*). Both the receptor and the ligand are missing in the Aphidoidea (infraorder of Sternorrhyncha, aphid; Kotwica-Rolinska et al. 2020). The absence of only the ligand in Fulgoromorpha may suggest that the early steps of signaling loss are captured in this species (Kotwica-Rolinska et al. 2020). The absence of ligands and receptors has also been observed in other hemimetabolous insects (Blattodea and Thysanoptera) and holometabolous insects (Kotwica-Rolinska et al. 2020). The presence of more than

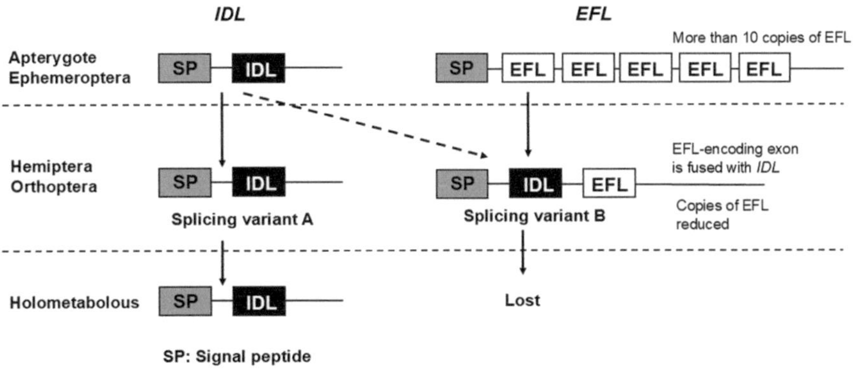

Fig. 3.5 Evolution of the EFLamide gene. SP signal peptide, IDL IDL-like peptide, EFLa EFLamide

ten paracopies of mature EFL in the precursor may suggest that EFL has some physiological functions in apterygote insects and Empheroptera. The number of paracopies in the precursor dramatically decreased to only one or two in Hemiptera and Orthoptera. Homozygous null *EFL* mutants do not show any visible defects in development, reproduction, lifespan, diapause induction, or circadian rhythmicity in the linden bug *Pyrrhocoris apterus* (Kotwica-Rolinska et al. 2020). In locusts, *EFL* is expressed in a few neurons, and EFLR can be activated by nanomolar concentrations of two predicted EFL peptides (Veenstra and Šimo 2020). These findings suggest that the EFL signaling system is still functional in Orthoptera.

The EFL signaling in insects may represent the process of gene loss during evolution. First, the number of paracopy genes in the precursor dramatically decreases, after which the ligand gene is lost, and finally, the receptor gene is lost. The function of EFL might become superfluous as insects evolve, regardless of its original function as an arthropod neurohormone (Veenstra and Šimo 2020) (Fig. 3.5).

3.4 Future Perspective

Currently, gene analysis is being conducted on many insect species, and the diversity of insect PHs/NPs is gradually being elucidated. However, many unknown aspects exist regarding the relationship between the loss/gain of PHs/NPs in individual species and their adaptation to their living environment. In the future, it will be necessary to elucidate the mechanisms of action of each neuropeptide and clarify how these actions contribute to survival to understand the evolutionary significance of neuropeptide loss/gain. In *D. melanogaster*, various genetic manipulation techniques have been developed, significantly contributing to the comprehensive understanding of the neuropeptide network. However, insects have developed diverse regulatory mechanisms to adapt to different habitats, and understanding the entire insect species cannot be achieved solely through *D. melanogaster*. Therefore, developing various genetic manipulation techniques for species other than *D. melanogaster* in the future will be necessary.

References

Abdel-Latif M, Hoffmann KH (2010) Neuropeptide regulators of the juvenile hormone biosynthesis (*in vitro*) in the beetle, *Tenebrio molitor* (Coleoptera, Tenebrionidae). Arch Insect Biochem Physiol 74:135–146. https://doi.org/10.1002/arch.20359

Adams MD, Celniker SE, Holt RA, Evans CA, Gocayne JD, Amanatides PG et al (2000) The genome sequence of *Drosophila melanogaster*. Science 287:2185–2195. https://doi.org/10.1126/science.287.5461.2185

Afifi S, Wahedi A, Paluzzi JP (2023) Functional insight and cell-specific expression of the adipokinetic hormone/corazonin-related peptide in the human disease vector mosquito, *Aedes aegypti*. Gen Comp Endocrinol 330:114145. https://doi.org/10.1016/j.ygcen.2022.114145

Ahn SJ, Choi MY (2018) Identification and characterization of capa and pyrokinin genes in the brown marmorated stink bug, *Halyomorpha halys* (Hemiptera): gene structure, immunocytochemistry, and differential expression. Arch Insect Biochem Physiol 99:e21500. https://doi.org/10.1002/arch.21500

Al-Anzi B, Armand E, Nagamei P, Olszewski M, Sapin V, Waters C et al (2010) The leucokinin pathway and its neurons regulate meal size in Drosophila. Curr Biol 20:969–978. https://doi.org/10.1016/j.cub.2010.04.039

Audsley N, Vandersmissen HP, Weaver R, Dani P, Matthews J, Down R et al (2013) Characterisation and tissue distribution of the PISCF allatostatin receptor in the red flour beetle, *Tribolium castaneum*. Insect Biochem Mol Biol 43:65–74. https://doi.org/10.1016/j.ibmb.2012.09.007

Barberà M, Martínez-Torres D (2017) Identification of the prothoracicotropic hormone (Ptth) coding gene and localization of its site of expression in the pea aphid *Acyrthosiphon pisum*. Insect Mol Biol 26:654–664. https://doi.org/10.1111/imb.12326

Bauknecht P, Jékely G (2015) Large-scale combinatorial deorphanization of platynereis neuropeptide GPCRs. Cell Rep 12:684–693. https://doi.org/10.1016/j.celrep.2015.06.052

Bläser M, Predel R (2020) Evolution of neuropeptide precursors in Polyneoptera (Insecta). Front Endocrinol (Lausanne) 11:197. https://doi.org/10.3389/fendo.2020.00197

Blomquist GJ, Dillwith JW, Adams TS (1987) Biosynthesis and endocrine regulation of sex pheromone production in Diptera. In: Prestwich GD, Blomquist GJ (eds) Pheromone biochemistry. Academic, Orlando, pp 217–250

Boerjan B, Cardoen D, Bogaerts A, Landuyt B, Schoofs L, Verleyen P (2010) Mass spectrometric profiling of (neuro)-peptides in the worker honeybee, *Apis mellifera*. Neuropharmacology 58: 248–258. https://doi.org/10.1016/j.neuropharm.2009.06.026

Brown MR, Graf R, Swiderek KM, Fendley D, Stracker TH, Champagne DE et al (1998) Identification of a steroidogenic neurohormone in female mosquitoes. J Biol Chem 273:3967–3971. https://doi.org/10.1074/jbc.273.7.3967

Christie AE (2020) Assessment of midgut enteroendocrine peptide complement in the honey bee, *Apis mellifera*. Insect Biochem Mol Biol 116:103257. https://doi.org/10.1016/j.ibmb.2019.103257

Costa CP, Okamoto N, Orr M, Yamanaka N, Woodard SH (2022) Convergent loss of prothoracicotropic hormone, a canonical regulator of development, in social bee evolution. Front Physiol 13:831928. https://doi.org/10.3389/fphys.2022.831928

Daubnerová I, Roller L, Satake H, Zhang C, Kim YJ, Žitňan D (2021) Identification and function of ETH receptor networks in the silkworm *Bombyx mori*. Sci Rep 11:11693. https://doi.org/10.1038/s41598-021-91022-8

Derst C, Dircksen H, Meusemann K, Zhou X, Liu S, Predel R (2016) Evolution of neuropeptides in non-pterygote hexapods. BMC Evol Biol 16:51. https://doi.org/10.1186/s12862-016-0621-4

Dickinson PS, Armstrong MK, Dickinson ES, Fernandez R, Miller A, Pong S et al (2018) Three members of a peptide family are differentially distributed and elicit differential state-dependent responses in a pattern generator-effector system. J Neurophysiol 119:1767–1781. https://doi.org/10.1152/jn.00850.2017

Downer KE, Haselton AT, Nachman RJ, Stoffolano JJG (2007) Insect satiety: sulfakinin localization and the effect of drosulfakinin on protein and carbohydrate ingestion in the blow fly, *Phormia regina* (Diptera: Calliphoridae). J Insect Physiol 53:106–112. https://doi.org/10.1016/j.jinsphys.2006.10.013

Duan Şahbaz B, Birgül Iyison N (2019) Prediction and expression analysis of G protein-coupled receptors in the laboratory stick insect, *Carausius morosus*. Turk J Biol 43:77–88. https://doi.org/10.3906/biy-1809-27

Farris SM (2023) Insect PRXamides: evolutionary divergence, novelty, and loss in a conserved neuropeptide system. J Insect Sci 23:3. https://doi.org/10.1093/jisesa/ieac079

Fiandra L, Caccia S, Giordana B, Casartelli M (2010) Leucine transport by the larval midgut of the parasitoid *Aphidius ervi* (Hymenoptera). J Insect Physiol 56:165–169. https://doi.org/10.1016/j.jinsphys.2009.09.015

Gao H, Li Y, Zhang X, Zhang H, Tian Y, Li B (2024) Unraveling the G protein-coupled receptor superfamily in aphids: contractions and duplications linked to phloem feeding. Gen Comp Endocrinol 347:114435. https://doi.org/10.1016/j.ygcen.2023.114435

Grmai L, Michaca M, Lackner E, Nampoothiri VPN, Vasudevan D (2024) Integrated stress response signaling acts as a metabolic sensor in fat tissues to regulate oocyte maturation and ovulation. Cell Rep 43:113863. https://doi.org/10.1016/j.celrep.2024.113863

Habenstein J, Schmitt F, Liessem S, Ly A, Trede D, Wegener C et al (2021) Transcriptomic, peptidomic, and mass spectrometry imaging analysis of the brain in the ant *Cataglyphis nodus*. J Neurochem 158:391–412. https://doi.org/10.1111/jnc.15346

Hansen KK, Stafflinger E, Schneider M, Hauser F, Cazzamali G, Williamson M et al (2010) Discovery of a novel insect neuropeptide signaling system closely related to the insect adipokinetic hormone and corazonin hormonal systems. J Biol Chem 285:10736–10747. https://doi.org/10.1074/jbc.M109.045369

Hao K, Tu X, Ullah H, McNeill MR, Zhang Z (2019) Novel *Lom-dh* genes play potential role in promoting egg Diapause of *Locusta migratoria* L. Front Physiol 10:767. https://doi.org/10.3389/fphys.2019.00767

Hauser F, Cazzamali G, Williamson M, Park Y, Li B, Tanaka Y et al (2008) A genome-wide inventory of neurohormone GPCRs in the red flour beetle *Tribolium castaneum*. J Front Neuroendocrinol 29:142–165. https://doi.org/10.1016/j.yfrne.2007.10.003

Hauser F, Neupert S, Williamson M, Predel R, Tanaka Y, Grimmelikhuijzen CJ (2010) Genomics and peptidomics of neuropeptides and protein hormones present in the parasitic wasp *Nasonia vitripennis*. J Proteome Res 9:5296–5310. https://doi.org/10.1021/pr100570j

Hawlitschek O, Sadílek D, Dey LS, Buchholz K, Noori S, Baez IL et al (2023) New estimates of genome size in Orthoptera and their evolutionary implications. PLoS One 18:e0275551. https://doi.org/10.1371/journal.pone.0275551

Hernández-Martínez S, Sánchez-Zavaleta M, Brito K, Herrera-Ortiz A, Ons S, Noriega FG (2017) Allatotropin: a pleiotropic neuropeptide that elicits mosquito immune responses. PLoS One 12:e0175759. https://doi.org/10.1371/journal.pone.0175759

Hewes RS, Taghert PH (2001) Neuropeptides and neuropeptide receptors in the *Drosophila melanogaster* genome. Genome Res 11:1126–1142. https://doi.org/10.1101/gr.169901

Huang C, Dong X, Yang X, Zou J, Yang M, Wang X, Li W, He Y (2024) Identification of neuropeptides and their G protein-coupled receptors in the predatory stink bug, *Arma custos*

(Hemiptera: Pentatomidae). Arch Insect Biochem Physiol 115:e22094. https://doi.org/10.1002/arch.22094

Hummon AB, Richmond TA, Verleyen P, Baggerman G, Huybrechts J, Ewing MA et al (2006) From the genome to the proteome: uncovering peptides in the Apis brain. Science 314:647–649. https://doi.org/10.1126/science.1124128

Huybrechts J, Bonhomme J, Minoli S, Prunier-Leterme N, Dombrovsky A, Abdel-Latief M et al (2010) Neuropeptide and neurohormone precursors in the pea aphid, *Acyrthosiphon pisum*. Insect Mol Biol 19(Suppl 2):87–95. https://doi.org/10.1111/j.1365-2583.2009.00951.x

Işbilir A, Duan Sahbaz B, Tuncgenc G, Bünemann M, Lohse MJ, Birgül-Iyison N (2020) Pharmacological characterization of the stick insect *Carausius morosus* Allatostatin-C receptor with its endogenous agonist. ACS Omega 5:32183–32194. https://doi.org/10.1021/acsomega.0c03382

Jin X, Tian Y, Zhang ZC, Gu P, Liu C, Han J (2021) A subset of DN1p neurons integrates thermosensory inputs to promote wakefulness via CNMa signaling. Curr Biol 31:2075–2087.e6. https://doi.org/10.1016/j.cub.2021.02.048

Jindal V, Li D, Rault LC, Fatehi S, Singh R, Mating M et al (2022) Bee-safe peptidomimetic acaricides achieved by comparative genomics. Sci Rep 12:17263. https://doi.org/10.1038/s41598-022-20110-0

Jung SH, Lee JH, Chae HS, Seong JY, Park Y, Park ZY et al (2014) Identification of a novel insect neuropeptide, CNMa and its receptor. FEBS Lett 588:2037–2041. https://doi.org/10.1016/j.febslet.2014.04.028

Jurenka R (2015) The PRXamide neuropeptide signaling system: conserved in animals. In: Advances in insect physiology, vol 49. Elsevier, pp 123–170

Kaneko Y, Hiruma K (2014) Short neuropeptide F (sNPF) is a stage-specific suppressor for juvenile hormone biosynthesis by corpora allata, and a critical factor for the initiation of insect metamorphosis. Dev Biol 393:312–319. https://doi.org/10.1016/j.ydbio.2014.07.014

Kotwica-Rolinska J, Krištofová L, Chvalová D, Pauchová L, Provazník J, Hejníková M et al (2020) Functional analysis and localisation of a thyrotropin-releasing hormone-type neuropeptide (EFL) in hemipteran insects. Insect Biochem Mol Biol 122:103376. https://doi.org/10.1016/j.ibmb.2020.103376

Kramer SJ, Toschi A, Miller CA, Kataoka H, Quistad GB, Li JP et al (1991) Identification of an allatostatin from the tobacco hornworm *Manduca sexta*. Proc Natl Acad Sci USA 88:9458–9462. https://doi.org/10.1073/pnas.88.21.9458

Krupp JJ, Billeter JC, Wong A, Choi C, Nitabach MN, Levine JD (2013) Pigment-dispersing factor modulates pheromone production in clock cells that influence mating in drosophila. Neuron 79: 54–68. https://doi.org/10.1016/j.neuron.2013.05.019

Leyria J, Philip R, Orchard I, Lange AB (2022) Gonadulin: a newly discovered insulin-like peptide involved in ovulation and oviposition in *Rhodnius prolixus*, a vector of Chagas disease. Insect Biochem Mol Biol 150:103848. https://doi.org/10.1016/j.ibmb.2022.103848

Li Y, Hernandez-Martinez S, Fernandez F, Mayoral JG, Topalis P, Priestap H et al (2006) Biochemical, molecular, and functional characterization of PISCF-allatostatin, a regulator of juvenile hormone biosynthesis in the mosquito *Aedes aegypti*. J Biol Chem 281:34048–34055. https://doi.org/10.1074/jbc.M606341200

Li B, Predel R, Neupert S, Hauser F, Tanaka Y, Cazzamali G et al (2008) Genomics, transcriptomics, and peptidomics of neuropeptides and protein hormones in the red flour beetle *Tribolium castaneum*. Genome Res 18:113–122. https://doi.org/10.1101/gr.6714008

Li C, Song X, Chen X, Liu X, Sang M, Wu W et al (2014) Identification and comparative analysis of G protein-coupled receptors in *Pediculus humanus humanus*. Genomics 104:58–67. https://doi.org/10.1016/j.ygeno.2014.06.002

Liessem S, Ragionieri L, Neupert S, Büschges A, Predel R (2018) Transcriptomic and neuropeptidomic analysis of the stick insect, *Carausius morosus*. J Proteome Res 17:2192–2204. https://doi.org/10.1021/acs.jproteome.8b00155

Luo LL, Lin Y, Linghu JH, Gong W, Luo YH, Liu M et al (2024) Genomics, transcriptomics, and peptidomics of the greater wax moth Galleria mellonella neuropeptides and their expression in response to lead stress. Insect Sci 31:773–791. https://doi.org/10.1111/1744-7917.13264

Matsumoto K, Suetsugu Y, Tanaka Y, Kotaki T, Goto SG, Shinoda T et al (2017) Identification of allatostatins in the brown-winged green bug *Plautia stali*. J Insect Physiol 96:21–28. https://doi.org/10.1016/j.jinsphys.2016.10.005

Misof B, Liu S, Meusemann K, Peters RS, Donath A, Mayer C et al (2014) Phylogenomics resolves the timing and pattern of insect evolution. Science 346:763–767. https://doi.org/10.1126/science.1257570

Mizoguchi A, Okamoto N (2013) Insulin-like and IGF-like peptides in the silkmoth *Bombyx mori*: discovery, structure, secretion, and function. Front Physiol 4:217. https://doi.org/10.3389/fphys.2013.00217

Mochizuki T, Sakamoto M, Tanizawa Y, Seike H, Zhu Z, Zhou YJ et al (2023) Best practices for comprehensive annotation of neuropeptides of *Gryllus bimaculatus*. Insects 14:121. https://doi.org/10.3390/insects14020121

Muratspahić E, Monjon E, Duerrauer L, Rogers SM, Cullen DA, Vanden Broeck J et al (2020) Oxytocin/vasopressin-like neuropeptide signaling in insects. Vitam Horm 113:29–53. https://doi.org/10.1016/bs.vh.2019.08.011

Myers EW, Sutton GG, Delcher AL, Dew IM, Fasulo DP, Flanigan MJ et al (2000) A whole-genome assembly of Drosophila. Science 287:2196–2204. https://doi.org/10.1126/science.287.5461.2196

Nässel DR, Wegener CA (2011) Comparative review of short and long neuropeptide F signaling in invertebrates: any similarities to vertebrate neuropeptide Y signaling? Peptides 32:1335–1355. https://doi.org/10.1016/j.peptides.2011.03.013

Ons S, Richter F, Urlaub H, Pomar RR (2009) The neuropeptidome of *Rhodnius prolixus* brain. Proteomics 9:788–792. https://doi.org/10.1002/pmic.200800499

Ormerod KG, LePine OK, Bhutta MS, Jung J, Tattersall GJ, Mercier AJ (2016) Characterizing the physiological and behavioral roles of proctolin in *Drosophila melanogaster*. J Neurophysiol 115:568–580. https://doi.org/10.1152/jn.00606.2015

Paluzzi JP, Vanderveken M, O'Donnell MJ (2014) The heterodimeric glycoprotein hormone, GPA2/GPB5, regulates ion transport across the hindgut of the adult mosquito, *Aedes aegypti*. PLoS One 9:e86386. https://doi.org/10.1371/journal.pone.0086386

Peng L, Yin HY, Huang C (2021) CNMa-CNMa receptor at microbiome-gut-brain axis: novel target to regulate feeding decision. Signal Transduct Target Ther 6:283. https://doi.org/10.1038/s41392-021-00708-y

Predel R, Nachman RJ, Gäde G (2001) Myostimulatory neuropeptides in cockroaches: structures, distribution, pharmacological activities, and mimetic analogs. J Insect Physiol 47:311–324. https://doi.org/10.1016/s0022-1910(00)00129-3

Predel R, Wegener C, Russell WK, Tichy SE, Russell DH, Nachman RJ (2004) Peptidomics of CNS-associated neurohemal systems of adult *Drosophila melanogaster*: a mass spectrometric survey of peptides from individual flies. J Comp Neurol 474:379–392. https://doi.org/10.1002/cne.20145

Ragionieri L, Verdonck R, Verlinden H, Marchal E, Vanden Broeck J, Predel R (2022) Schistocerca neuropeptides – an update. J Insect Physiol 136:104326. https://doi.org/10.1016/j.jinsphys.2021.104326

Redeker J, Bläser M, Neupert S, Predel R (2017) Identification and distribution of products from novel tryptopyrokinin genes in the locust, *Locusta migratoria*. Biochem Biophys Res Commun 486:70–75. https://doi.org/10.1016/j.bbrc.2017.02.135

Riehle MA, Garczynski SF, Crim JW, Hill CA, Brown MR (2002) Neuropeptides and peptide hormones in *Anopheles gambiae*. Science 298:172–175. https://doi.org/10.1126/science.1076827

Roller L, Yamanaka N, Watanabe K, Daubnerová I, Zitnan D, Kataoka H et al (2008) The unique evolution of neuropeptide genes in the silkworm *Bombyx mori*. Insect Biochem Mol Biol 38: 1147–1157. https://doi.org/10.1016/j.ibmb.2008.04.009

Roller L, Čižmár D, Gáliková Z, Bednár B, Daubnerová I, Žitňan D (2016) Molecular cloning, expression and identification of the promoter regulatory region for the neuropeptide trissin in the nervous system of the silkmoth *Bombyx mori*. Cell Tissue Res 364:499–512. https://doi.org/10.1007/s00441-015-2352-z

Roller L, Beke G, Stano M, Daubnerova I, Mizoguchi A, Satake H et al (2017) BomBase – web atlas of the neuroendocrine system in *Bombyx mori*. The online database resource. http://bombase.org/

Sanger F, Coulson AR (1975) A rapid method for determining sequences in DNA by primed synthesis with DNA polymerase. J Mol Biol 94:441–448. https://doi.org/10.1016/0022-2836(75)90213-2

Schoofs L, Nachman RJ (2006) Chapter 28—Sulfakinins. In: Kastin AJ (ed) Handbook of biologically active peptides. Academic, Burlington, pp 183–187

Schooley DA, Horodyski FM, Coast GM (2012) Hormones controlling homeostasis in insects. Insect Endocrinol:366–429. https://doi.org/10.1016/b978-0-12-384749-2.10009-3

Sekiguchi M, Katoh S, Yokosako T, Saito A, Sakai M, Fukuda A et al (2024) Biochem Biophys Res Commun 704:149705. https://doi.org/10.1016/j.bbrc.2024.149705

Spiegel CN, Dias DBS, Araki AS, Hamilton JGC, Brazil RP, Jones TM (2016) The Lutzomyia longipalpis complex: a brief natural history of aggregation-sex pheromone communication. Parasites Vectors 9:580. https://doi.org/10.1186/s13071-016-1866-x

Stay B, Tobe SS (2007) The role of allatostatins in juvenile hormone synthesis in insects and crustaceans. Annu Rev Entomol 52:277e299. https://doi.org/10.1146/annurev.ento.51.110104.151050

Sterkel M, Urlaub H, Rivera-Pomar R, Ons S (2011) Functional proteomics of neuropeptidome dynamics during the feeding process of *Rhodnius prolixus*. J Proteome Res 10:3363–3371. https://doi.org/10.1021/pr2001012

Stork NE (2018) How many species of insects and other terrestrial arthropods are there on Earth? Annu Rev Entomol 63:31–45. https://doi.org/10.1146/annurev-ento-020117-043348

Sturm S, Ramesh D, Brockmann A, Neupert S, Predel R (2016) Agatoxin-like peptides in the neuroendocrine system of the honey bee and other insects. J Proteome 132:77–84. https://doi.org/10.1016/j.jprot.2015.11.021

Tanaka Y, Suetsugu Y, Yamamoto K, Noda H, Shinoda T (2014) Transcriptome analysis of neuropeptides and G-protein coupled receptors (GPCRs) for neuropeptides in the brown planthopper *Nilaparvata lugens*. Peptides 53:125–133. https://doi.org/10.1016/j.peptides.2013.07.027

Tang J, Yu R, Zhang Y, Xie J, Song X, Feng F et al (2022) Molecular and functional analysis of eclosion hormone-like gene involved in post-eclosion behavior in a beetle. J Insect Physiol 142: 104429. https://doi.org/10.1016/j.jinsphys.2022.104429

Terhzaz S, Rosay P, Goodwin SF, Veenstra JA (2007) The neuropeptide SIFamide modulates sexual behavior in Drosophila. Biochem Biophys Res Commun 352:305–310. https://doi.org/10.1016/j.bbrc.2006.11.030

Urlacher E, Soustelle L, Parmentier ML, Verlinden H, Gherardi MJ, Fourmy D et al (2016) Honey bee allatostatins target Galanin/Somatostatin-like receptors and modulate learning: a conserved function? PLoS One 11:e0146248. https://doi.org/10.1371/journal.pone.0146248

Van Loy T, Vandersmissen HP, Van Hiel MB, Poels J, Verlinden H, Badisco L et al (2007) Comparative genomics of leucine-rich repeats containing G protein-coupled receptors and their ligands. Gen Comp Endocrinol 155:14–21. https://doi.org/10.1016/j.ygcen.2007.06.022

Van Sinay E, Mirabeau O, Depuydt G, Van Hiel MB, Peymen K, Watteyne J et al (2017) Evolutionarily conserved TRH neuropeptide pathway regulates growth in *Caenorhabditis elegans*. Proc Natl Acad Sci USA 114:E4065–E4074. https://doi.org/10.1073/pnas.1617392114

Veenstra JA (2010) What the loss of the hormone neuroparsin in the melanogaster subgroup of Drosophila can tell us about its function. Insect Biochem Mol Biol 40:354–361. https://doi.org/10.1016/j.ibmb.2010.03.001

Veenstra JA (2014) The contribution of the genomes of a termite and a locust to our understanding of insect neuropeptides and neurohormones. Front Physiol 5:454. https://doi.org/10.3389/fphys.2014.00454

Veenstra JA (2016) Allatostatins C, double C and triple C, the result of a local gene triplication in an ancestral arthropod. Gen Comp Endocrinol 230–231:153–157. https://doi.org/10.1016/j.ygcen.2016.04.013

Veenstra JA (2020) Gonadulins, the fourth type of insulin-related peptides in decapods. Gen Comp Endocrinol 296:113528. https://doi.org/10.1016/j.ygcen.2020.113528

Veenstra JA (2023) Differential expression of some termite neuropeptides and insulin/IGF-related hormones and their plausible functions in growth, reproduction and caste determination. PeerJ 11:e15259. https://doi.org/10.7717/peerj.15259

Veenstra JA, Šimo L (2020) The TRH-ortholog EFLamide in the migratory locust. Insect Biochem Mol Biol 116:103281. https://doi.org/10.1016/j.ibmb.2019.103281

Verlinden H, Sterck L, Li J, Li Z, Yssel A, Gansemans Y et al (2020) First draft genome assembly of the desert locust, *Schistocerca gregaria*. F1000Res 9:775. https://doi.org/10.12688/f1000research

Wang X, Fang X, Yang P, Jiang X, Jiang F, Zhao D et al (2014) The locust genome provides insight into swarm formation and long-distance flight. Nat Commun 5:2957. https://doi.org/10.1038/ncomms3957

Wang Y, Wang M, Yin S, Jang R, Wang J, Xue Z et al (2015) NeuroPep: a comprehensive resource of neuropeptides. Database 2015:bav038. https://doi.org/10.1093/database/bav038. Print 2015

Wyder S, Kriventseva EV, Schröder R, Kadowaki T, Zdobnov EM (2007) Quantification of ortholog losses in insects and vertebrates. Genome Biol 8:R242. https://doi.org/10.1186/gb-2007-8-11-r242

Xie J, Sang M, Song X, Zhang S, Kim D, Veenstra JA et al (2020) A new neuropeptide insect parathyroid hormone iPTH in the red flour beetle *Tribolium castaneum*. PLoS Genet 16:e1008772. https://doi.org/10.1371/journal.pgen.1008772

Xu G, Gu GX, Teng ZW, Wu SF, Huang J, Song QS et al (2016) Identification and expression profiles of neuropeptides and their G protein-coupled receptors in the rice stem borer *Chilo suppressalis*. Sci Rep 6:28976. https://doi.org/10.1038/srep28976

Yeoh JGC, Pandit AA, Zandawala M, Nässel DR, Davies SA, Dow JAT (2017) DINeR: database for insect neuropeptide research. Insect Biochem Mol Biol 86:9–19. https://doi.org/10.1016/j.ibmb.2017.05.001

Yu B, Li DT, Wang SL, Xu HJ, Bao YY, Zhang CX (2016) Ion transport peptide (ITP) regulates wing expansion and cuticle melanism in the brown planthopper, *Nilaparvata lugens*. Insect Mol Biol 25:778–787. https://doi.org/10.1111/imb.12262

Zandawala M (2012) Calcitonin-like diuretic hormones in insects. Insect Biochem Mol Biol 42:816–825. https://doi.org/10.1016/j.ibmb.2012.06.006

Chapter 4
Characteristic Features of Insect Peptide Hormones and Endocrine Systems: Comparison with Vertebrate Hormones and Endocrine Systems

Akira Mizoguchi

Abstract Although the insect endocrine system shares many features with those of vertebrates, some features are unique to insects. This chapter summarizes characteristics of insect peptide hormones (PHs) and endocrine systems compared to those of vertebrates. Whereas vertebrates have many endocrine glands that produce and release PHs, most of the insect PHs are produced by neurosecretory cells in the central nervous system and released from neurohemal organs or areas. The circulatory systems also differ: insects have an open circulatory system, while vertebrates have closed one. This difference may be related to the differences in the structure and function of endocrine systems. In insects and some other invertebrates, precursor proteins sometimes generate multiple copies of small PHs or neuropeptides with similar sequences, a feature not seen in vertebrates.

Keywords Insect · Peptide hormones · Neuropeptides · Endocrine system · Central nervous system · Brain · Neurosecretory cells · Neurohemal organs · Corpora cardiaca (CC) · Corpora allata (CA)

4.1 Insect Endocrine System

4.1.1 What Are Peptide Hormones and Neuropeptides?

According to the classical definition, hormones are biologically active substances secreted by specific cells that circulate in the blood and modify the function of specific cells or tissues in minute amounts. Organs that secrete hormones are called endocrine organs or glands. However, this concept of hormones has changed significantly over the past century due to the development of endocrinology, the field of research that deals with hormones. For example, it has been shown that many

---

A. Mizoguchi (✉)
Nagoya, Aichi, Japan
e-mail: mizoamizo@icloud.com

A. Mizoguchi, N. Okamoto (eds.), *Insect Peptide Hormones and Neuropeptides*, Entomology Monographs, https://doi.org/10.1007/978-981-95-7510-7_4

hormones are produced by various tissues other than known endocrine organs and that many hormones act not only through the circulatory system, in an endocrine fashion, but also locally, in a paracrine fashion. Therefore, currently, paracrine substances are also called hormones. In addition, the boundaries between the nervous and endocrine systems have been blurred by the discovery that many neurons secrete peptides. Peptides secreted by neurons are called neuropeptides (NPs), some of which are released into the circulation and function as hormones. In this regard, the brain is also an important endocrine gland. Vasopressin and oxytocin in vertebrates are produced by neurosecretory cells (NSCs) in the hypothalamus and transported via axons to and released into the blood from the posterior pituitary as hormones. In addition, several peptide hormones (PHs), called releasing hormones and release-inhibiting hormones, are produced by hypothalamic NSCs and released in the median eminence of the hypothalamus. The phenomenon in which neurons secrete hormones is called neurosecretion, and the hormones that are secreted by neurons are called neurosecretory hormones or neurohormones. Neurosecretory hormones are mostly peptides or proteins, but some biogenic amines are also secreted into the blood as hormones by neurons. The NSCs are neurons that are specialized to secrete hormones and typically have larger cell bodies than usual neurons. The NSCs are also characterized by their axon terminal structures that do not form synapses and have large cell surfaces facing bloodstream. However, it has recently been discovered that many normal neurons also produce and release various types of NPs. The NPs are secreted from varicosities (lump-like bulges) on the axons or dendrites, modulating the function or activity of other neurons. This effect of NPs is called neuromodulation, and NPs with this effect are called neuromodulators. The action of neuromodulators can be considered a type of paracrine action. Some NPs are secreted within synapses and function as neurotransmitters. Neurotransmitters are distinguished from neuromodulators by their different mechanisms of action: neurotransmitters are secreted within the synapses and activate ionotropic receptors on post-synaptic neurons or muscle cells, eliciting post-synaptic potentials, whereas neuromodulators bind to metabotropic receptors, usually G protein-coupled receptors, on other neurons to modulate their function or activity.

This book provides an overview of the synthesis, secretion, and physiological functions of insect PHs and NPs. However, as discussed above, it is often not easy to determine whether a peptide of interest is a hormone, neuromodulator, or neurotransmitter. This is because the actual secretion pathway and mode of action are poorly understood for many PHs/NPs. Therefore, PHs and NPs are not strictly discriminated in this book. Some peptides are called by the names with the word "hormone" in them, such as eclosion hormone (EH). However, this does not necessarily indicate that those peptides always act as hormones. Indeed, EH has been suggested to also act as a neuromodulator (Gong et al. 2019). Allatostatins, for example, have traditionally been treated as one of the representative insect hormones. In fact, however, these peptides are now thought to act as neuromodulators or neurotransmitters.

4.1.2 Insect Endocrine System

In vertebrates, many types of PHs are produced by various endocrine glands and released into the bloodstream, whereas neurosecretory hormones are limited to those produced in the hypothalamus. In contrast, in insects, most PHs are neurosecretory hormones, and only a few types of PHs are secreted by non-neural endocrine cells. Thus, the most dominant endocrine organ of insects is the central nervous system (CNS). The CNS produces a variety of NPs. The CNS of arthropods, including insects, is a ventral nerve cord, a chain of many ganglia located along the ventral center of the body. A relatively large ganglion situated at the anterior end of the nerve cord is called the brain. This is typically followed by the subesophageal ganglion, three thoracic ganglia, and a variable number of abdominal ganglia. However, the number of ganglia varies depending on the species and developmental stage (larva, pupa, or adult) due to ganglion fusion. In extreme cases, as seen in the fruit fly *Drosophila melanogaster*, thoracic and abdominal ganglia are fused into one large structure, which is called the "ventral nerve cord."

On the other hand, the best known hormones in insects, ecdysteroids (ECDs) and juvenile hormones (JHs), are produced by the prothoracic glands (PGs) and corpora allata (CA; singular is corpus allatum), respectively. The PGs are paired amorphous endocrine organs composed of ECD-producing cells, typically located around the base of the trachea in the prothorax. However, their shape and location vary significantly depending on the species. The CA are a pair of small, usually spherical endocrine organs composed of JH-producing cells and connected by nerves to the corpora cardiaca (CC; singular is corpus cardiacum). The size of the CA varies greatly depending on JH synthesis activity.

The ECD and JH are not PHs but steroid and terpenoid hormones, respectively. These lipophilic hormones play a central role in the regulation of many aspects of insect life, including development, metabolism, reproduction, and behavior. However, this book does not cover these hormones in detail, as excellent reviews on these hormones have already been published (Gilbert 2012; Saleuddin et al. 2020). Nonetheless, both hormones frequently appear in this book because many types of PHs/NPs regulate the synthesis of ECD and JH. Therefore, in this book, readers should be able to learn about various functions of ECD and JH through the effects of PHs/NPs on the PGs and CA.

Another important endocrine organ in insects is the CC, which secrete adipokinetic hormone (AKH), an important hormone involved in metabolic regulation. Within the CC, the AKH-producing cells are often intermingled with the neurosecretory axons coming from the brain. However, in some insects, such as locusts, they are completely separated from the axons to form the glandular lobe of the CC. In this case, the remaining part of the CC is called the storage lobe. The CC is connected by nerves to the brain on one side and to the CA on the other side (Fig. 4.3). The size and shape of the CC vary remarkably between insect groups: they are small and paired in moths, whereas they are large and single by fusion in locusts. Furthermore, the CC of flies, together with the CA and the PGs, form the ring gland by fusion. As the CC and CA are located at the back of the brain, they are collectively called the retrocerebral complex.

In addition to the CNS, PGs, and retrocerebral complex mentioned above, the insect endocrine system includes peripheral endocrine cells in the digestive tract, ovary, and epitracheal glands. There are many endocrine cells within the epithelial tissue of the intestinal tract, and they are called enteroendocrine cells. Detections by immunohistochemistry and/or in situ hybridization have revealed that a wide variety of NPs are produced by enteroendocrine cells. Ovaries are known to produce ECDs and express some PH genes. In lepidopteran insects, epitracheal glands contain Inka cells that produce ecdysis-triggering hormone (see Sect. 2.4.1). Epitracheal glands are located at the base of the trachea in each segment of the larval body. However, in many other orders of insects, Inka cells are widely distributed throughout the body along the trachea (Zitňan and Adams 2012).

4.1.3 Open Circulatory System

Whereas vertebrates have a closed circulatory system, insects, like other arthropods and most mollusks, have an open circulatory system. This difference may be reflected in the differences in the endocrine systems between vertebrates and insects. The circulatory system of insects is composed of the hemolymph, dorsal vessel, and hemocoel (Hillyer and Pass 2020). In this system, blood and interstitial fluid cannot be distinguished. Therefore, both fluids are collectively called hemolymph. The dorsal vessel is usually divided into the heart in the abdomen and the aorta in the thorax, with the heart having a wider luminal diameter, incurrent ostia, alary muscles, and thicker walls. As the heart beats, in most cases, hemolymph is drawn in through the posterior open end and pairs of lateral openings called ostia of the dorsal vessel and then extruded through its anterior open end in the head. In this way, the heart functions as a pump that generates a hemolymph flow within the hemocoel. However, hemolymph does not diffuse freely throughout the hemocoel, but instead flows along channel-like routes that are created by the arrangement of internal organs and by diaphragms (Hillyer and Pass 2020). Thus, it is clear that this type of circulatory system is less efficient in circulating hormones within the body compared to the closed circulatory system in vertebrates. However, we can find that the insect endocrine system has various structural ingenuity to transport hormones efficiently through this inefficient circulatory system.

For example, it is noteworthy that the main endocrine organs of insects, such as the brain, CC, CA, and PGs, are located in the head and prothorax. This is because hemolymph extruded from the front end of the aorta flows from the head to the rest of the body, allowing it to efficiently transport hormones secreted by these endocrine organs. It is also interesting to note that many hormones are produced not only in the brain but in the thoracic and abdominal ganglia (see Chap. 5). In the open circulatory system, secreted hormones are rapidly and significantly diluted due to the abundance of hemolymph. In the larva of the silkworm *Bombyx mori*, approximately a quarter of the body weight is hemolymph (Horie et al. 1971). Therefore, it may be rational to produce and release hormones in various parts of the body in order to distribute them efficiently throughout the body. The location of prothoracicotropic hormone (PTTH)-release site and its target organ, PG, is also noteworthy. As described in Chap. 1, *Bombyx* PTTH is produced in the brain and released from the CA to stimulate ECD synthesis by the PG. Interestingly, in *B. mori*, the anterior end of the PG extends anteriorly and is located in close proximity to the CA. There is no doubt that the proximity of hormone release site and its target organ is advantageous for signal transmission. Furthermore, axonal projections of hormone secretory neurons to their target tissues may be the most reliable means for signal transmission. Orcokinin is known to be another PG-stimulatory NP in *B. mori*. The PGs are innervated by a pair of orcokinin-producing neurons originating from the prothoracic ganglion, providing a good example of direct control of target tissues by NPs through innervation (see Fig. 5.2 in Chap. 5).

4.2 Insect Neuroendocrine System

4.2.1 Neurosecretory Cells

Neurosecretory hormones are produced by NSCs. Many NSCs are distributed in the cortex of the brain and other ganglia and are most abundant in the brain, especially in the medial part of the protocerebrum. The distribution of NSCs/neurons that produce PHs/NPs has been comprehensively investigated in *D. melanogaster* and *B. mori*, and maps of these cells have been published (Park 2008; Nëssel and Zandawala 2019; Roller et al. 2017). The NSCs in the brain are largely divided into medial and lateral NSCs, depending on their location (medial or lateral) within the protocerebrum. The medial and lateral regions of the protocerebrum are also called pars intercerebralis and pars lateralis, respectively. Each region has several cell populations consisting of one to several (more than ten in some cases) NSCs per hemisphere that produce the same PH/NP (Fig. 4.1). However, the number of NSCs varies widely between species and even between developmental stages. The NSCs are also present in the tritocerebrum and other parts of the brain. Outside the brain, NSC populations are primarily distributed in each ganglion along the ventral nerve cord. They are fewer in number and more widely distributed than those of the brain (Raabe 1989).

The NSCs are composed of relatively large but variable-sized cell bodies, long axons with collaterals, and branched and swollen axon terminals (Fig. 4.2). The collateral is branched repeatedly, forming a dendritic field within the brain or other ganglia. It is thought that NSCs may communicate with each other and/or other neurons in this area. The cell body has well-developed rough endoplasmic reticulum (rER) and Golgi apparatus and abundant secretory granules, features similar to PH-producing endocrine cells. Secretory granules are transported along actin

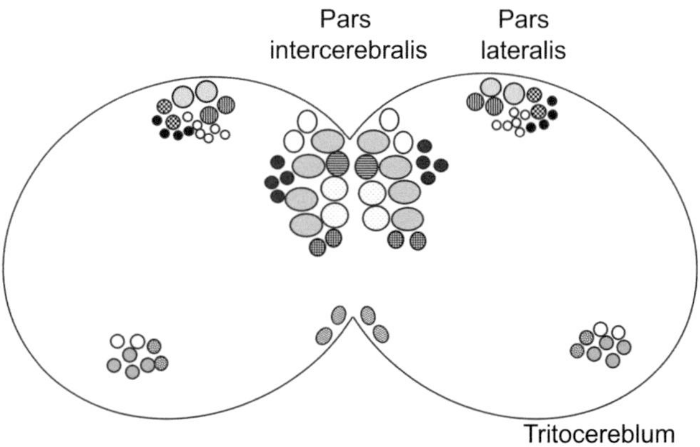

Fig. 4.1 Schematic representation of the NSC distribution in the insect brain

Fig. 4.2 Schematic drawing of a NSC

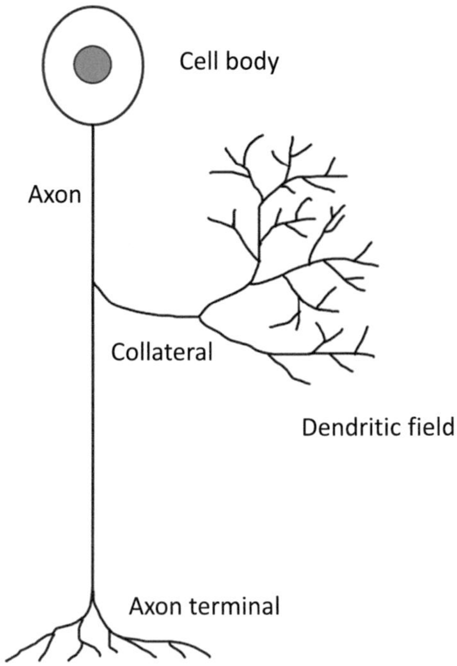

filaments within the axon to the axon terminal and released when the cell (neuron) is excited. The mechanism by which the neurosecretory material is released by neuronal excitation in insects is thought to be the same as that in vertebrates. Axon terminals are located in the neurohemal organ or area, where they are in contact with a thin connective tissue sheath that separates them from the hemolymph.

4.2.2 Neurosecretory Pathways

Well-known neurohemal organs in insects are the CC and CA, both of which are also endocrine organs as discussed above. The CC are connected to the brain by three pairs of nerves, nervi corporis cardiaci (NCC) I, II, and III (Fig. 4.3). Axons from NSCs in the pars intercerebralis, pars lateralis, and tritocerebrum reach the CC through NCC I, II, and III, respectively. In some insects, NCC I and II fuse to form NCC I + II. Some neurosecretory axons project to the ipsilateral CC, while others traverse the contralateral lobe of the brain, projecting to the contralateral CC. Some of the neurosecretory axons from the brain pass through the CC to reach the CA via nervi corporis allati (NCA) I (Fig. 4.3). The CA are also connected to the subesophageal ganglion by NCA II. The relative importance of the CC and CA as

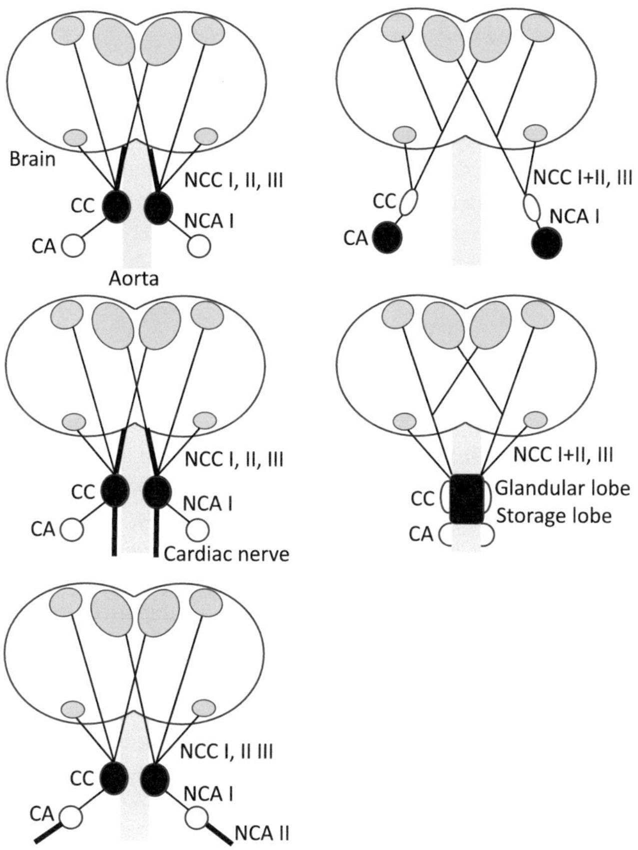

Fig. 4.3 Various arrangements of neurohemal organs and areas in insects. The gray ellipses in the brain represent the populations of medial, lateral, or tritocerebral NSCs. The black ellipses and thick lines represent neurohemal organs and areas, respectively. See text for abbreviations. (Modified from Raabe (1989))

neurohemal organs varies among insect groups. For example, in locusts, most neurosecretory axons terminate in the CC, whereas in moths, some brain NSCs have their axon terminals in the CA. Neurosecretory pathways from the brain NSCs have also been found in the circumesophageal connectives and the connectives of the ventral nerve cord in many insects (Raabe 1989).

4.2.3 Release Sites

Neurosecretory hormones are released into the hemolymph from neurohemal organs or areas. The neurohemal areas refer to more diffuse structures compared to neurohemal organs. The main neurohemal organ for brain-derived neurohormones is the CC in many insects, but in some insects, including lepidopterans, it is the CA. The CC often fuse with the lateroventral wall of the aorta, but may only partly fuse or may not connect with the aorta at all. The CC are sometimes associated to the sympathetic hypocerebral ganglion that innervates the gut. The neurosecretory axons entering the CC form a large bundle in the center of the organ and then send out numerous branches to reach the connective membrane or its invaginations, where the neurosecretory products are released into the hemolymph (Raabe 1989). Some part of the axons entering the CC pass through them and enter the NCA I to reach the CA or enter the cardiac or recurrent nerves, depending on the species. Some neurosecretory products are released from the cardiac nerves. Thus, these nerves serve as neurohemal areas. The CA also function as neurohemal organs in some insects, but in some other insects, the axons from the CC pass through the CA to reach the NCA II, where neurosecretory products are released. Therefore, in the latter case, these nerves act as neurohemal areas. In some species, the CC primarily contain the glandular intrinsic NSCs, and neurosecretory axons from the brain terminate in the aortic wall to which the CC are fused. In such species, the aortic wall serves as a neurohemal area. In several insects, neurosecretory axon terminals are found in contact with the sheath of the NCC, NCA, or esophageal nerves, indicating that these nerves can also act as neurohemal areas. Various arrangements of release sites for brain neurosecretory products are schematically shown in Fig. 4.3. However, some other nerves or tissues could also act as neurohemal areas in some insects.

The neurohemal organs for neurosecretory peptides produced by NSCs in other ganglia are the perisympathetic organs (POs or PSOs), also called the periventral organs. Their name derives from their location on the unpaired sympathetic nerves. The POs show metameric distribution and significant morphological variation. They have been classified into three types according to their anatomical location: median, transverse, and mediotransverse (Raabe 1989). The PO contains neurosecretory axons that repeatedly branch and form numerous terminals. These axons are surrounded by a large number of glial cells and are entirely covered by a thin neural lamella. The neural lamella may invaginate to form sinuses, where neurosecretory material is released through the neural lamella. The neurosecretory materials contained in the PO typically originate from the posterior ganglion. However, in the tobacco hornworm *Manduca sexta*, specific axons in the PO have been shown to originate from the anterior ganglion. In some insects, link nerves have been suggested to serve as a neurohemal area for neurosecretory materials produced in the ventral ganglia (Raabe 1989).

4.2.4 Production and Release of NPs

The mechanisms by which insect PHs/NPs are produced and released are thought to be essentially the same as those for vertebrate PHs. Here, we review the mechanisms known for vertebrate PHs.

The PHs/NPs are produced by the expression of the genes encoding them. The first product of gene expression is the preprohormone, which consists of a signal peptide and a prohormone. The signal peptide consists of approximately 20 hydrophobic amino acid residues. This hormone precursor protein is produced, processed, and released by pathways common to secretory proteins. When the ribosome starts to translate the mRNA for a hormone, the N-terminal signal peptide is bound to the signal recognition particle (SRP), which guides the signal peptide and ribosome to the SRP receptor on the rER. The SRP is a complex of RNA and several proteins. The SRP receptor transfers the ribosome to the Sec translocon on the rER, and then the SRP and its receptor are released from the ribosome. The ribosome continues to synthesize the preprohormone on the rER, and the protein being synthesized enters the lumen of the rER guided by the signal peptide. The signal peptide of the preprohormone is cut off within the rER by signal peptidase to generate the prohormone. In the rER, n-linked glycosylation of the prohormone may begin to occur. These and other modifications of precursor proteins (preprohormones) are called processing. Prohormones are concentrated within the rER and taken up by transport vesicles generated by rER budding. The vesicles are transported to the cis face of the Golgi apparatus, where the vesicle contents (prohormones) are transferred to the Golgi apparatus by membrane fusion. The transport of materials by transport vesicle within the cytoplasm is called vesicle transport. The Golgi apparatus consists of several membrane-enclosed flattened disks called cisternae. Prohormones taken up at the cis face of the Golgi are transferred between cisternae by vesicle transport, toward the trans face of the Golgi, where a trans-Golgi network (TGN) is developed. Each Golgi cisterna contains specific enzymes involved in the modification of prohormones, and prohormones are processed one after another while they travel between cisternae toward the TGN. Processed prohormones are packed into secretory vesicles or granules by budding at the TGN. Processed prohormones may be further processed in secretory granules as well. Processing of prohormones occurring in the Golgi apparatus and secretory granules includes cleavage of the proteins at dibasic cleavage sites, removal of the C-terminal basic amino acid residues after protein cleavage, formation of disulfide bridges between cysteine residues, C-terminal amidation and N-terminal pyroglutamylation of the peptides, and

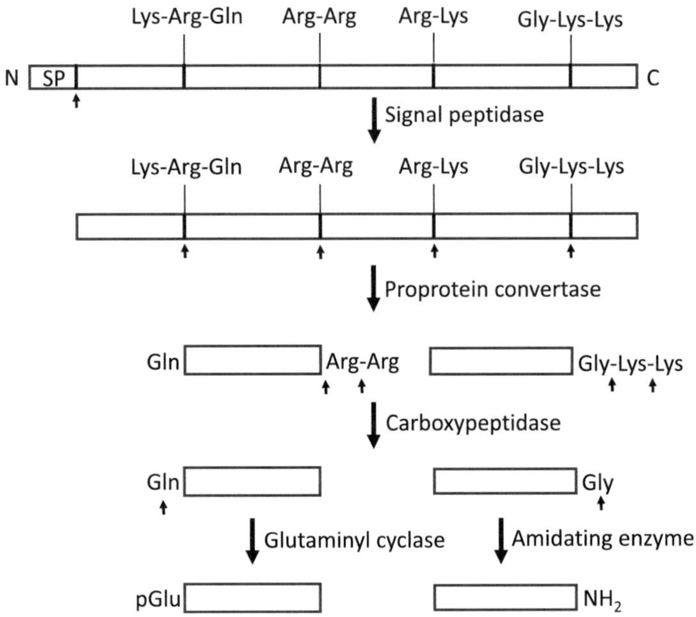

Fig. 4.4 Processing of a preprohormone. The large arrows denote the actions of the indicated enzymes. The small arrows indicate the targets of the enzymes. SP, signal peptide. N and C, N- and C-termini of the preprohormone, respectively

sulfation and glycosylation of the peptides. Mature hormones resulting from the processing of prohormones are stored in secretory granules. The NPs are also generated from their precursors in the same way. Figure 4.4 schematically shows the stepwise processing of preprohormones.

Dibasic cleavage site refers to a motif consisting of two basic amino acid residues, arginine and lysine residues, in any combination or order. This site is cleaved at its C-terminus by prohormone convertases (PCs). However, the dibasic sites are not always cleaved by these enzymes. On the other hand, monobasic sites may also be cleaved in some cases. The basic amino acid residues remaining at the cleavage site are removed by carboxypeptidase E. If the C-terminus of the peptide resulting from prohormone cleavage is a glycine residue, the amidating enzyme removes the glycine residue, leaving an amino group. If the N-terminus of the generated peptide is glutamic acid or glutamine, it is converted to pyroglutamic acid through cyclization by the enzyme glutaminyl cyclase. C-terminal amidation and N-terminal pyroglutamylation of peptides are effective in protecting them from degradation.

Mature PHs/NPs contained in secretory granules are released into the blood, extracellular fluid, or synaptic cleft by exocytosis, a secretory process that utilizes membrane fusion between secretory granules and the cell membrane.

4.3 Characteristic Features of Insect PHs/NPs

Insect PH/NP molecules themselves do not have any distinctive features unique to insects. However, when insect PHs/NPs are compared with vertebrate PHs/NPs, some notable features can be found. Such features may not be unique to insect PHs/NPs but rather to those of arthropods or protostomes. Here, we focus on the characteristics of insect PHs/NPs noticeable in comparison to vertebrate PHs/NPs.

4.3.1 Conserved and Differential Evolution of Ancestral Signaling Systems

As discussed in Chap. 2, a large number of homologous peptide signaling systems, which consist of homologous ligands and homologous receptors, exist between insects and vertebrates. However, the physiological functions of these homologous systems often differ remarkably between them. For example, gonadotropin-releasing hormone in vertebrates acts on the anterior pituitary to stimulate the release of gonadotropins (in mammals) or other hormones (in teleosts), whereas its insect homolog, AKH, acts on the fat body to mobilize carbohydrates and lipids (see Sect. 9.2). Calcitonin in vertebrates acts on osteoclasts to suppress their activity, thereby lowering blood calcium levels, whereas its insect homolog, calcitonin-like diuretic hormone, acts on Malpighian tubules to promote diuresis (see Sect. 9.3). The functions of these signaling systems in the common ancestor of vertebrates and insects are unknown. Therefore, it is unclear which of these different functions of homologous signaling systems reflects the function of the ancestral signaling system, both, or neither. In general, however, the functions of each signaling system appear to change between animal lineages through evolution, depending on the animal's lifestyle, including morphology, development, physiology, and environmental adaptation.

On the other hand, some peptide signaling systems are involved in similar functions in both vertebrates and insects. For example, insulin in vertebrates and insulin-like peptide in insects both regulate growth and metabolism (see Chaps. 7 and 9). Moreover, the secretion of both hormones is stimulated by food intake, suggesting that this type of hormone plays an essential and universal role in linking animal growth with an animal's nutritional status. Somatostatin in vertebrates functions as a hormone that inhibits the release of some other hormones, various neuronal activities, gastrointestinal function, etc., while its insect homolog, allatostatin C, inhibits JH biosynthesis in lepidopterans and inhibits muscle contractions in most insects (see Sect. 2.3.3). Therefore, this type of hormone appears to act as an inhibiting hormone to various biological processes in both animal groups. The NP Y in vertebrates and NP F in insects both modulate feeding behaviors (see Chap. 11), although each hormone is also involved in other functions. Similar functions of homologous signaling systems in vertebrates and insects are thought to derive from the functions of the corresponding signaling systems in their common ancestor. These functions have likely been conserved across metazoans over hundreds of millions of years of evolutionary history because they were universal and essential for animal survival.

4.3.2 Insects Have Fewer PHs/NPs Than Vertebrates

Although many peptide signaling systems are conserved across metazoans, they represent only a portion of the total repertoire of PHs/NPs in vertebrates and insects, respectively. When the total numbers of PHs/NPs are compared between vertebrates and insects, we notice that vertebrates have many more PHs/NPs (Takei et al. 2015). One reason for this may simply be the difference in the number of researchers engaging in endocrinological studies of each animal group. It is evident that far more researchers are involved in the studies of vertebrates, especially humans or mammals. However, there may be some more fundamental reasons for this. The difference in the number of PHs/NPs between vertebrates and insects may reflect the differences in the complexity of the body construction and nervous system between them. It is reasonable to assume that controlling a more complex body requires more signaling systems. However, there is another reason for the abundance of PHs/NPs in vertebrates. It is known that three rounds of whole-genome duplication have occurred during the evolution of vertebrates. The first one occurred in the common ancestor of all vertebrates and the second in jawed vertebrates after the divergence of jawless (agnathan) and jawed (gnathostome) vertebrates. The third one occurred only in the lineage of teleosts. Therefore, most vertebrates, including humans, potentially harbor at least four sets of ancestral genomes, although many of them may have been lost during evolution of each animal. As a result, vertebrates have many paralogous genes. This also applies to the genes encoding PHs/NPs. Indeed, there are many paralogous PHs/NPs in vertebrates. Although some of them have similar functions, others have different functions. For example, vasopressin and oxytocin are paralogous hormones but have different functions. Thus, this may be another reason why vertebrates have many more PHs/NPs than insects. From another perspective, however, the duplication and diversification of hormones through whole-genome duplications enabled the creation of complex bodies and the development of complex nervous and endocrine systems in vertebrates.

4.3.3 Most PHs Are Neurosecretory Hormones

Vertebrates have many PHs produced by glandular endocrine organs. However, as seen in Chap. 2, many of their homologs are present even in protostomes, indicating that the origin of vertebrate PHs can be traced back to the common ancestor of bilaterians. Interestingly, invertebrate homologs of vertebrate PHs are all produced by neurons. Therefore, the origin of vertebrate PHs is thought to be NPs used in the common ancestor of bilaterians. In insects, most PHs are produced by neurons, similar to those in other invertebrates. In other words, most insect PHs are neurosecretory hormones. Because of their small size, invertebrates are thought not to have needed to develop glandular endocrine organs necessary to produce a large amount of PHs.

4.3.4 Many Similar Peptides Are Generated from a Single Precursor

The allatostatin A (AST-A) precursor of the cockroach *Diploptera punctata* generates as many as 13 AST-A peptides through processing. The number of AST-A peptides derived from the precursor protein is 2–14 in insects and up to 35 in crustaceans (Stay and Tobe 2007). The myoinhibiting peptide (MIP) precursor of the kissing bug *Rhodnius prolixus* generates 12 copies of MIPs. Other insects also produce multiple MIP peptides from a single precursor (Lange et al. 2012). FMRFamide (FMRFa) was first discovered in mollusks, and later N-terminally extended forms of this peptide have been identified in insects. In *D. melanogaster*, the FMRFa precursor contains ten copies of FMRFa and FMRFa-related peptides (Nëssel and Winther 2010). The leucokinin (LK) precursor of *R. prolixus* contains 12 LK-type peptides, although *D. melanogaster* has only 1 LK in its precursor. As these examples demonstrate, in insects, many copies of PHs/NPs are often produced from a single precursor protein. The production of multiple copies of PHs/NPs from a precursor protein is not limited to insects, but is also widespread in other arthropods and other invertebrates, including mollusks and annelids. In contrast, in vertebrates, multiple copies of a PH/NP are rarely generated from a PH/NP precursor, and the copy number is smaller. This mechanism may have evolved in protostomes to produce the same PHs/NPs with high efficiency.

4.3.5 There Are Many Myo-Regulatory Peptides in Insects

When you look at the list of insect PHs/NPs (Table 1.2), you will notice that there are a number of peptides with names containing "myo" or "kinin" that suggest they act on muscle. This fact may reflect the history of the discovery of insect hormones. In the early stages of insect endocrinology, the hormones targeted for purification were those involved in the regulation of specific processes in insect development, metabolism, reproduction, and behavior, such as PTTH, which induces metamorphosis; AKH, which mobilizes lipids from the fat body; ovary ecdysteroidogenic hormone, which promotes ovary development; and EH, which induces eclosion. These hormones were purified using a combination of biochemical techniques, while monitoring their presence with bioassays that can specifically test their effects. However, the number of hormones that could be purified using this method was limited. In the 1980s, with the widespread use of HPLC for peptide purification, there was growing momentum to efficiently and comprehensively identify insect hormones. At that time, it was known that many of the signal molecules that work in peripheral tissues also work in the CNS. Thus, the gut and heart were often used as bioassay systems to screen signal molecules in the CNS, especially in locusts and cockroaches. Consequently, many PHs that stimulate or inhibit muscle contractions in the gut or heart were purified and sequenced. These peptides are now known to have pleiotropic effects. For example, orcokinin regulates circadian locomotor activity in the cockroach *Leucophaea maderae* (Wei and Stengl 2011) and ecdysteroidogenesis in *B. mori* (Yamanaka et al. 2011), and leucokinin stimulates fluid secretion from the Malpighian tubules in the moth *Heliothis zea* (Torfs et al. 1999) and regulates multiple aspects of physiology and behavior in *D. melanogaster* (Nëssel 2021). Thus, the research to identify novel signal molecules using bioassays with the gut or heart has greatly contributed to the development of insect endocrinology.

4.4 Conclusion

Although the endocrine systems of insects appear to be somewhat different from those of vertebrates, these differences reflect the differences in their lifestyles, such as body plan, body size, patterns of growth and development, physiology, and reproduction, and the fundamental mechanisms of intercellular and interorgan signaling via hormones and NPs are the same. Insects have small body sizes, ventral nerve cords, and open vascular systems and have developed sophisticated endocrine systems to match these characteristics. One relevant significant difference between insect and vertebrate hormones appears to be that most insect PHs are produced by neurons. The PHs produced by neurons are sent through axons either to the neurohemal organs/areas, where they are released into the hemolymph, or directly to target tissues.

Insect development and reproduction are directly controlled by the major insect hormones ECDs and JHs, but their secretion is regulated by various types of PHs/NPs, whose secretion is modulated by a variety of internal and external factors, allowing growth-dependent and environmentally adapted development and reproduction. Insect PHs/NPs are also deeply involved in regulating growth, metabolism, water and ion balance, behavior, body color changes, circadian rhythms, and more. The secretion of each PH/NP is interconnected, enabling the integrated regulation of insect growth, homeostasis, and environmental adaptation.

References

Gilbert LI (ed) (2012) Insect endocrinology. Academic Press, San Diego

Gong C, Ouyang Z, Zhao W, Wang J, Li K, Zhou P et al (2019) A neuronal pathway that commands deceleration in Drosophila larval light-avoidance. Neurosci Bull 35:959–968. https://doi.org/10.1007/s12264-019-00349-w

Hillyer JF, Pass G (2020) The insect circulatory system: structure, function, and evolution. Annu Rev Entomol 65:121–143. https://doi.org/10.1146/annurev-ento-011019-025003

Horie Y, Iguchi T, Watanabe K, Yanagisawa H (1971) On the haemolymph volume of the silkworm, Bombyx mori. J Ser Sci Jap 40:330–334. https://doi.org/10.11416/kontyushigen1930.40.330

Lange AB, Alim U, Vandersmissen HP, Mizoguchi A, Broeck JV, Orchard I (2012) The distribution and physiological effects of the myoinhibiting peptides in the kissing bug, Rhodnius prolixus. Front Neurosci 6:00098. https://doi.org/10.3389/fnins.2012.00098

Nässel DR (2021) Leucokinin and associated neuropeptides regulate multiple aspects of physiology and behavior in Drosophila. Int J Mol Sci 22:1940. https://doi.org/10.3390/ijms22041940

Nässel DR, Winther AME (2010) Drosophila neuropeptides in regulation of physiology and behavior. Prog Neurobiol 92:42–104. https://doi.org/10.1016/j.pneurobio.2010.04.010

Nässel DR, Zandawala M (2019) Recent advances in neuropeptide signaling in Drosophila, from genes to physiology and behavior. Prog Neurobiol 179:101607. https://doi.org/10.1016/j.pneurobio.2019.02.003

Park D, Veenstra JA, Park JH, Paul H, Taghert PH (2008) Mapping peptidergic cells in Drosophila: where DIMM fits in. PLoS One 3:e1896. https://doi.org/10.1371/journal.pone.0001896

Raabe M (1989) Recent developments in insect neurohormones. Plenum Press, New York

Roller L, Beke G, Stano M, Daubnerova I, Mizoguchi A, Satake H et al (2017) BomBase—web atlas of the neuroendocrine system in Bombyx mori. Online database resource. http://bombase.org/

Saleuddin S, Lange AB, Orchard I (2020) Advances in invertebrate (neuro)endocrinology, Vol 2, Arthropoda. Apple Academic Press, Burlington

Stay B, Tobe SS (2007) The role of allatostatins in juvenile hormone synthesis in insects and crustaceans. Annu Rev Entomol 52:277–299. https://doi.org/10.1146/annurev.ento.51.110104.151050

Takei Y, Ando H, Tsutsui K (eds) (2015) Handbook of hormones: comparative endocrinology for basic and clinical research, 1st edn. Academic Press, London. https://doi.org/10.1016/C2013-0-15395-0

Torfs P, Nieto J, Veelaert D, Boon D, van de Water G, Waelkens E et al (1999) The kinin peptide family in invertebrates. Ann N Y Acad Sci 897:361–375. https://doi.org/10.1111/j.1749-6632.1999.tb07906.x

Wei H, Stengl M (2011) Light affects the branching pattern of peptidergic circadian pacemaker neurons in the brain of the cockroach Leucophaea maderae. J Biol Rhythm 26:507–517. https://doi.org/10.1177/0748730411419968

Yamanaka N, Roller L, Zitňan D, Satake H, Mizoguchi A, Kataoka H et al (2011) Bombyx orcokinins are brain-gut peptides involved in the neuronal regulation of ecdysteroidogenesis. J Comp Neurol 519:238–246. https://doi.org/10.1002/cne.22517

Zitňan D, Adams M (2012) Neuroendocrine regulation of Ecdysis. In: Gilbert LI (ed) Insect endocrinology. Academic Press, San Diego

Chapter 5
Localization of Peptide Hormone/Neuropeptide-Producing Cells in Insects

Yoshiaki Tanaka

Abstract Peptide hormone/neuropeptide (PH/NP)-producing cells are widely distributed throughout the central and peripheral nervous systems. Technological progress in immunohistochemistry and in situ hybridization has significantly contributed to our understanding of PH/NP-producing cells. Advanced techniques such as double-label immunochemistry, in situ hybridization, and the use of reporter proteins such as green fluorescent protein have contributed to the construction of PH/NP atlases in several insects. The atlas can provide comprehensive views of neuropeptide distribution and signaling networks. Most neuropeptide genes are expressed in the central and peripheral nervous systems. Specific neuropeptides are produced in the corpora cardiaca, subesophageal ganglion, frontal ganglion, thoracic ganglia, and abdominal ganglia and regulate development and behavior. Peripheral cells are also a source of neuropeptides. In non-nervous organs, enteroendocrine cells in the midgut release multiple neuropeptides to regulate feeding activity and respond to physiological changes in the intestine. These distributed control systems may be beneficial for quick and flexible responses of insects to environmental changes and may contribute to insect survival and prosperity.

Keywords Central nervous system (CNS) · Corpora cardiaca (CC) · Ecdysis-triggering hormone (ETH) · Frontal ganglion (FG) · Immunohistochemistry (IHC) · In situ hybridization · Neurosecretory cells (NSCs) · Peripheral organ · Subesophageal ganglion (SG) · Reporter gene

Peptide hormones/neuropeptides (PHs/NPs) represent a diverse class of signaling molecules that play pivotal roles in modulating various physiological and behavioral processes in insects. These peptides are produced by specific cells and exert their functions through interactions with receptors on target cells, influencing different aspects of insect life, including development, reproduction, feeding, and response to stress. The study of PHs/NPs and their producing cells in insects has garnered

Y. Tanaka (✉)
Institute of Agrobiological Sciences, NARO, Tsukuba, Japan
e-mail: tanaka.yoshiaki660@naro.go.jp

A. Mizoguchi, N. Okamoto (eds.), *Insect Peptide Hormones and Neuropeptides*, Entomology Monographs, https://doi.org/10.1007/978-981-95-7510-7_5

considerable attention, as understanding their localization and function can reveal fundamental aspects of insect biology and offer potential targets for pest management strategies.

The PH/NP-producing cells are distributed throughout the central nervous system (CNS) and peripheral nervous system. These include neurosecretory cells (NSCs) in the brain that project to neurohemal organs such as the corpora cardiaca (CC) and corpora allata (CA). Some PHs/NPs are synthesized by intrinsic cells of these organs or peripheral endocrine cells, whereas others are produced by NSCs located elsewhere in the CNS or peripheral nervous system.

The localization of PHs/NPs within insects is critical for understanding their function. The PHs/NPs are often produced in specific neurons within the brain and other parts of the nervous system, from which they can be axonally transported to different release sites. The precise mapping of these producing cells and their projection patterns provides invaluable clues about the potential roles of PHs/NPs.

5.1 Technological Advancements in the Localization Analysis of PH/NP-Producing Cells in Insects

Technological advancements in immunohistochemistry and in situ hybridization have significantly contributed to our understanding of PH/NP-producing cells. Proctolin (PRT, H-Arg-Tyr-Leu-Pro-Thr-OH) is a pentapeptide first extracted from cockroaches (Staratt et al. 1975). This peptide is known to have many neurohormonal effects and has been associated with specific, identified cockroach neurons. The PRT-producing cells in the cockroach CNS were determined by immunohistochemistry with rabbit antisera against synthetic PRT peptide (Bishop et al. 1981). This was the first identification of PH/NP-producing cells in insects.

A monoclonal antibody was produced by immunizing mice with a synthetic decapeptide corresponding to the N-terminal portion of the A-chain of bombyxin, a peptide from the silkworm *Bombyx mori* (Mizoguchi et al. 1987). Immunohistochemistry with this antibody revealed that bombyxin is produced by four pairs of mid-dorsal NSCs of the brain and is transferred to and released from the CA of *B. mori.*

Localization studies using in situ hybridization across various insect models have shed light on PH/NP expression in the CNS, linking mRNA expression to specific biological functions and phenotypes and thus enhancing our understanding of the roles of PHs/NPs in regulatory and physiological processes in insects (Normand and Bloch 1991). Eclosion hormone (EH) is a neurosecretory hormone that initiates the ecdysis behavior of insects. In situ hybridization revealed that only two pairs of large NSCs in the ventromedial region of the larval brain contained mRNA in the hawkmoth *Manduca sexta*. These ventromedial cells are also immunopositive with an antiserum against EH, confirming that *EH* is expressed only in these cells (Horodyski et al. 1989).

Various techniques, such as double staining with immunohistochemistry and in situ hybridization or double-labeled in situ hybridization, are frequently used to reveal peptide colocalization and visualize similarities and differences in gene expression patterns within tissues. Pigment-dispersing factor (PDF) regulates the activity rhythm (see Sect. 13.3.1), and FMRFamide (FMRF) was proposed as an additional candidate regulatory factor because double immunolabeling suggested that PDF and FMRF coexist in the same cells of several insects. However, double-labeled in situ hybridization demonstrated that *PDF* and *FMRF* are not coexpressed in the same cells in the *housefly Musca domestica* and the fruitfly *Drosophila melanogaster*. This study suggested that FMRF may have roles in circadian functions that are independent of PDF, even if it is involved in circadian rhythms (Matsushima et al. 2007). Using these techniques, we can discriminate gene expressions at single-cell resolution.

Cell labeling using reporter genes has been established to identify the expression of a gene of interest. The genes encoding fluorescent and luminescent proteins and beta-galactosidase of *Escherichia coli* (LacZ) are commonly used as reporters to induce visually identifiable characteristics. DNA constructs of the reporter gene fused with the regulatory sequence of another gene are transformed into the insect to express the reporter protein. This technology is potentially helpful for examining cell-specific gene expression patterns when the PH/NP or mRNA levels are relatively low. In *D. melanogaster*, an 8 kb *FMRF* DNA fragment (including 5 kb of the 5′ upstream sequence) fused with *LacZ* (reporter) was introduced into the *Drosophila* genome using P-element transformation (Schneider et al. 1993). In this transformant, *LacZ* expression mimics nearly all spatial aspects of the normal expression pattern of the FMRF gene. This result indicates that the cell-specific gene expression pattern of the FMRF gene is generated by transcriptional regulation resulting from multiple promoter/enhancer regions and that the multiple control regions are included in the 8 kb *FMRF* DNA fragment.

By using recombinant *Autographa californica* multiple nucleopolyhedrovirus-mediated gene transfer, cell labeling is also applicable in Lepidoptera (Shiomi et al. 2003; Daubnerová et al. 2009). The trissin (TR) gene is expressed in the frontal ganglion (FG). Nevertheless, the nerve projection of TR-producing cells could not be explored by immunohistochemistry, as an antibody against TR has not been developed. Enhanced green fluorescent protein expression driven by the TR promoter revealed the projection of TR-producing neurons (Roller et al. 2016). This result shows that this system is also helpful for exploring the nervous network of peptidergic neurons.

Mass spectrometric methods are increasingly used to analyze the peptidome of the CNS, even at the single-cell level (Hummon et al. 2006). In this method, tissues or cells of interest need to be identified and isolated, but immunostained neurons yield negative results, as the antibody might suppress ion signals in mass spectra (Neupert and Predel 2005). Reporter genes also identify and isolate specific neurons in *D. melanogaster* (Neupert et al. 2007). When a Gal4 promoter line is used to drive a gene for fluorescent protein under UAS control, the fluorescent protein is expressed in single neurons of interest. The neurons can be isolated singly under a fluorescence microscope and then subjected to MALDI-TOF mass spectrometry. The MALDI-TOF mass spectrometry can detect ion signals in isolated neurons, as the fluorescent protein expressed in these cells does not disturb ion signals in mass spectra. The isolation of single neurons using the Gal4/UAS system benefits the single-cell peptidome.

A PH/NP atlas is a powerful tool for identifying and visualizing the distribution of PHs/NPs in the CNS. By combining the various techniques mentioned above, PH/NP atlases have been reported in the adult brain of *D. melanogaster* (Diesner et al. 2018), in the CNS of *B. mori* (Roller et al. 2017; Liu et al. 2023), in the brain of the honey bee (Kaiser et al. 2022), in the larval ventral nerve cord of the mosquito *Aedes aegypti* (Yin et al. 2024), and in the brains of two ant species, the leafcutter ant *Atta sexdens* and the black garden ant *Lasius niger* (Geiber et al. 2023). In *A. aegypti*, the expression of PHs/NPs, neurotransmitters, and their receptors has been analyzed using single-cell RNA sequencing technologies to dissect gene expression at single-cell resolution. The data indicates the organs where each PH/NP is expressed at some stage from larva to adult. This approach provides a comprehensive view of the cellular composition and possible peptidergic signaling networks involved. For example, the effects of extensive crosstalk among calcitonin-like diuretic hormone (CT-DH), CCHamide (CCH) 2, and allatotropin (AT) on larval growth have been explored from the coexpression of receptors and PHs/NPs (Yin et al. 2024).

Connectomics, which uses transmission electron microscopy to provide high-resolution reconstructions of neural circuits, is a powerful tool for unraveling the intricate PH/NP network in *D. melanogaster*. This approach helps elucidate how neurons are interconnected and how these interactions regulate behavior and cognition. For example, a recent study by Hückesfeld et al. (2021) identified a novel carbon dioxide-responsive sensory pathway that targets specific NSCs, including those that express corazonin (CRZ) and corticotropin-releasing factor-like diuretic hormone (CRF-DH).

5.2 Localization and Function of PHs/NPs in the CNS

Table 5.1 shows the expression of PH/NP genes based on Bombase data (Roller et al. 2017). The data indicates the organs where each PH/NP is expressed at some stage from larva to adult. However, the distribution of PHs/NPs varies depending on the developmental stage. For more details, refer to Bombase.

5.2.1 Brain

The NSCs of the insect brain were first described in the pars intercerebralis of the honeybee *Apis mellifera* (Weyer 1935). The NSCs are characterized by electron-dense granules containing PH/NP (Panov 1980). Some PHs/NPs are produced in

Table 5.1 List of organs or cells that express PHs/NPs

Peptide precursor genes (abbreviation)	Organs or cells expressing the genes for PH/NP									
	Brain	CC	SG	FG	TG	AG	EEC	L1	EPG	Other non-neural organ
Adipokinetic hormone/corazonin-related peptide (ACP)	++			+		+				
Adipokinetic hormones (AKH)		++								
Agatoxin-like peptide/ LQDVamide (ALP)	+	+	+	+	+	++	+			
Allatostatin A (AST-A)	++			+	+	+	+			
Allatostatin C (AST-C)	+		+		+	++	+			
Allatostatin CC (AST-CC)	+		++		++	++	+			
Allatostatin CCC (AST-CCC)[1]							+			
Allatotropin (AT)	+		+	+	+	+	+			
Bursicon α subunit (Burs)			++		++	++				
Capability/ cardioacceleratory peptide 2b/periviscerokinin (CAPA/PK)[2]	+		+		+	++				
CCHamide 1 (CCH1)	+		+		+	+	+			
CCHamide 2 (CCH2)	++		+		++	+	+			
CNMamide (CNM)[3]	+				+	+	+			
Corazonin (CRZ)	++		+		+	+				
Corticotropin-releasing factor (CRF)-like diuretic hormone (CRF-DH)	++		+	+	+	++				
Crustacean cardioactive peptide (CCAP)	+		++		++	++		+		
Calcitonin-like diuretic hormone/ DH31 (CT-DH)	++			+	++	+	+			
Ecdysis-triggering hormone (ETH)										+
Eclosion hormone (EH)	++									
EFLamide (EFL)[4]	+									
Elevenin (ELE)[5]	+		+		+					+

(continued)

Table 5.1 (continued)

Peptide precursor genes (abbreviation)	Organs or cells expressing the genes for PH/NP									
	Brain	CC	SG	FG	TG	AG	EEC	L1	EPG	Other non-neural organ
FMRFamide (FMRF)	+		+		++	+				
Glycoprotein hormone alpha 2 (GPA2)						++				
Glycoprotein hormone beta 5 (GPB5)						++				
Gonadulin/ILP 8 (ILP8)[6]										+
Hansolin[7]										+
IDL-like (IDL)	+			+[14]		+[14]				
IMFamide (IMF)	+									
Insulin-like peptide (ILP)	++					+				+
Ion-transport peptide (ITP)	++									
Ion-transport peptide-like (ITPL)	++				+	+		+		
Inotocin (ITC)[8]			+							
Insect parathyroid hormone (iPTH)[9]	+		+		+	+	+			
ITG-like (ITG)	+		+	+	+	+				
K5 (K5)	+	+		+	+	+	+			
Kinin (K)	+		+		+	++				
Limostatin (LST)[10]		++								
Myoinhibitory peptide (MIP)	++		+	+	+	+			+	
Myosuppressin (MS)	+		+		++	+	+	+[15]		
Natalisin (NTL)	+		+		+	+				
Neuroparsin/ovary ecdysteroidogenic hormone (OEH)	++[13]					+				
Neuropeptide F (NPF)	+		+	+	+	++	+			
Neuropeptide-like precursor 1 (NPLP1)	+		+		+	+				
NVP-like (NVP)	?									
Orcokinin (OK)	+		+		++	++				
Partner of bursicon/bursicon β subunit (Pburs)		++	++		++	++				
Pyrokinin/pheromone biosynthesis-activating neuropeptide (PK/PBAN)	+		+		+					

(continued)

Table 5.1 (continued)

Peptide precursor genes (abbreviation)	Organs or cells expressing the genes for PH/NP									
	Brain	CC	SG	FG	TG	AG	EEC	L1	EPG	Other non-neural organ
Pigment-dispersing factor (PDF)	+					+	+			
Proctolin (PRT)[11]	+		+		+	+				
Prothoracicotropic hormone (PTTH)	++									
RFLamide (RFL)[12]										+
RYamide (RY)	+					+	+			
Short neuropeptide F (sNPF)	+	+	+	+	+	+	+			
SIFamide (SIF)	+									
Sulfakinin (SK)	+				+	+	+			
Tachykinin-related peptide (TRP)	+		+		+	+	+			
Trissin (TR)	+			+						

Most of the data are taken from *B. mori* (Bombase), with the following data taken from other insects: (1) *A.mellifera*, (2), (15) *M.sexta*, (3), (10), (11) *D.melanogaster* (4) *P. apterus*, (5) *N. lugens*, (6) *R.prolixus* (7), (12) *P. simplex*, (8), (9) *T.castaneum*, (13) *A.aegypti*, (14) *B.terrestris* (Ragionieri 2025). +; expression detected. ++; expression in the NSC

NSCs, but others are also expressed in intrinsic or extrinsic neurons of the brain and act as neuromodulators or neurotransmitters. Most PHs/NPs are expressed in both the brain and other nervous tissues. However, the genes for EFLamide (EFL), EH, IMFamide (IMF), prothoracicotropic hormone (PTTH), and SIFamide (SIF) are exclusively expressed in the brain.

Ion-transport peptide (ITP) and longer ITP-like (ITPL) are alternatively spliced multifunctional PH/NP involved in the regulation of water and ion homeostasis, metabolism, cuticle expansion, melanization, and reproduction (Dircksen 2009; Yu et al. 2016; Gáliková et al. 2018). *ITP* and *ITPL* are expressed in different tissues. In *B. mori*, *ITP* is exclusively expressed in the NSCs of the brain and exceptionally expressed in the male-specific neurons of the abdominal neuromere 9 (MAN9) of pharate adults. In contrast, *ITPL* is expressed in numerous central and peripheral neurons (Klöcklerová et al. 2023). The predominant expression of *ITP* in the brain, in contrast to the diverse neuronal expression of *ITPL*, has also been reported in other insects (Sajadi and Paluzzi 2024).

The PTTH stimulates both peak (molting period) and basal levels of ecdysteroid biosynthesis (feeding period). Several PHs/NPs regulate the synthesis and secretion of PTTH through the receptors expressed on PTTH neurons. In *D. melanogaster*, the expression of *PTTH* is negatively controlled by PDF neurons (McBrayer et al. 2007), and the timing of PTTH secretion for peak ecdysteroid biosynthesis is regulated by both allatostatin A (AST-A) neurons and leucine-rich repeat-containing G-protein-coupled receptor 3 (LGR3)-positive neurons. The AST-A signaling promotes PTTH secretion (Deveci et al. 2019). LGR3 is a receptor for *Drosophila* insulin-like peptide 8 (DILP8), and DILP8/LGR3 signaling suppresses PTTH secretion (Vallejo et al. 2015). Basal levels of ecdysteroid in hemolymph negatively affect the larval growth rate during the feeding period without affecting pupariation timing (see Sect. 7.4.1), and PTTH stimulates basal levels of ecdysteroid biosynthesis in *D. melanogaster*. *CRZ* is expressed in the NSCs of the brain and negatively controls PTTH neurons only during the middle third-instar stage to ensure the proper size of the body (Imura et al. 2020). CRZ signaling is involved in PTTH accumulation via posttranscriptional regulation rather than via transcriptional regulation of PTTH.

5.2.2 Corpora Cardiaca

The CC are the major neurohemal organ in insects, but they also contain intrinsic NSCs (see Sect. 4.1.2). Adipokinetic hormone (AKH) is the first insect hormone to be identified (Stone et al. 1976) and mediates the mobilization of energy substrates in many insects. In the migratory locust *Locusta migratoria*, AKH is synthesized in the cell bodies of the glandular lobes (composed of intrinsic cells) of the CC (Bogerd et al. 1995). The AKH-producing cells (APCs) of the CC possess intrinsic nutrient sensors, which become active during low energy (high metabolic demand and/or low nutrient availability) in *D. melanogaster*. The ACPs release AKH in response to low energy conditions to raise circulating levels of carbohydrate and lipid energy reserves. This mechanism is functionally similar to alpha cells in the mammalian pancreas (Nelson et al. 2021). In addition to AKH, APCs are suggested to produce several PHs/NPs. Limostatin suppresses insulin secretion under poor nutrient conditions to prevent the unnecessary consumption of energy in *D. melanogaster* (Alfa et al. 2015).

Partner of bursicon (*Pburs*) is coexpressed with *bursicon alpha subunit* (*Burs*) in the same NSCs and forms a heterodimeric protein hormone bursicon that initiates the tanning of the cuticle. In the APCs of Lepidoptera, only *Pburs* is expressed, where it forms a homodimer (Dai et al. 2008). Pburs homodimer in the APCs is unique to Lepidoptera, but the physiological function remains unknown. Agatoxin-like peptide/LQDVamide (ALP, Sturm et al. 2016; Roller et al. 2017) and K5 (Mitsumasu et al. 2009) are also expressed in intrinsic cells of the CC, but their functions are unknown in any insect.

Juvenile hormone (JH) biosynthesis in the CA is thought to be stimulated by allatotropin (AT). However, AT has no allatotropic effect on JH synthesis in the CC–CA complex of larvae in *B. mori* (Kaneko and Hiruma 2014). In addition, the AT receptor is not highly expressed in the CA; instead, the receptor for short neuropeptide F (sNPF) is predominantly transcribed in the CA of *B. mori* (Yamanaka et al. 2008). The sNPF is a multifunctional group of neurohormones that regulates various physiological processes in insects. In *B. mori*, sNPF is produced in AT receptor-expressing cells of the CC. As sNPFs have apparent allatostatic activity (Kaneko and Hiruma 2014), AT may inhibit the synthesis or release of sNPFs in the CC, thereby exerting its indirect allatotropic effect (Yamanaka et al. 2008).

5.2.3 Subesophageal Ganglion

The subesophageal ganglion (SG) of insects is located below the esophagus inside the head. It comprises three pairs of fused ganglia: the mandibular, maxillary, and labial ganglia. The SG is a crucial PH/NP-producing organ that induces embryonic diapause in *B. mori* and pheromone biosynthesis in Lepidoptera.

The pyrokinin (PK)/pheromone biosynthesis-activating neuropeptide (PBAN) peptide family is a major group of insect PH/NP with a conserved C-terminal domain of FXPRLamide. In *B. mori*, the PK/PBAN gene is called the diapause hormone (DH)-PBAN gene. The DH is released in the diapause-destined pupae of females and induces embryonic diapause in the eggs of *B. mori* (see Sect. 8.2.1). The PBAN is released into the hemolymph of female adults and activates the biosynthesis of pheromones in moths. *DH-PBAN* encodes five neuropeptides: DH, PBAN, SG neuropeptide (SGNP)-α, SGNP-β, and SGNP-γ (Sato et al. 1993). *DH-PBAN* is expressed in three clusters of NSCs—mandibular, maxillary, and labial cells (Sato et al. 1994)—and the neuropeptides are transported to the CC and released into the hemolymph (Yamashita et al. 1998). However, ablation experiments and the firing activity of NSCs suggest that functional differentiation occurs in these cells (Fig. 5.1). Labial cells are responsible for diapause induction (Ichikawa 2003), whereas maxillary cells are primarily responsible for pheromone production (Ichikawa and Kamito 2003). As the cell- or tissue-specific processing mechanisms of precursor polyproteins are well understood in vertebrates (Douglass et al. 1984), the selective production/inactivation of PH/NP may occur in these cells.

The cyclic neuropeptide inotocin (ITC) is an insect homolog of oxytocin/vasopressin (see Sect. 2.3.10). It is expressed only in a pair of cells in the SG in beetles,

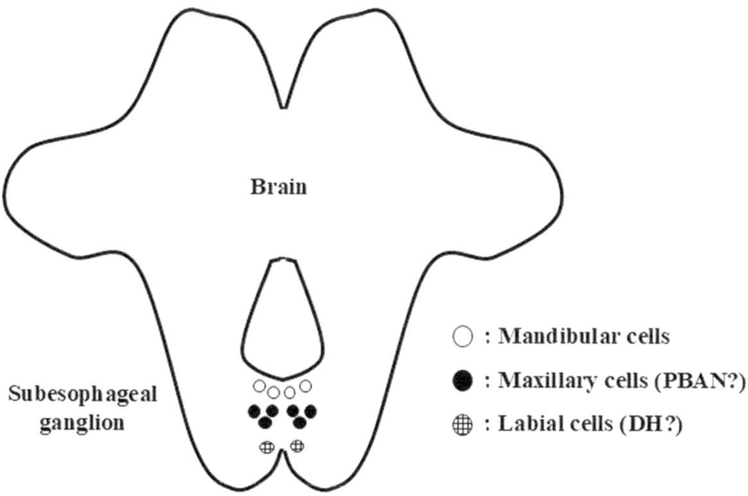

Fig. 5.1 Schematic drawing of NSCs in the SG of *B. mori*. PBAN, pheromone biosynthesis-activating neuropeptide; DH, diapause hormone

cockroaches, mantids, grasshoppers, locusts, and ants (Aikins et al. 2008; Fetter-Pruneda et al. 2021). In the red flour beetle *Tribolium castaneum*, ITC neurons have axons that project symmetrically on both sides of the body. These projections extend forward toward the brain and backward toward the terminal abdominal ganglion. Each axon forms many bulging structures (varicosities) along its length (Aikins et al. 2008). The ITC receptor is expressed in the SG and throughout the brain, ventral nerve cord, Malpighian tubules, gut, oenocytes, and ovaries, which is consistent with the diverse roles of ITC in regulating behavior and physiology. In the ant *Camponotus fellah*, ITC regulates the expression of cytochrome P450 4G1 (CYP4G1) and the synthesis of cuticular hydrocarbons, which play essential roles in desiccation resistance once workers initiate foraging (Fetter-Pruneda et al. 2021).

5.2.4 Frontal Ganglion

The FG is responsible for foregut motor activity and, hence, the passage of food through the gut (Audsley and Weaver 2008). Six neurons of very similar size (frontal neurons (FN)-I, FN-II, FN-III) and a single larger cell (FN-IV) of the FG are identified by peptidergic contents and axonal projections in *B. mori* (Roller et al. 2016). FN-I produces AT, AST-A, CT-DH, and myoinhibitory peptide (MIP). FN-II is paired cells that produce TR, CT-DH, MIP, and sNPF. Paired FN-III samples presented strong TR and weak sNPF expressions. FN-IV is a larger single cell with TR expression (Table 5.2). *ALP* and *K5* are also expressed in the FG of *B. mori*, but the neurons that express these genes have not been described (Mitsumasu et al. 2009; Roller et al. 2017). The FN-I bears short arborizations within the neuropil of the FG and a single posterior axonal projection innervating the dorsal surface of the crop to terminate at the posterior end of the foregut. The FN-II projects single lateral branches via each frontal connective into the tritocerebrum, where they arborize. Another posterior branch along the recurrent and esophageal nerves arborizes on the stomodeal valve, forming a branching axonal ring over the foregut–midgut boundary. The FN-III neurons probably also project axonal branches via frontal connections into the brain, whereas the FN-IV neurons project only along the recurrent nerve on the foregut–midgut surface.

Several peptides produced in the FG affect feeding activity (Table 5.2). The CT-DH increases the time to the initiation of feeding in *B. mori* larvae (Nagata and Nagasawa 2017). The AT is a feeding inhibitory peptide (Matsumoto et al. 2019), whereas sNPF positively modulates feeding motivation by reducing the inhibitory effects of AT in the midgut (Matsumoto et al. 2019). The MIP pathway has a switch-like role in regulating feeding satiety in *D. melanogaster* (Min and Chung 2016). The AST-A reduces food intake (Aguilar et al. 2003). The roles of ALP, K5, and TR in feeding have yet to be discovered.

Table 5.2 Neurons in the FG of *B. mori*

Neuron	Innervate to:	Peptides	Function
FN-I	FGforegut	AT	Inhibit feeding
		AST-A	Reduce food intake
		CT-DH	Increase the time to the initiation of feeding
		MIP	Induce satiety (stop feeding)
FN-II	Brainforegut-midgut boundary	TR	Unknown
		CT-DH	Increase the time to the initiation of feeding
		MIP	Induce satiety (stop feeding)
		sNPF	Activate feeding motivation
FN-III	Brain	TR	Unknown
		sNPF (weakly)	Activate feeding motivation
FN-IV	Foregut-midgut surface	TR	Unknown

5.2.5 *Thoracic Ganglia*

The thoracic ganglia (TGs) in insects are principally involved in the control of locomotion. These controls are mediated through efferent neurons and hormone release from the NSCs of the ganglion. The PRT is known to act as a neuromodulator and neurotransmitter in various insects. In the TG, PRT is produced in skeletal motor neurons that are involved in modulating the rhythmic muscle contractions necessary for flight and locomotion (Keshishian and O'Shea 1985). The PHs/NPs produced in the NSCs of the TGs are released into the hemolymph through various neurohemal organs, such as the CC, CA, and transverse nerves (see Fig. 5.3), depending on the insect species (Veenstra 1987).

Bombyx FMRF (BRF) gene encodes four FMRF-related peptides (FaRPs). All the FaRPs have static effects on ecdysteroid biosynthesis by the prothoracic gland (PG) in *B. mori*. The *BRF* is predominantly expressed in the NSCs of the TG, and the *BRF*-expressing neurons of the prothoracic ganglion innervate the surface of the PG and supply all FaRPs (Yamanaka et al. 2006). Ecdysteroid biosynthesis by the PG was previously thought to be regulated only by humoral factors such as PTTH. This study provides the first report of peptides controlling ecdysteroid biosynthesis by direct innervation (Fig. 5.2). One of the two pairs of *BRF*-expressing NSCs of the prothoracic ganglion coexpressed the orcokinin (OK) A gene (splicing variant of *OK*) in *B. mori*. The OKA gene encodes five OKA-related peptides, and the two peptides (Bommo-OKA-I and Bommo-OKA-II) containing the conserved N-terminal OKA sequence (NFDEIDRS-) have a tropic effect on ecdysteroid biosynthesis by the PG (Yamanaka et al. 2011). These results indicate that both tropic and static factors are delivered to the PG by direct innervation. A possible role of the complex features of PG-innervating neurons is to maintain ecdysteroid titer at a low

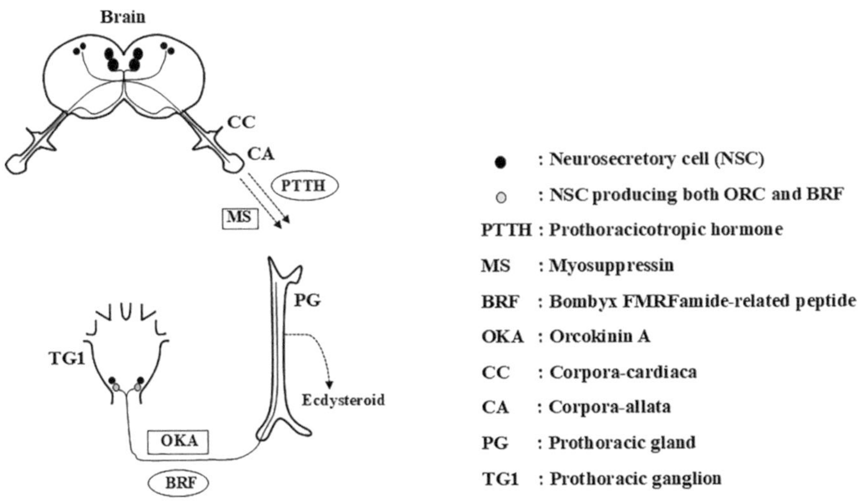

Fig. 5.2 Schematic drawing of neural and humoral regulation of ecdysteroid biosynthesis by the PG. Solid line, neural innervation; dotted line, humoral regulation; circle, tropic PH/NP (peptide hormone/neuropeptide) on ecdysteroid biosynthesis; square, static PH/NP on ecdysteroid biosynthesis

level during the intermolt period, as low ecdysteroid levels have a significant effect on controlling tissue growth and proliferation (Yamanaka et al. 2011).

5.2.6 Abdominal Ganglia

The abdominal ganglia (AGs) oversee various vital functions, from reproduction to digestion. Different types of NSCs are present in the AGs, and these cells vary in size, staining properties, and secretory activity (Aydemir and Ergen 2001). The PHs/NPs are released into the hemolymph through neurohemal sites such as the perisympathetic organ and the transverse nerve. *Burs* are coexpressed with *Pburs* in four bilateral neurons in the thoracic and AGs of the third-instar larva of *D. melanogaster*, and the somata in the posterior AGs expresses only *Burs* (Luo et al. 2005).

Heterodimers of glycoprotein hormone alpha 2 and glycoprotein hormone beta 5 (GPA2/GPB5) play critical roles in development, diuresis, and reproduction. The GPA2/GPB5 signaling acts to delay egg production in adult females of the blood-sucking bug *Rhodnius prolixus* (Al-Dailami et al. 2022). Both GPA2 and GPB5 are produced in the bilateral pairs of NSCs of the AGs. These cells are distinct from the NSCs that produce crustacean cardioactive peptide (CCAP), Burs/Pburs, kinin, and CT-DH in two dipteran insects (Sellami et al. 2011; Rocco and Paluzzi et al. 2020). GPA2/GPB5 produced in two pairs of lateral NSCs of each AG appears to be released from the transverse nerve in *B. mori* (Roller et al. 2017).

MAN9 neurons are male-specific and are missing in females of *B. mori* (Čižmár et al. 2019). MAN9 starts to differentiate in spinning fifth-instar larvae, and their number increases to ~20 in pharate adult males. The AT, AST-C, CT-DH, MIP, and ITP are produced in MAN9. This neuron innervates the accessory glands and sperm ducts. The CT-DH and AT stimulate spontaneous seminal vesicle and accessory gland contractions, whereas AST-C and MIP cause dose-dependent inhibition of contractions (Čižmár et al. 2019). The function of ITP produced in MAN9 has not yet been explored.

5.2.7 PH/NP-Producing Cells Regulated by Ecdysis-Triggering Hormone Signaling

The ETH produced by endocrine Inka cells in the epitracheal glands (Fig. 5.3) of insects is a crucial signal for regulating the ecdysis sequence in insects. The ETH directly targets specific peptidergic and aminergic neurons in the CNS that express ETH receptors. By activating these neurons, ETH initiates different behaviors during the ecdysis process: pre-ecdysis, ecdysis, and post-ecdysis (Daubnerová et al. 2021). These primary target neurons of ETH express distinct splicing variants of the ETH receptor (ETHR-A or ETHR-B). Neurons in the CNS that express the ETHR gene produce a wide variety of PHs/NPs (see Figs. 6 and 10 in Daubnerová et al. 2021). Neurons expressing the ETHR-A subtype (ETHR-A neurons) generate many peptides, such as kinin, CRF-DH, CT-DH, AST-A, AST-CC, EH, FMRF, CCAP, CCH1, MIP, bursicon, neuropeptide F (NPF), and sNPFs (Kim et al. 2006; Daubnerová et al. 2021). In contrast, neurons expressing ETHR-B produce fewer peptides, such as MIPs, AST-A, PDF, and RFamide-like peptides. Some neurons, such as DLT cells of TG1–3 and PM9 neurons of the terminal abdominal ganglion (TAG), express both ETHR-A and ETHR-B. These specific cells produce NPF, CCH1, CT-DH, and MIPs.

The roles of different PHs/NPs in ETHR neurons of *B. mori* are as follows: (1) kinin and CRF-DH control pre-ecdysis I; (2) PDF is involved in pre-ecdysis II; (3) AST-CC, bursicon, CCAP, and MIPs initiate ecdysis behavior; (4) CCH1, CCAP, and MIPs in PM9 cells of the TAG regulate contractions essential for shedding of the cuticle lining the hindgut; (5) MIPs and sNPFs in VL8 cells of the TAG terminate ecdysis; and (6) bursicon is involved in post-ecdysis behavior.

In this way, ETH acts as a command factor to activate downstream regulatory PHs/NPs within the CNS, controlling a behavioral sequence critical for successfully shedding the old cuticle (Daubnerová et al. 2021).

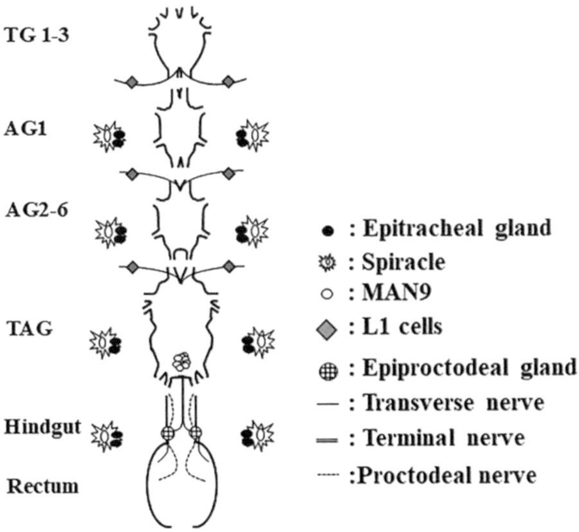

Fig. 5.3 Schematic drawing of L1 cells, epiproctodeal gland, epitracheal gland, and MAN9

5.3 Localization and Function of PHs/NPs in Peripheral Organs

5.3.1 Enteroendocrine Cells

Enteroendocrine cells (EECs) in the insect midgut respond to physiological changes in the intestine by releasing multiple peptides to control food intake, gastrointestinal activity, and systemic metabolism. The EECs in *Drosophila* were classified into ten subtypes based on single-cell RNA sequencing analysis. These EECs collectively express genes encoding at least 14 different peptide hormones, including AST-A, AST-C, tachykinin-related peptide (TRP), NPF, sNPF, CT-DH, CCH1, CCH2, MIP, neuropeptide-like precursor 2 (NPLP2), GPB5, ITPL, CCAP, and OK (Guo et al. 2022). In *B. mori*, AT, RYa, sulfakinin (SK), PDF, K5, AST-CC, and ALP are additionally expressed in EECs, but NPLP2 (absent in *B. mori*), GPB5, and CCAP are not expressed in EECs (Roller et al. 2022). Each subtype of EEC occupies a specific midgut region and expresses different PHs/NPs. Some gut PHs/NPs and gustatory receptors are coexpressed in EECs. In *D. melanogaster*, 12 gustatory receptors were coexpressed in EECs expressing gut PHs/NPs (Park and Kwon 2011). Therefore, gustatory receptors are expected to play a role in controlling the secretion of gut PHs/NPs. For example, EECs sense dietary L-glutamate through the metabotropic glutamate receptor, which then reduces the secretion of NPF from EECs, thereby promoting feeding in mated females of *D. melanogaster* (Gao et al. 2024).

5.3.2 Link Neurons 1 (L1 Cells)

Link neurons 1 (L1 cells) are located on the peripheral nerves in a variety of insect species (Fig. 5.3 and Finlayson and Osborne 1968; Wall and Taghert 1991a, b). The L1 cells in all thoracic and abdominal segments have a central projection into transverse nerves and distal projections that contact the heart (Wall and Taghert 1991a, b). *ITPL* is expressed in L1 cells of both the thoracic and the abdominal segments, whereas L1 cells of the abdominal segments coexpress *CCAP* with *ITPL* (Dai et al. 2007; Klöcklerová et al. 2023). ITPL in L1 cells may integrate signals from sensory systems monitoring body homeostasis or movements, as proposed in *D. melanogaster* (Dircksen 2009). Myosuppressin gene is strongly expressed in L1 cells of segments T2–A1 in *M. sexta* (Lu et al. 2002), but its function is unclear.

5.3.3 Epiproctodeal Glands

The epiproctodeal glands are located on each proctodeal nerve, just anterior to the rectum (Fig. 5.3 in this chapter and Davis et al. 2003). In fourth-instar larvae of *M. sexta*, the epiproctodeal glands exhibit a distinct cycle of synthesis and sudden release of MIP that coincides with a rapid decline in the ecdysteroid titer. The MIP was identified as a prothoracicostatic peptide in *B. mori* and inhibits ecdysteroid biosynthesis (Hua et al. 1999). The function of epiproctodeal glands is thought to be the timely release of MIP during the molting cycle to inhibit the PGs and thus facilitate the sudden decline in the ecdysteroid titer (Davis et al. 2003).

5.4 Future Prospects

One of the main characteristics of insects is that they have an open vascular system. This system offers many benefits, including efficient energy usage, flexible substance transport, a lightweight structure, and environmental adaptivity. An important aspect of this system is the precise timing of receptor expression in specific cells, as well as the timely production and secretion of ligands to ensure proper binding to their receptors. This coordination is critical for maintaining the open vascular system properly.

In addition, insects have a distributed control system, which means that many PHs/NPs are produced in various organs. This allows each thoracic and abdominal ganglion to independently control basic functions such as respiration, circulation, heartbeat, diuresis, and gut motility through the action of PHs and NPs, without waiting for direct commands from the brain (Nässel 1996). This enables quick and flexible responses of insects to sudden environmental changes. These systems have contributed to insect survival and prosperity.

Regulatory mechanism via PHs/NPs plays a crucial role in adapting to various environmental conditions. Research in this field, particularly in *Drosophila*, has provided excellent insights into how insects adapt to environmental conditions. However, these findings offer only a partial view. Comparative studies with other insect species are essential for a comprehensive understanding of the diverse regulatory networks that insects have evolved to survive under dynamic environmental change.

References

Aguilar R, Maestro JL, Vilaplana L, Pascual N, Piulachs MD, Bellés X (2003) Allatostatin gene expression in brain and midgut, and activity of synthetic allatostatins on feeding-related processes in the cockroach Blattella germanica. Regul Pept 115:171–177. https://doi.org/10.1016/s0167-0115(03)00165-4

Aikins MJ, Schooley DA, Begum K, Detheux M, Beeman RW, Park Y (2008) Vasopressin-like peptide and its receptor function in an indirect diuretic signaling pathway in the red flour beetle. Insect Biochem Mol Biol 38:740–748. https://doi.org/10.1016/j.ibmb.2008.04.006

Al-Dailami AN, Orchard I, Lange AB (2022) Investigating the role of glycoprotein hormone GPA2/GPB5 signaling in reproduction in adult female Rhodnius prolixus. Front Insect Sci 2(2022): 1096089. https://doi.org/10.3389/finsc.2022.1096089

Alfa RW, Park S, Skelly KR, Poffenberger G, Jain N, Gu X (2015) Suppression of insulin production and secretion by a decretin hormone. Cell Metab 21:323–334. https://doi.org/10.1016/j.cmet.2015.01.006

Audsley N, Weaver RJ (2008) Neuropeptides associated with the regulation of feeding in insects. Gen Comp Endocrinol 162:93–104. https://doi.org/10.1016/j.ygcen.2008.08.003

Aydemir EM, Ergen G (2001) Histological investigations on the neurosecretory cells in thoracic and abdominal ganglia of Melanogryllus desertus (Pallas, 1771), (Othoptera: Gryllidae). Turk J Zool 25:83–86

Bishop CA, O'Shea M, Miller RJ (1981) Neuropeptide proctolin (H-Arg-Tyr-Leu-pro-Thr-OH): immunological detection and neuronal localization in insect central nervous system. Proc Natl Acad Sci USA 78:5899–5902. https://doi.org/10.1073/pnas.78.9.5899

Bogerd J, Kooiman FP, Pijnenburg MA, Hekking LH, Oudejans RC, Van der Horst DJ (1995) Molecular cloning of three distinct cDNAs, each encoding a different adipokinetic hormone precursor, of the migratory locust, Locusta migratoria. Differential expression of the distinct adipokinetic hormone precursor genes during flight activity. J Biol Chem 270:23038–23043. https://doi.org/10.1074/jbc.270.39.23038

Čižmár D, Roller L, Pillerová M, Sláma K, Žitňan D (2019) Multiple neuropeptides produced by sex-specific neurons control activity of the male accessory glands and gonoducts in the silkworm Bombyx mori. Sci Rep 9:2253. https://doi.org/10.1038/s41598-019-38761-x

Dai L, Zitnan D, Adams ME (2007) Strategic expression of ion transport peptide gene products in central and peripheral neurons of insects. J Comp Neurol 500:353–367. https://doi.org/10.1002/cne.21192

Dai L, Dewey EM, Zitnan D, Luo CW, Honegger HW, Adams ME (2008) Identification, developmental expression, and functions of bursicon in the tobacco hawkmoth, Manduca sexta. J Comp Neurol 506:759–774. https://doi.org/10.1002/cne.21575

Daubnerová I, Roller L, Žitňan D (2009) Transgenesis approaches for functional analysis of peptidergic cells in the silkworm Bombyx mori. Gen Comp Endocrinol 162:36–42. https://doi.org/10.1016/j.ygcen.2008.11.028

Daubnerová I, Roller L, Satake H, Zhang C, Kim YJ, Žitňan D (2021) Identification and function of ETH receptor networks in the silkworm Bombyx mori. Sci Rep 11:11693. https://doi.org/10.1038/s41598-021-91022-8

Davis NT, Blackburn MB, Golubeva EG, Hildebrand JG (2003) Localization of myoinhibitory peptide immunoreactivity in Manduca sexta and Bombyx mori, with indications that the peptide has a role in molting and ecdysis. J Exp Biol 206(Pt 9):1449–1460. https://doi.org/10.1242/jeb.00234

Deveci D, Martin FA, Leopold P, Romero NM (2019) AstA signaling functions as an evolutionary conserved mechanism timing juvenile to adult transition. Curr Biol 29:813–822.e4. https://doi.org/10.1016/j.cub.2019.01.053

Diesner M, Predel R, Neupert S (2018) Neuropeptide mapping of dimmed cells of adult drosophila brain. J Am Soc Mass Spectrom 29:890–902. https://doi.org/10.1007/s13361-017-1870-1

Dircksen H (2009) Insect ion transport peptides are derived from alternatively spliced genes and differentially expressed in the central and peripheral nervous system. J Exp Biol 212(Pt 3): 401–412. https://doi.org/10.1242/jeb.026112

Douglass J, Civelli O, Herbert E (1984) Polyprotein gene expression: generation of diversity of neuroendocrine peptides. Annu Rev Biochem 53:665–715. https://doi.org/10.1146/annurev.bi.53.070184.003313

Fetter-Pruneda I, Hart T, Ulrich Y, Gal A, Oxley PR, Olivos-Cisneros L et al (2021) PLoS Biol 19: e3001305. https://doi.org/10.1371/journal.pbio.3001305

Finlayson LH, Osborne MP (1968) Peripheral neurosecretory cells in the stick insect (Carausius morosus) and the blowfly larva (Phormia terraenovae). J Insect Physiol 14:1793–1801. https://doi.org/10.1016/0022-1910(68)90211-4

Gáliková M, Dircksen H, Nässel DR (2018) The thirsty fly: ion transport peptide (ITP) is a novel endocrine regulator of water homeostasis in Drosophila. PLoS Genet 14:e1007618. https://doi.org/10.1371/journal.pgen.1007618

Gao J, Zhang S, Deng P, Wu Z, Lemaitre B, Zhai Z et al (2024) Dietary L-Glu sensing by enteroendocrine cells adjusts food intake via modulating gut PYY/NPF secretion. Nat Commun 15:3514. https://doi.org/10.1038/s41467-024-47465-4

Geier B, Gil-Mansilla E, Liutkevičiūtė Z, Hellinger R, Vanden Broeck J, Oetjen J (2023) Multiplexed neuropeptide mapping in ant brains integrating microtomography and three-dimensional mass spectrometry imaging. PNAS Nexus 2:pgad144. https://doi.org/10.1093/pnasnexus/pgad144

Guo X, Lv J, Xi R (2022) The specification and function of enteroendocrine cells in drosophila and mammals: a comparative review. FEBS J 289:4773–4796. https://doi.org/10.1111/febs.16067

Horodyski FM, Riddiford LM, Truman JW (1989) Isolation and expression of the eclosion hormone gene from the tobacco hornworm, Manduca sexta. Proc Natl Acad Sci USA 86: 8123–8127. https://doi.org/10.1073/pnas.86.20.8123

Hua YJ, Tanaka Y, Nakamura K, Sakakibara M, Nagata S, Kataoka H (1999) Identification of a prothoracicostatic peptide in the larval brain of the silkworm, Bombyx mori. J Biol Chem 274: 31169–31173. https://doi.org/10.1074/jbc.274.44.31169

Hückesfeld S, Schlegel P, Miroschnikow A, Schoofs A, Zinke I, Haubrich AN et al (2021) Unveiling the sensory and interneuronal pathways of the neuroendocrine connectome in drosophila. elife 10:e65745. https://doi.org/10.7554/eLife.65745

Hummon AB, Amare A, Sweedler JV (2006) Discovering new invertebrate neuropeptides using mass spectrometry. Mass Spectrom Rev 25:77–98. https://doi.org/10.1002/mas.20055

Ichikawa T (2003) Firing activities of neurosecretory cells producing diapause hormone and its related peptides in the female silkmoth, Bombyx mori. I labial cells. Zoolog Sci 20:971–978. https://doi.org/10.2108/zsj.20.971

Ichikawa T, Kamimoto S (2003) Firing activities of neurosecretory cells producing diapause hormone and its related peptides in the female silkmoth, Bombyx mori. II. Mandibular and maxillary cells. Zool Sci 20:979–983. https://doi.org/10.2108/zsj.20.979

Imura E, Shimada-Niwa Y, Nishimura T, Hückesfeld S, Schlegel P, Ohhara Y et al (2020) The Corazonin-PTTH neuronal Axis controls systemic body growth by regulating basal Ecdysteroid biosynthesis in Drosophila melanogaster. Curr Biol 30:2156–2165.e5. https://doi.org/10.1016/j.cub.2020.03.050

Kaiser A, Hensgen R, Tschirner K, Beetz E, Wüstenberg H, Pfaff M et al (2022) A three-dimensional atlas of the honeybee central complex, associated neuropils and peptidergic layers of the central body. J Comp Neurol 530:2416–2438. https://doi.org/10.1002/cne.25339

Kaneko Y, Hiruma K (2014) Short neuropeptide F (sNPF) is a stage-specific suppressor for juvenile hormone biosynthesis by corpora allata, and a critical factor for the initiation of insect metamorphosis. Dev Biol 393:312–319. https://doi.org/10.1016/j.ydbio.2014.07.014

Keshishian H, O'Shea M (1985) The distribution of a peptide neurotransmitter in the postembryonic grasshopper central nervous system. J Neurosci 5:992–1004. https://doi.org/10.1523/JNEUROSCI.05-04-00992.1985

Kim YJ, Zitnan D, Galizia CG, Cho KH, Adams ME (2006) A command chemical triggers an innate behavior by sequential activation of multiple peptidergic ensembles. Curr Biol 16:1395–1407. https://doi.org/10.1016/j.cub.2006.06.027

Klöcklerová V, Gáliková Z, Roller L, Žitňan D (2023) Differential expression of ITP and ITPL indicate multiple functions in the silkworm Bombyx mori. Cell Tissue Res 392:715–731. https://doi.org/10.1007/s00441-023-03752-y

Liu X, Zhang Z, Hu B, Chen K, Yu Y, Xiang H et al (2023) Single-cell transcriptomes provide insights into expansion of glial cells in Bombyx mori. Insect Sci 10:1–14. https://doi.org/10.1111/1744-7917.13294

Lu D, Lee KY, Horodyski FM, Witten JL (2002) Molecular characterization and cell-specific expression of a Manduca sexta FLRFamide gene. J Comp Neurol 446:377–396. https://doi.org/10.1002/cne.10205

Luo CW, Dewey EM, Sudo S, Ewer J, Hsu SY, Honegger HW et al (2005) Bursicon, the insect cuticle-hardening hormone, is a heterodimeric cystine knot protein that activates G protein-coupled receptor LGR2. Proc Natl Acad Sci USA 102:2820–2825. https://doi.org/10.1073/pnas.0409916102

Matsumoto S, Ohara A, Nagai-Okatani C, Zhou YJ, Fujinaga D, Seike H et al (2019) Antagonistic effect of short neuropeptide F on Allatotropin-inhibited feeding motivation of the silkworm larva, Bombyx mori. Zool Sci 36:58–67. https://doi.org/10.2108/zs180119

Matsushima A, Takano K, Yoshida T, Takeda Y, Yokotani S, Shimohigashi Y et al (2007) Double-labeled in situ hybridization reveals the lack of co-localization of mRNAs for the circadian neuropeptide PDF and FMRFamide in brains of the flies Musca domestica and Drosophila melanogaster. J Biochem 141:867–877. https://doi.org/10.1093/jb/mvm094

McBrayer Z, Ono H, Shimell M, Parvy JP, Beckstead RB, Warren JT (2007) Prothoracicotropic hormone regulates developmental timing and body size in Drosophila. Dev Cell 13:857–871. https://doi.org/10.1016/j.devcel.2007.11.003

Min S, Chung J (2016) Identification of a neural pathway governing satiety in drosophila. BMB Rep 49:137–138. https://doi.org/10.5483/bmbrep.2016.49.3.046

Mitsumasu K, Tanaka Y, Niimi T, Yamashita O, Yaginuma T (2009) Novel gene encoding precursor protein consisting of possible several neuropeptides expressed in brain and frontal ganglion of the silkworm, Bombyx mori. Peptides 30:1233–1240. https://doi.org/10.1016/j.peptides.2009.03.023

Mizoguchi A, Ishizaki H, Nagasawa H, Kataoka H, Isogai A, Tamura S (1987) A monoclonal antibody against a synthetic fragment of bombyxin (4K-prothoracicotropic hormone) from the silkmoth, Bombyx mori: characterization and immunohistochemistry. Mol Cell Endocrinol 51:227–235. https://doi.org/10.1016/0303-7207(87)90032-3

Nagata S, Nagasawa H (2017) Calcitonin-like peptide hormone (CT/DH) in the frontal ganglia as a feeding regulatory peptide of the silkworm, Bombyx mori. Peptides 98:23–28. https://doi.org/10.1016/j.peptides.2016.05.002

Nässel DR (1996) Neuropeptides, amines and amino acids in an elementary insect ganglion: functional and chemical anatomy of the unfused abdominal ganglion. Prog Neurobiol 48:325–420. https://doi.org/10.1016/0301-0082(95)00048-8

Nelson JM, Saunders CJ, Johnson EC (2021) The intrinsic nutrient sensing Adipokinetic hormone producing cells function in modulation of metabolism, activity, and stress. Int J Mol Sci 22: 7515. https://doi.org/10.3390/ijms22147515

Neupert S, Predel R (2005) Mass spectrometric analysis of single identified neurons of an insect. Biochem Biophys Res Commun 327:640–645. https://doi.org/10.1016/j.bbrc.2004.12.086

Neupert S, Johard HA, Nässel DR, Predel R (2007) Single-cell peptidomics of drosophila melanogaster neurons identified by Gal4-driven fluorescence. Anal Chem 79:3690–3694. https://doi.org/10.1021/ac062411p

Normand E, Bloch BJ (1991) Simultaneous detection of two messenger RNAs in the central nervous system: a simple two-step in situ hybridization procedure using a combination of radioactive and non-radioactive probes. Histochem Cytochem 39:1575–1578. https://doi.org/10.1177/39.11.1918932

Panov AA (1980) Demonstration of neurosecretory cells in the insect central nervous system. In: Strausfeld NJ, Miller TA (eds) Neuroanatomical techniques. Insect nervous system. Springer-Verlag, New York, Heidelberg, Berlin, pp 25–50

Park JH, Kwon JY (2011) Heterogeneous expression of Drosophila gustatory receptors in enteroendocrine cells. PLoS One 6:e29022. https://doi.org/10.1371/journal.pone.0029022

Ragionieri L (2025) Neuropeptide profiling of Bombus terrestris (Hymenoptera, Apidae) and comparative analysis across hymenopteran species. Insect Biochem Mol Biol 182:104347. https://doi.org/10.1016/j.ibmb.2025.104347

Rocco DA, Paluzzi JV (2020) Expression profiling, downstream signaling, and inter-subunit interactions of GPA2/GPB5 in the adult mosquito Aedes aegypti. Front Endocrinol (Lausanne) 11:158. https://doi.org/10.3389/fendo.2020.00158

Roller L, Čižmár D, Gáliková Z, Bednár B, Daubnerová I, Žitňan D (2016) Molecular cloning, expression and identification of the promoter regulatory region for the neuropeptide trissin in the nervous system of the silkmoth Bombyx mori. Cell Tissue Res 364:499–512. https://doi.org/10.1007/s00441-015-2352-z

Roller L, Beke G, Stano M, Daubnerova I, Mizoguchi A, Satake H et al (2017) BomBase—web atlas of the neuroendocrine system in Bombyx mori. Online Database Resource. http://bombase.org/

Roller L, Daubnerová I, Mizoguchi A, Satake H, Tanaka Y, Stano M et al (2022) Expression analysis of peptidergic enteroendocrine cells in the silkworm Bombyx mori. Cell Tissue Res 389:385–407. https://doi.org/10.1007/s00441-022-03666-1

Sajadi F, Paluzzi JV (2024) Molecular characterization, localization, and physiological roles of ITP and ITP-L in the mosquito, Aedes aegypti. Front Insect Sci 4:1374325. https://doi.org/10.3389/finsc.2024.1374325

Sato Y, Oguchi M, Menjo N, Imai K, Saito H, Ikeda M et al (1993) Precursor polyprotein for multiple neuropeptides secreted from the suboesophageal ganglion of the silkworm Bombyx mori: characterization of the cDNA encoding the diapause hormone precursor and identification of additional peptides. Proc Natl Acad Sci USA 90:3251–3255. https://doi.org/10.1073/pnas.90.8.3251

Sato Y, Ikeda M, Yamashita O (1994) Neurosecretory cells expressing the gene for common precursor for diapause hormone and pheromone biosynthesis-activating neuropeptide in the suboesophageal ganglion of the silkworm, Bombyx mori. Gen Comp Endocrinol 96:27–36. https://doi.org/10.1006/gcen.1994.1156

Schneider LE, Roberts MS, Taghert PH (1993) Cell type-specific transcriptional regulation of the Drosophila FMRFamide neuropeptide gene. Neuron 10:279–291. https://doi.org/10.1016/0896-6273(93)90318-l

Sellami A, Agricola HJ, Veenstra JA (2011) Neuroendocrine cells in Drosophila melanogaster producing GPA2/GPB5, a hormone with homology to LH, FSH and TSH. Gen Comp Endocrinol 170:582–588. https://doi.org/10.1016/j.ygcen.2010.11.015

Shiomi K, Kajiura Z, Nakagaki M, Yamashita O (2003) Baculovirus-mediated efficient gene transfer into the central nervous system of the silkworm, Bombyx mori. J Insect Biotechnol Sericol 72:149–155. https://doi.org/10.11416/jibs.72.149

Starratt AN, Brown BE (1975) Structure of the pentapeptide proctolin, a proposed neurotransmitter in insects. Life Sci 17:1253–1256. https://doi.org/10.1016/0024-3205(75)90134-4

Stone JV, Mordue W, Batley KE, Morris HR (1976) Structure of locust adipokinetic hormone, a neurohormone that regulates lipid utilization during flight. Nature 263:207–211. https://doi.org/10.1038/263207a0

Sturm S, Ramesh D, Brockmann A, Neupert S, Predel R (2016) Agatoxin-like peptides in the neuroendocrine system of the honey bee and other insects. J Proteome 132:77–84. https://doi.org/10.1016/j.jprot.2015.11.021

Vallejo DM, Juarez-Carreño S, Bolivar J, Morante J, Dominguez M (2015) A brain circuit that synchronizes growth and maturation revealed through Dilp8 binding to Lgr3. Science 350: aac6767. https://doi.org/10.1126/science.aac6767

Veenstra JA (1987) Diversity in neurohaemal organs for homologous neurosecretory cells in different insect species as demonstrated by immunocytochemistry with an antiserum to molluscan cardioexcitatory peptide. Neurosci Lett 73:33–37. https://doi.org/10.1016/0304-3940(87)90026-7

Wall JB, Taghert PH (1991a) Segment-specific modifications of a neuropeptide phenotype in embryonic neurons of the moth, Manduca sexta. J Comp Neurol 309:375–390. https://doi.org/10.1002/cne.903090307

Wall JB, Taghert PH (1991b) The timing of initial neuropeptide expression by an identified insect neuron does not depend on interactions with its normal peripheral target. J Neurobiol 22:935–956. https://doi.org/10.1002/neu.480220906

Weyer F (1935) ber driisenartige Nervenzellen im Gehirn der Honigbiene, Apis mellifera. Zool Anz 112:137–141

Yamanaka N, Zitnan D, Kim YJ, Adams ME, Hua YJ, Suzuki Y et al (2006) Regulation of insect steroid hormone biosynthesis by innervating peptidergic neurons. Proc Natl Acad Sci USA 103: 8622–8627. https://doi.org/10.1073/pnas.0511196103

Yamanaka N, Yamamoto S, Zitnan D, Watanabe K, Kawada T, Satake H et al (2008) Neuropeptide receptor transcriptome reveals unidentified neuroendocrine pathways. PLoS One 3:e3048. https://doi.org/10.1371/journal.pone.0003048

Yamanaka N, Roller L, Zitňan D, Satake H, Mizoguchi A, Kataoka H et al (2011) Bombyx orcokinins are brain-gut peptides involved in the neuronal regulation of ecdysteroidogenesis. J Comp Neurol 519:238–246. https://doi.org/10.1002/cne.22517

Yamashita O, Imai K, Saito H, Shiomi K, Sato Y (1998) Phe-X-Pro-Arg-Leu-NH(2) peptide producing cells in the central nervous system of the silkworm, Bombyx mori. J Insect Physiol 44:333–342. https://doi.org/10.1016/s0022-1910(97)00140-6

Yin C, Morita T, Parrish JZ (2024) A cell atlas of the larval Aedes aegypti ventral nerve cord. Neural Dev 19:2. https://doi.org/10.1186/s13064-023-00178-8

Yu B, Li DT, Wang SL, Xu HJ, Bao YY, Zhang CX (2016) Ion transport peptide (ITP) regulates wing expansion and cuticle melanism in the brown planthopper, Nilaparvata lugens. Insect Mol Biol 25:778–787. https://doi.org/10.1111/imb.12

Part II
Physiological Roles of Peptide Hormones and Neuropeptides

Chapter 6
Regulation of Insect Development

Shinji Nagata and Hitomi Seike

Abstract Insect postembryonic development, including molting, metamorphosis, body size control, and diapause, is hormonally regulated by the brain, prothoracic glands (PGs), and corpora allata (CA). The PGs and CA produce ecdysteroids and juvenile hormone, respectively, both of which are direct regulators of insect development. The brain secretes prothoracicotropic hormone (PTTH), the main regulator of PG activity. Hormone secretory activity of CA and PGs is positively and negatively regulated by a variety of peptide hormones and neuropeptides. This chapter reviews their effects on these endocrine organs and discusses the physiological significance of their actions in regulating insect development.

Keywords Development · Growth · Molting · Metamorphosis · Ecdysteroid · Juvenile hormone

6.1 Introduction

Insects display a remarkable diversity in their growth and development, which is essential for their survival and adaptation to various environments. Unlike vertebrates, insects undergo distinct developmental stages characterized by dramatic transformations known as metamorphosis. Understanding these processes—from repeated molting during their growth and metamorphosis to body size regulation—is crucial for grasping the complexity of insect postembryonic development.

Insect metamorphosis is generally classified into two significant types: complete (holometabolism) and incomplete (hemimetabolism) metamorphosis. Holometabolous insects undergo four distinct stages, egg, larva, pupa, and adult, each with its own unique functions and morphology. In contrast, hemimetabolous insects undergo three stages, egg, nymph, and adult, with nymphs resembling adults except for undeveloped wings and reproductive organs, which mature after the final molt.

S. Nagata (✉) · H. Seike
Department of Integrated Biosciences, Graduate School of Frontier Sciences, The University of Tokyo, Kashiwa, Japan
e-mail: shinjin@edu.k.u-tokyo.ac.jp; hitomis@edu.k.u-tokyo.ac.jp

A. Mizoguchi, N. Okamoto (eds.), *Insect Peptide Hormones and Neuropeptides*, Entomology Monographs, https://doi.org/10.1007/978-981-95-7510-7_6

Both holo- and hemimetabolous insects undergo a specific number of ecdyses during their life cycle, though the number varies among species. The final molt, or eclosion, marks the emergence of the adult with fully developed wings.

Another key developmental strategy is diapause, a period of suspended development that allows survival under adverse conditions such as cold or drought. Diapause is not merely a reduction in metabolism but a highly regulated process controlled by environmental cues through hormonal signaling, ensuring survival and precise developmental timing (see Chap. 8).

Molting, the periodic shedding of the exoskeleton, enables growth and stage transitions and is tightly controlled by hormones. Foundational studies identified key endocrine organs involved in this regulation. In 1917, Kopeć demonstrated the brain's essential role in molting and metamorphosis. Later, Wigglesworth (1936) and Williams (1948, 1950) showed that the corpora allata (CA) and the prothoracic glands (PGs) are also critical. Fukuda (1940) further revealed that the PGs secrete humoral factors that induce molting, while Williams (1946) demonstrated that brain-derived hormones stimulate the PGs to release metamorphosis-inducing hormones. These discoveries established the brain, CA, and PGs as the central endocrine organs regulating insect molting and metamorphosis.

As ecdysozoans, insects grow by repeatedly shedding their exoskeleton, making the timing of molting crucial for proper growth. Studies on the tobacco hornworm *Manduca sexta* by Nijhout and Williams (1974) revealed that larval development depends on reaching a "critical weight." According to their hypothesis, larvae can only proceed to the next stage (instar) once they attain this weight. The critical weight is influenced by the growth rate, which is modulated by nutrition, environmental conditions, and genetic factors—all contributing to final body size and morphology (Davidowitz et al. 2003; Nijhout 2003).

This chapter outlines the endocrine mechanisms that regulate insect growth and development and explains how hormonal control underlies their remarkable developmental success.

6.2 Growth and Developmental Modulation by JH and Ecdysteroids

As mentioned in Sect. 6.1, insect development is precisely regulated by endocrine factors that coordinate growth and molting. Among them, two major peripheral endocrine factors, juvenile hormone (JH) and ecdysteroids, play pivotal roles in controlling molting and metamorphosis (Truman 2005). The regulated secretion of these hormones enables insects to transition between developmental stages (Truman and Riddiford 2023). This section describes the endocrine mechanisms governing the secretion of JH and ecdysteroids.

6.2.1 Retrocerebral Complex and the PGs

JH and ecdysteroids are synthesized in specialized endocrine organs. JH is produced by the corpora allata (CA), a pair of glands located near the brain and rich in secretory cells (see Sect. 4.1.2). Another neural structure, the corpora cardiaca (CC), lies along the neural pathway between the brain and the CA. These organs are interconnected by nerves from the brain, forming the retrocerebral complex—a major neuroendocrine center that regulates hormone release. In larvae and pupae, ecdysteroids are produced by the PGs, which are typically located near the spiracles of the prothorax. In fact, the PGs produce ecdysone, one of ecdysteroids, but it is not a biologically active form of this hormone. The actual bioactive hormone is the one that is hydroxylated at the 20th carbon, which is called 20-hydroxyecdysone, in the peripheral tissues such as the fat body. In this chapter, ecdysone and its derivatives, including 20-hydroxyecdysone, are collectively referred to as ecdysteroids. "Ecdysone" is the name of a specific steroid molecule, but it is also sometimes used to mean a molting hormone, in which case ecdysone is synonymous with ecdysteroids.

6.2.2 Classical Scheme

In holometabolous insects, molting and metamorphosis are tightly regulated by JH and ecdysteroids, a framework known as "The Classical Scheme" of insect postembryonic development (De Loof et al. 2013) (Fig. 6.1). The primary hormone that triggers molting is ecdysteroids. The increased ecdysteroid titers in the hemolymph (insect blood) induces molting and metamorphosis.

The developmental outcome of molting depends on the presence or absence of JH. When JH levels remain high during an ecdysteroid surge, the insect undergoes a larval–larval molt and retains its larval form. Conversely, when JH levels are low or absent, ecdysteroid stimulation induces metamorphosis, leading to the pupal stage. This hormonal interplay between JH and ecdysteroids forms the foundation of endocrine control in holometabolous insects, orchestrating the transitions from larva to pupa and ultimately to adult.

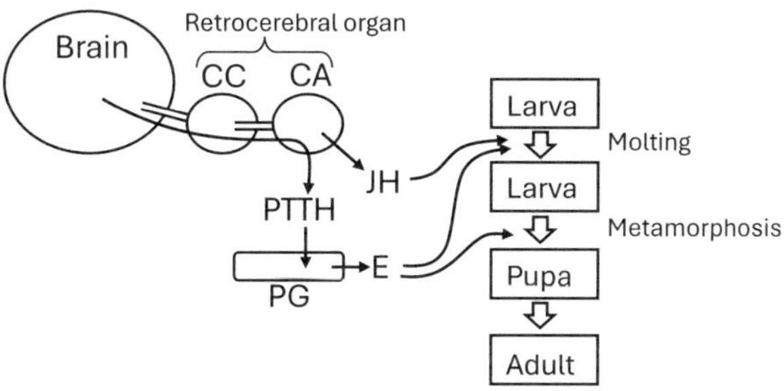

Fig. 6.1 Schematic diagram of the classical scheme of insect molting and metamorphosis. CC Corpora cardiaca, CA copora allata, JH juvenile hormone, E ecdysteroids, PG prothoracic glands

6.3 Regulation of Development by Endocrine Modulation on JH Production

Several neuropeptides have been identified as regulators of JH synthesis, thereby influencing insect development, metamorphosis, and reproduction. Since the CA is the site of JH production, allatoregulatory peptides such as allatotropin (AT) and allatostatins (ASTs) play central roles in JH control (Audsley et al. 2008). These peptides not only regulate JH biosynthesis but also act in various other physiological processes. In addition, insulin-like peptides (ILPs; see Sect. 2.2.1) have recently been characterized as allatotropic factors that stimulate JH synthesis in the CA (Leyria 2024) (Table 6.1).

6.3.1 Effects of AT and ILPs on JH Production

AT was first identified in *M. sexta* as a neuropeptide that stimulates JH biosynthesis in the CA in vitro (Kataoka et al. 1989; see Sect. 2.3.4). However, evidence for its stimulatory activity in vivo remains limited. In the adult *B. mori* and the fall armyworm *Spodoptera frugiperda*, AT instead exerts inhibitory effects on the adult CA (Kaneko and Hiruma 2015; Griebler et al. 2008). In *B. mori*, AT receptors are highly expressed in the CC rather than the CA, and AT receptor-expressing cells in the CC co-express short neuropeptide F (NPF; see Sect. 2.3.5). The CA, in turn, highly express sNPF receptors, and their JH synthesis activity in vitro is suppressed by sNPF. These finding suggest that AT may regulate the CA indirectly via sNPF produced by the CC (Yamanaka et al. 2008; Matsumoto et al. 2019).

In addition to its role in JH regulation, AT has diverse physiological roles (Verlinden et al. 2015). In *B. mori*, AT promotes visceral muscle contraction, suggesting a role in gut motility (Matsumoto et al. 2019). It has also been implicated in reproductive regulation and immune responses, highlighting its multifunctionality across various physiological systems (Verlinden et al. 2015).

In contrast, disruption of ILP signaling leads to reduced JH levels in several insects, including *D. melanogaster*, *A. aegypti*, and the German cockroach *Blattella germanica* (Tu et al. 2005; Pérez-Hedo et al. 2013; Abrisqueta et al. 2014). Therefore, ILPs exert allatotropic functions on the CA in modulating JH levels, eventually affecting vitellogenesis and reproduction (Leyria 2024).

Table 6.1 Amino acid sequences of *B. mori* allatoregulatory peptides

Allatoregulatory peptide	Sequence
Allatotropin	GFKNVEMMTARGFa
Allatostatin A (FGLa/AST)	SQYDFGLa
Allatostatin B (WX_6W/AST)	AWQDLNSAWa
Allatostatin C (PSCF/AST)	pEVRFRQCYFNPISCF
Allatostatin CC	GRVLRCFFNAVTCF

6.3.2 *Effects of ASTs on JH Production*

ASTs are a group of neuropeptides that inhibit JH biosynthesis in the CA and are classified into three families: AST-A (see Sect. 2.3.2), AST-B (see Sect. 2.4.4), and AST-C (see Sect. 2.3.3).

AST-A, first identified in the cockroach *Diploptera punctata* (Woodhead et al. 1989), suppresses JH biosynthesis in some groups of insects (see Sect. 2.3.2). In many species, the AST-A receptor (AST-AR) is expressed in the CA. In the dipteran species including *D. melanogaster*, two types of AST-ARs, AST-AR1 and AST-AR2, are expressed in the CA (Birgül et al. 1999), while the other insects have only one AST-AR (Wegener and Chen 2022). However, only limited studies on several insects such as the brown planthopper *Nilaparvata lugens* have demonstrated that AST-A can inhibit biosynthetic pathway of JH in the CA (Gao et al. 2024). In *D. melanogaster*, AST-A also regulates prothoracicotropic hormone (PTTH)–producing cells and ILP-producing cells at specific developmental stages, resulting in growth perturbation (Deveci et al. 2019).

AST-B was originally identified as a factor that inhibits JH production in crickets (see Sect. 2.4.4). As its amino acid sequence is identical to that of myoinhibiting peptide (MIP), the terms AST-B and MIP are used interchangeably (see Sect. 2.4.4). JH production activity by the CA isolated from the adult cricket *Gryllus bimaculatus* is inhibited by AST-B (Lorenz et al. 1995). Similar activity of AST-B can be observed in the beetle *Tribolium castaneum* (Abdel-latief and Hoffmann 2014). However, evidence for the expression of AST-B/MIP receptor in the CA remains lacking or inconclusive in insects, although AST-neurons reaching the CA have been observed and the receptor transcript have been detected in the whole CC-CA complex (Oh et al. 2014; Tsukamoto and Nagata 2016; Lange et al. 2012).

AST-C was first identified from *M. sexta* as an inhibiting factor on JH production by the CA (Kramer et al. 1991). Subsequent investigations on AST-C in the black-back prominent moth *Clostera anastomosis* and Chinese white pine beetle *Dendroctonus armandi* have revealed that AST-C modulates JH production by transcriptional alteration of JH acid methyl transferase (JHAMT), a key enzyme in JH synthesis (Dong et al. 2017; Sun et al. 2022). Expression of the AST-C receptor in the CA has been reported in several species, including *D. melanogaster* and *Aedes aegypti* (Wang et al. 2012; Mayoral et al. 2010).

6.4 Hormones Affecting the PGs in Relation to Ecdysteroid Production

Insect molting and metamorphosis are controlled by a finely tuned endocrine system. Central to this process is the PGs (see Sect. 4.1.2), an endocrine organ responsible for synthesizing and secreting the molting hormone ecdysteroids (Yamanaka et al. 2013; Lafont et al. 2012). Ecdysteroids play pivotal roles in initiating molting and coordinating developmental transitions (Niwa and Niwa 2016). Their production and release are regulated by a complex network of peptide hormones and neuropeptides that ensure these critical events occur at the proper developmental stages.

6.4.1 Prothoracicotropic Hormone

The principal regulator of ecdysteroid secretion is the brain-derived neuropeptide, PTTH (see Sects. 1.1.6 and 2.6.1), which is released from neurosecretory cells in the brain. In *B. mori*, PTTH is produced by two pairs of neurosecretory cells in the brain, transported axonally to the CA, and secreted into the hemolymph. *B. mori* PTTH is a homodimer of a polypeptide consisting of 109 amino acid residues linked via a disulfide bond (Kataoka et al. 1989). An *N*-linked sugar chain is attached to the 41st asparagine residue possibly to reinforce the stability of its structure. PTTH orthologs are conserved among insect species with low sequence similarity but having conserved cysteine residues required to generate the bioactive form. Reflecting the low sequence similarity, PTTH is highly species specific (Nagata et al. 2005). Several factors are suggested to regulate PTTH secretion, including JH, serotonin, and light-dark cycles (Smith and Rybczynski 2012). PTTH acts directly on the PGs by binding to its receptor tyrosine kinase, Torso, which is specifically expressed in the PGs (Rewitz et al. 2009; Yamanaka et al. 2013). Changes in hemolymph PTTH titers are strongly correlated with changes in ecdysteroid levels, supporting the concept that PTTH regulates ecdysteroid secretion (Mizoguchi et al. 2001, 2013). In fact, PTTH acts on the ecdysteroidogenic enzymes in the PGs such as CYP302a1/Disembodied in *B. mori* (Niwa et al. 2005). Therefore, PTTH is a pivotal factor contributing to the normal development by modulating the ecdysteroid levels in the hemolymph. Loss of PTTH function in *D. melanogaster* causes a significant delay in metamorphosis, indicating that PTTH is not essential for PG activation but is critical for the precise timing of ecdysteroid release (McBrayer et al. 2018; Shimell et al. 2018). *PTTH* knockout in *B. mori* results in various developmental defects, including developmental arrest and precocious metamorphosis that skips the final larval molt (Uchibori-Asano et al. 2017), demonstrating its essential role in normal development.

PTTH also plays a crucial role in regulating pupal diapause. In the cabbage army moth *Mamestra brassicae*, PTTH secretion is shut down after pupation, resulting in PG inactivation and pupal diapause (Mizoguchi et al. 2013).

Recent studies revealed that, in addition to PTTH, many other peptides and regulatory factors act directly or indirectly act on the PGs to fine-tune ecdysteroid biosynthesis (Yamanaka et al. 2013; Niwa and Niwa 2016; Kannangara et al. 2021) (Fig. 6.2). The major factors are listed below.

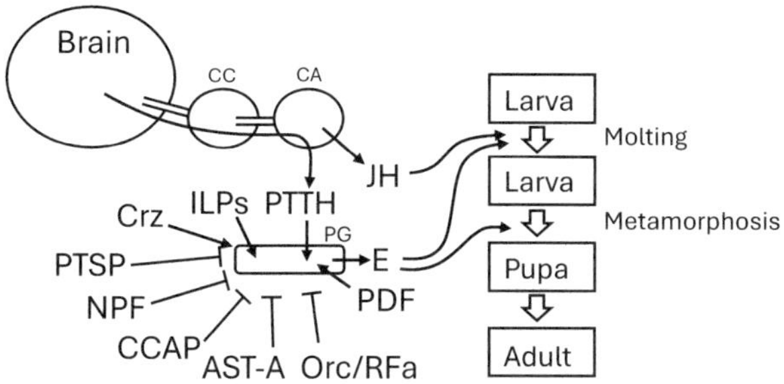

Fig. 6.2 Schematic diagram of the orchestration of neuropeptides modulating ecdysteroidogenesis in the prothoracic glands. ILPs Insulin-like peptides, Crz corazonin, PTSP prothoracicostatic peptide, NPF neuropeptide F, CCAP crustacean cardioacceleratory peptide, AST-A allatostatin-A, Orc orcokinin, PDF pigment dispersing factor

6.4.2 Other PG-Regulatory Peptides

ILPs The first ILP identified in insects was bombyxin, purified from *B. mori* heads as a peptide that stimulates the PGs of the saturniid moth *Samia cynthia ricini* (see Sect. 1.1). This peptide did not show PG-stimulating activity in *B. mori* from which it was purified in any bioassays conducted at the time. However, subsequent studies demonstrated that bombyxin stimulates *B. mori* PGs under a long-term in vitro culture (Gu et al. 2015). Thus, both bombyxin and PTTH—two hormones purified by Ishizaki, Suzuki, and colleagues (Nagasawa et al. 1984; 1986; Kawakami et al. 1990; Kataoka et al. 1991)—are now recognized as bona fide regulators of ecdysteroidogenesis.

In *B. mori*, four pairs of large neurosecretary cells in the dorsomedial part of the brain produce bombyxin, which is transferred to the CA via axons and released into the hemolymph. As demonstrated using the PGs cultured in vitro, the brain-derived ILPs directly act on the PGs to stimulate ecdysteroid production. Several genetic studies in 2005 revealed that *Drosophila* ILPs (DILPs) possess ecdysteroidogenic activity (Caldwell et al. 2005; Colombani et al. 2005; Mirth et al. 2005). Specific activation of phosphoinositide 3-kinase (PI3K) in the PG enhanced the transcription of ecdysteroidogenic "Halloween" genes and accelerated pupariation, whereas PI3K inhibition suppressed ecdysteroid production and delayed development. Since PI3K is a central effector of insulin/IGF signaling (IIS), these studies demonstrated that DILPs act on the PG to regulate developmental timing by promoting ecdysteroid biosynthesis. ILPs function as a nutrient-sensitive factor, like vertebrate insulin. Nutritional conditions strongly influence IIS activity in the PG (Koyama et al. 2013), and IIS also promotes endoreplication in PG cells, a process essential for ecdysone biosynthesis (Ohhara et al. 2017).

Tissue damage to larval imaginal discs delays metamorphosis via the secrction of DILP8, a relaxin-like peptide, from regenerating tissues (Colombani et al. 2012; Garelli et al. 2012). DILP8 acts on its receptor LGR3, expressed in brain neurons that directly contact PTTH-producing and DILP-secreting cells, thereby inhibiting both PTTH and DILP release (Colombani et al. 2015; Garelli et al. 2015; Vallejo et al. 2015). This DILP8–LGR3 signaling forms a systemic feedback loop coordinating regeneration with developmental timing.

Neuropeptide F Neueropeptide F (NPF; 2.3.11), best known for its role in behavior (see Chap. 11), also suppresses ecdysteroid levels and steroidogenic gene expression in the PGs of *D. melanogaster* (Kannangara et al. 2021). NPF is expressed broadly in the central nervous system and enteroendocrine cells in the gut.

Prothoracicostatic Peptide Prothoracicostatic peptide (PTSP; see Sect. 2.4.4) was identified in *B. mori* as a factor to counterbalance PTTH's stimulation of ecdysteroid synthesis in the PGs in vitro (Hua et al. 1999; Yamanaka et al. 2006). The structure of PTSP is identical to that of AST-B/MIP. In *B. mori*, PTSP is released from the epiproctodeal gland (EPG) during larval and pupal moltings. *PTSP* expression in the EPG is regulated by the hemolymph ecdysteroid titers (Yamanaka et al. 2010). Its receptor is highly expressed in the PGs, particularly at the end of each larval instar (Yamanaka et al. 2010).

Myosuppressin Myosuppressin (see Sect. 2.4.5), originally identified as a myoinhibiting peptide, also acts as a potent prothoracicostatic factor in *B. mori*, inhibiting ecdysteroid synthesis at concentrations lower than PTSP (Yamanaka et al. 2005). In *B. mori*, myosuppressin is produced in the brain neurosecretory cells projecting neurohemal organs, including the CC. Its receptor is highly expressed in the PGs especially during the feeding period of each larval instar (Yamanaka et al. 2010). In *M. brassicae*, myosuppressin has been suggested to contribute to PG inactivation during the onset of pupal diapause (Yamada et al. 2017).

Crustacean Cardioactive Peptide Crustacean cardioactive peptide (CCAP) has diverse physiological roles (see Sect. 2.3.7). In the desert locust *Schistocerca gregaria*, CCAP directly suppresses ecdysteroid production via receptors expressed in the PGs. Knockdown of CCAP or its receptor increases expression of ecdysteroid biosynthetic enzymes and elevates ecdysteroid levels (Verbakel et al. 2021).

Allatostatins Although originally identified as inhibitors of JH biosynthesis, some ASTs also suppress ecdysteroid production in the PGs of specific species (Bendena et al. 2020; Yamanaka et al. 2013). AST-A inhibits the activity of PTTH neurons and eventually suppresses ecdysteroid secretion by the PGs in *D. melanogaster* (Kannangara et al. 2021). AST-C inhibits ecdysteroidogenesis in crustaceans (Liu et al. 2021; Hoppes et al. 2025). In insects, few studies have been conducted on the action of AST-C on the PGs. However, since the AST-C receptor is expressed in the PGs of *B. mori*, it is possible that AST-C may also suppress PG activity in insects (Liu et al. 2021).

Pigment Dispersing Factor Pigment dispersing factor (PDF; see Sect. 2.4.6), well known for its role in circadian rhythm regulation (see Chap. 13), also stimulates ecdysteroid biosynthesis in *B. mori* PGs (Iga et al. 2014). The PDF receptor gene is expressed in the PGs in addition to the central site for regulation of circadian rhythms, such as the medulla and the interface between the optic lobe and central brain (protocerebrum) (Ikeno et al. 2014; Abdelsalam et al. 2008; Colizzi et al. 2023).

Orcokinin In *B. mori*, orcokinin (see Sect. 2.3.12) directly stimulates ecdysteroid production in the PGs in vitro (Yamanaka et al. 2005). Orcokinin-expressing neurons in the prothoracic ganglion project through the transverse nerve to the PG surface, suggesting neuronal regulation of PG activity. The molecular identity of the receptor for orcokinin remains to be clarified.

Corazonin The corazonin (see Sect. 2.3.1)–PTTH neuronal axis modulates larval growth rate and final body size in *D. melanogaster* by regulating basal ecdysteroid synthesis in the PGs. Octopaminergic neurons upstream of this axis convey environmental and nutritional cues to adjust developmental timing and growth rate (Imura et al. 2020). This coordinated neuroendocrine system ensures adaptive developmental progression under varying conditions.

Taken together, the interplay of these stimulatory and inhibitory peptides underscores the complexity of the endocrine network controlling PG activity (Yamanaka et al. 2013; Niwa and Niwa 2016). Multiple neuropeptides, octopamine, and environmental inputs converge on the PGs to achieve precisely timed developmental transitions. Recent studies also highlight how metabolic and environmental signals—such as nutrient availability and stress—are integrated into this peptide regulatory network (Colombani et al. 2005; Yamanaka et al. 2013; Ohhara et al. 2015; Kannangara et al. 2021).

6.5 Peptide Hormones Regulating Insect Growth and Development Independently of Ecdysteroid and JH

Insects adjust growth and development in response to nutrition. Nutrient availability strongly influences developmental timing and body size (see Chaps. 7, 9, and 11).

ILPs serve as central regulators of cellular and systemic growth, developmental progression, and metabolic activity in insects (Okamoto and Yamanaka 2015; Chowański et al. 2021). In model species such as *D. melanogaster* and *B. mori*, multiple ILP genes exhibit tissue- and stage-specific expression patterns, enabling fine-tuned responses to nutritional input.

Basically, under nutrient-rich conditions, ILP secretion is upregulated in the brain, promoting rapid growth and developmental progression (Koyama et al. 2013; Ohhara and Yamanaka 2022). Conversely, starvation suppresses ILP production, leading to developmental delay and reduced adult body size (Ikeya et al. 2002; Okamoto and Yamanaka 2015). Loss-of-function mutants for brain-derived ILPs in *D. melanogaster* display reduced body size; impaired carbohydrate and lipid metabolism, eventually causing delayed development; and extended lifespan (Grönke et al. 2010). This fact illustrates pleiotropic functions of ILPs in insects. In *A. aegypti*,

ILPs also regulate post-blood-meal vitellogenesis, underscoring their role in reproductive as well as somatic development (Brown et al. 2008; Gulia-Nuss et al. 2011). This indicates that lifelong growth and development of insects are modulated by ILPs along with the nutritional states. The molecular and physiological mechanisms of ILP-mediated growth regulation are discussed in detail in Chap. 7.

6.6 Conclusions and Perspectives

As discussed above, insect growth and development are orchestrated by a dynamic network of peptide hormones that act through reciprocal regulation among multiple signaling pathways and their target tissues. This intricate system enables the integration of nutritional status, environmental stresses, tissue integrity, and internal developmental programs with remarkable flexibility and precision (Nagata and Zhou 2019; Nässel and Zandawala 2020).

Beyond the framework of classical hormones, the ever-expanding repertoire of insect peptide hormones constitutes a highly sophisticated regulatory network that ensures both developmental precision and adaptive plasticity. The diversity of their mechanisms across insect taxa, along with their context-dependent modulation, highlights an evolutionary refinement in physiological control. This complexity is particularly evident in the regulation of molting and metamorphosis, where key neuropeptides coordinate energy balance, feeding behavior, nutrient allocation, and systemic physiological processes. Furthermore, insulin-like peptides (ILPs) play a central role in mediating growth control in response to nutritional cues, as discussed in detail in Chap. 7. Together, these studies underscore that the temporal and spatial orchestration of neuropeptidergic signaling not only ensures successful developmental transitions but also underlies the extraordinary ecological and evolutionary success of insects.

References

Abdel-latief M, Hoffmann KH (2014) Functional activity of allatotropin and allatostatin in the pupal stage of a holometablous insect, *Tribolium castaneum* (Coleoptera, Tenebrionidae). Peptides 53:172–184. https://doi.org/10.1016/j.peptides.2013.10.007

Abdelsalam S, Uemura H, Umezaki Y, Saifullah AS, Shimohigashi M, Tomioka K (2008) Characterization of PDF-immunoreactive neurons in the optic lobe and cerebral lobe of the cricket, *Gryllus bimaculatus*. J Insect Physiol 54(7):1205–1212. https://doi.org/10.1016/j.jinsphys.2008.05.001

Abrisqueta M, Süren-Castillo S, Maestro JL (2014) Insulin receptor-mediated nutritional signalling regulates juvenile hormone biosynthesis and vitellogenin production in the German cockroach. Insect Biochem Mol Biol 49:14–23. https://doi.org/10.1016/j.ibmb.2014.03.005

Audsley N, Matthews HJ, Price NR, Weaver RJ (2008) Allatoregulatory peptides in Lepidoptera, structures, distribution and functions. J Insect Physiol 54(6):969–980. https://doi.org/10.1016/j.jinsphys.2008.01.012

Bendena WG, Hui JHL, Chin-Sang I, Tobe SS (2020) Neuropeptide and microRNA regulators of juvenile hormone production. Gen Comp Endocrinol 295:113507. https://doi.org/10.1016/j.ygcen.2020.113507

Birgül N, Weise C, Kreienkamp HJ, Richter D (1999) Reverse physiology in *Drosophila*: identification of a novel allatostatin-like neuropeptide and its cognate receptor structurally related to the mammalian somatostatin/galanin/opioid receptor family. EMBO J 18:5892–5900. https://doi.org/10.1093/emboj/18.21.5892

Brown MR, Clark KD, Gulia M, Zhao Z, Garczynski SF, Crim JW et al (2008) An insulin-like peptide regulates egg maturation and metabolism in the mosquito *Aedes aegypti*. Proc Natl Acad Sci USA 105(15):5716–5721. https://doi.org/10.1073/pnas.0800478105

Caldwell PE, Walkiewicz M, Stern M (2005) Ras activity in the Drosophila prothoracic gland regulates body size and developmental rate via ecdysone release. Curr Biol 15(20):1785–1795. https://doi.org/10.1016/j.cub.2005.09.011

Chowański S, Spochacz M, Szymczak M, Kudlewska M, Tomasik M, Kozłowski M et al (2021) Insulin-like peptides and cross-talk with other factors in regulation of insect metabolism. Front Physiol 12:701203. https://doi.org/10.3389/fphys.2021.701203

Colizzi FS, Veenstra JA, Rezende GL, Helfrich-Förster C, Martínez-Torres D (2023) Pigment-dispersing factor is present in circadian clock neurons of pea aphids and may mediate photoperiodic signalling to insulin-producing cells. Open Biol 13(6):230090. https://doi.org/10.1098/rsob.230090

Colombani J, Bianchini L, Layalle S, Pondeville E, Dauphin-Villemant C, Antoniewski C et al (2005) Antagonistic actions of ecdysone and insulins determine final size in *Drosophila*. Science 310(5748):667–670. https://doi.org/10.1126/science.1119432

Colombani J, Andersen DS, Léopold P (2012) Secreted peptide Dilp8 coordinates Drosophila tissue growth with developmental timing. Science 336(6081):582–585. https://doi.org/10.1126/science.1216689

Colombani J, Andersen DS, Boulan L, Boone E, Romero N, Virolle V et al (2015) Drosophila Lgr3 couples organ growth with maturation and ensures developmental stability. Curr Biol 25 (20):2723–2729. https://doi.org/10.1016/j.cub.2015.09.020

Davidowitz G, D'Amico LJ, Nijhout HF (2003) Critical weight in the development of insect body size. Evol Dev 5(2):188–197. https://doi.org/10.1046/j.1525-142x.2003.03026.x

De Loof A, Boerjan B, Ernst UR, Schoofs L (2013) The mode of action of juvenile hormone and ecdysone: towards an epi-endocrinological paradigm? Gen Comp Endocrinol 188:35–45. https://doi.org/10.1016/j.ygcen.2013.02.004

Deveci D, Martin FA, Leopold P, Romero NM (2019) AstA signaling functions as an evolutionary conserved mechanism timing juvenile to adult transition. Curr Biol 29:813–822. e4. https://doi.org/10.1016/j.cub.2019.01.053

Dong YQ, Wang ZY, Jing TZ (2017) Functional characterization of ASTC (allatostatin C) and ASTCC (allatostatin double C) in *Clostera anastomosis* (Lepidoptera: Notodontidae). Gene 598:1–8. https://doi.org/10.1016/j.gene.2016.10.038

Fukuda S (1940) Induction of pupation in silkworm by transplanting the prothoracic gland. Proc Imper Aca 16(8):414–416

Gao J, Zhang S, Deng P, Wu Z, Lemaitre B, Zhai Z et al (2024) Dietary L-Glu sensing by enteroendocrine cells adjusts food intake via modulating gut PYY/NPF secretion. Nat Commun 15(1):3514. https://doi.org/10.1038/s41467-024-47465-4

Garelli A, Gontijo AM, Miguela V, Caparros E, Dominguez M (2012) Imaginal discs secrete insulin-like peptide 8 to mediate plasticity of growth and maturation. Science 336(6081):579–582. https://doi.org/10.1126/science.1216735

Garelli A, Heredia F, Casimiro AP, Macedo A, Nunes C, Garcez M et al (2015) Dilp8 requires the neuronal relaxin receptor Lgr3 to couple growth to developmental timing. Nat Commun 6:8732. https://doi.org/10.1038/ncomms9732

Griebler M, Westerlund SA, Hoffmann KH, Meyering-Vos M (2008) RNA interference with the allatoregulating neuropeptide genes from the fall armyworm *Spodoptera frugiperda* and its effects on the JH titer in the hemolymph. J Insect Physiol 54(6):997–1007. https://doi.org/10.1016/j.jinsphys.2008.04.019

Grönke S, Clarke DF, Broughton S, Andrews TD, Partridge L (2010) Molecular evolution and functional characterization of *Drosophila* insulin-like peptides. PLoS Genet 6(2):e1000857. https://doi.org/10.1371/journal.pgen.1000857

Gu SH, Chen CH, Hsieh YC, Lin PL, Young SC (2015) Modulatory effects of bombyxin on ecdysteroidogenesis in *Bombyx mori* prothoracic glands. J Insect Physiol 72:61–69. https://doi.org/10.1016/j.jinsphys.2014.11.007

Gulia-Nuss M, Robertson AE, Brown MR, Strand MR (2011) Insulin-like peptides and the target of rapamycin pathway coordinately regulate blood digestion and egg maturation in the mosquito *Aedes aegypti*. PLoS One 6(5):e20401. https://doi.org/10.1371/journal.pone.0020401

Hoppes JL, Wilcockson DC, Webster SG (2025) Allatostatin-C signaling in the crab *Carcinus maenas* is implicated in the ecdysis program. J Exp Biol 228(5):JEB249929. https://doi.org/10.1242/jeb.249929

Hua YJ, Tanaka Y, Nakamura K, Sakakibara M, Nagata S, Kataoka H (1999) Identification of a prothoracicostatic peptide in the larval brain of the silkworm, *Bombyx mori*. J Biol Chem 274(44):31169–31173. https://doi.org/10.1074/jbc.274.44.31169

Iga M, Nakaoka T, Suzuki Y, Kataoka H (2014) Pigment dispersing factor regulates ecdysone biosynthesis via *Bombyx* neuropeptide G protein coupled receptor-B2 in the prothoracic glands of *Bombyx mori*. PLoS One 9(7):e103239. https://doi.org/10.1371/journal.pone.0103239

Ikeno T, Numata H, Goto SG, Shiga S (2014) Involvement of the brain region containing pigment-dispersing factor-immunoreactive neurons in the photoperiodic response of the bean bug, *Riptortus pedestris*. J Exp Biol 217(Pt 3):453–462. https://doi.org/10.1242/jeb.091801

Ikeya T, Galic M, Belawat P, Nairz K, Hafen E (2002) Nutrient-dependent expression of insulin-like peptides from neuroendocrine cells in the CNS contributes to growth regulation in *Drosophila*. Curr Biol 12(15):1293–1300. https://doi.org/10.1016/s0960-9822(02)01043-6

Imura E, Shimada-Niwa Y, Nishimura T, Hückesfeld S, Schlegel P, Ohhara Y et al (2020) The corazonin-PTTH neuronal axis controls systemic body growth by regulating basal ecdysteroid biosynthesis in *Drosophila melanogaster*. Curr Biol 30(11):2156–2165.e5. https://doi.org/10.1016/j.cub.2020.03.050

Kaneko Y, Hiruma K (2015) Allatotropin inhibits juvenile hormone biosynthesis by the corpora allata of adult *Bombyx mori*. J Insect Physiol 80:15–21. https://doi.org/10.1016/j.jinsphys.2015.01.005

Kannangara JR, Mirth CK, Warr CG (2021) Regulation of ecdysone production in *Drosophila* by neuropeptides and peptide hormones. Open Biol 11(2):200373. https://doi.org/10.1098/rsob.200373. PMID: 33593157

Kataoka H, Toschi A, Li JP, Carney RL, Schooley DA, Kramer SJ (1989) Identification of an allatotropin from adult *Manduca sexta*. Science 243(4897):1481–1483. https://doi.org/10.1126/science.243.4897.1481

Kataoka H, Nagasawa H, Isogai A, Ishizaki H, Suzuki A (1991) Prothoracicotropic hormone of the silkworm, Bombyx mori: amino acid sequence and dimeric structure. Agric Biol Chem 55(1):73–86. https://doi.org/10.1080/00021369.1991.10870535

Kawakami A, Kataoka H, Oka T, Mizoguchi A, Kimura-Kawakami M, Adachi T et al (1990) Molecular cloning of the Bombyx mori prothoracicotropic hormone. Science 247(4948):1333–1335. https://doi.org/10.1126/science.2315701

Kopeć S (1917) Experiments on metamorphosis of insects. Bull Int Acad Sci Cracovie B:57–60

Koyama T, Mendes CC, Mirth CK (2013) Mechanisms regulating nutrition-dependent developmental plasticity through organ-specific effects in insects. Front Physiol 4:263. https://doi.org/10.3389/fphys.2013.00263

Kramer SJ, Toschi A, Miller A, Kataoka H, Quistad GB, Li JP et al (1991) Identification of an allatostatin from the tobacco hornworm *Manduca sexta*. Proc Natl Acad Sci USA 88(21):9458–9462. https://doi.org/10.1073/pnas.88.21.9458

Lafont R, Dauphin-Villemant C, Warren JT, Rees HH (2012) Ecdysteroid chemistry and biochemistry. In: Gilbert LI (ed) Insect endocrinology. Academic, pp 106–176

Lange AB, Alim U, Vandersmissen HP, Mizoguchi A, Vanden Broeck J, Orchard I (2012) The distribution and physiological effects of the myoinhibiting peptides in the kissing bug, *Rhodnius prolixus*. Front Neurosci 6:98. https://doi.org/10.3389/fnins.2012.00098

Leyria J (2024) Endocrine factors modulating vitellogenesis and oogenesis in insects: an update. Mol Cell Endocrinol 587:112211. https://doi.org/10.1016/j.mce.2024.112211

Liu A, Shi W, Lin D, Ye H (2021) A possible role of allatostatin C in inhibiting ecdysone biosynthesis revealed in the mud crab *Scylla paramamosain*. Front Mar Sci 8:740251. https://doi.org/10.3389/fmars.2021.740251

Lorenz MW, Kellner R, Hoffmann KH (1995) A family of neuropeptides that inhibit juvenile hormone biosynthesis in the cricket, *Gryllus bimaculatus*. J Biol Chem 270(36):21103–21108. https://doi.org/10.1074/jbc.270.36.21103

Matsumoto S, Kutsuna N, Daubnerová I, Roller L, Žitňan D, Nagasawa H et al (2019) Enteroendocrine peptides regulate feeding behavior via controlling intestinal contraction of the silkworm *Bombyx mori*. PLoS One 14(7):e0219050. https://doi.org/10.1371/journal.pone.0219050

Mayoral JG, Nouzova M, Brockhoff A, Goodwin M, Hernandez-Martinez S, Richter D et al (2010) Allatostatin-C receptors in mosquitoes. Peptides 31(3):442–450. https://doi.org/10.1016/j.peptides.2009.04.013

McBrayer Z, Ono H, Shimell MJ, Parvy JP Beckstead RB, Warren JT et al (2018) Prothoracicotropic hormone regulates developmental timing and body size in *Drosophila*. Dev Cell 13(6):857–871. https://doi.org/10.1016/j.devcel.2007.11.003

Mirth C, Truman JW, Riddiford LM (2005) The role of the prothoracic gland in determining critical weight for metamorphosis in *Drosophila melanogaster*. Curr Biol 15(20):1796–1807. https://doi.org/10.1016/j.cub.2005.09.017

Mizoguchi A, Ohashi Y, Hosoda K, Ishibashi J, Kataoka H (2001) Developmental profile of the changes in the prothoracicotropic hormone titer in hemolymph of the silkworm *Bombyx mori*: correlation with ecdysteroid secretion. Insect Biochem Physiol 31:349–358. https://doi.org/10.1016/s0965-1748(00)00127-2

Mizoguchi A, Ohsumi S, Kobayashi K, Okamoto N, Yamada N, Tateishi K et al (2013) Prothoracicotropic hormone acts as a neuroendocrine switch between pupal diapause and adult development. PLoS One 8:e60824. https://doi.org/10.1371/journal.pone.0060824

Nagasawa H, Kataoka H, Hori Y, Isogai A, Tamura S, Suzuki A et al (1984) Isolation and some characterization of the prothoracicotropic hormone from Bombyx mori. Gen Comp Endocrinol 53(1):143–152. https://doi.org/10.1016/0016-6480(84)90234-x

Nagata S, Zhou YJ (2019) Feeding-modulating neuropeptides and peptide hormones in insects. Adv Insect Physiol 57:137–172. https://doi.org/10.1016/bs.aiip.2019.10.002

Nagata S, Kataoka H, Suzuki a (2005) Silk moth neuropeptide hormone: prothoracicotropic hormones and others. Ann N Y Acad Sci 1040:38–52. https://doi.org/10.1196/annals.1327.004

Nässel DR, Zandawala M (2020) Hormonal axes in *Drosophila*: regulation of hormone release and multiplicity of actions. Cell Tissue Res 382(2):233–266. https://doi.org/10.1007/s00441-020-03264-z

Nijhout HF (2003) The control of body size in insects. Dev Biol 261(1):1–9. https://doi.org/10.1016/s0012-1606(03)00276-8

Nijhout HF, Williams CM (1974) Control of moulting and metamorphosis in the tobacco hornworm, *Manduca sexta* (L.): growth of the last-instar larva and the decision to pupate. J Exp Biol 61(2):481–491. https://doi.org/10.1242/jeb.61.2.493

Niwa YS, Niwa R (2016) Transcriptional regulation of insect steroid hormone biosynthesis and its role in controlling timing of molting and metamorphosis. Develop Growth Differ 58(1):94–105. https://doi.org/10.1111/dgd.12248

Niwa R, Sakudoh T, Namiki T, Saida K, Fujimoto Y, Kataoka H (2005) The ecdysteroidogenic P450 Cyp302a1/disembodied from the silkworm, *Bombyx mori*, is transcriptionally regulated by prothoracicotropic hormone. Insect Mol Biol 14(5):563–571. https://doi.org/10.1111/j.1365-2583.2005.00587.x

Oh Y, Yoon SE, Zhang Q, Chae HS, Daubnerová I, Shafer OT et al (2014) A homeostatic sleep-stabilizing pathway in *Drosophila* composed of the sex peptide receptor and its ligand, the myoinhibitory peptide. PLoS Biol 12(10):e1001974. https://doi.org/10.1371/journal.pbio.1001974

Ohhara Y, Yamanaka N (2022) Internal sensory neurons regulate stage-specific growth in *Drosophila*. Development 149(21):dev200440. https://doi.org/10.1242/dev.200440

Ohhara Y, Niwa YS, Niwa R, Kayashima Y, Hayashi Y, Akagi K et al (2015) Autocrine regulation of ecdysone synthesis by β3-octopamine receptor in the prothoracic gland is essential for *Drosophila* metamorphosis. Proc Natl Acad Sci USA 112(5):1452–1457. https://doi.org/10.1073/pnas.1414966112. PMID: 25605909

Ohhara Y, Kobayashi S, Yamanaka N (2017) Nutrient-dependent endocycling in steroidogenic tissue dictates timing of metamorphosis in *Drosophila melanogaster*. PLoS Genet 13(1):e1006583. https://doi.org/10.1371/journal.pgen.1006583

Okamoto N, Yamanaka N (2015) Nutrition-dependent control of insect development by insulin-like peptides. Curr Opin Insect Sci 11:21–30. https://doi.org/10.1016/j.cois.2015.08.001

Pérez-Hedo M, Rivera-Perez C, Noriega FG (2013) The insulin/TOR signal transduction pathway is involved in the nutritional regulation of juvenile hormone synthesis in *Aedes aegypti*. Insect Biochem Mol Biol 43(6):495–500. https://doi.org/10.1016/j.ibmb.2013.03.008

Rewitz KF, Yamanaka N, Gilbert LI, O'Connor MB (2009) The insect neuropeptide PTTH activates receptor tyrosine kinase torso to initiate metamorphosis. Science 326(5958):1403–1405. https://doi.org/10.1126/science.1176450

Shimell M, Pan X, Martin FA, Ghosh A, Leopold P, O'Connor MB (2018) Prothoracicotropic hormone modulates environmental adaptive plasticity through the control of developmental timing. Development 145(6):dev159699. https://doi.org/10.1242/dev.159699

Smith W, Rybczynski R (2012) Prothoracicotropic hormone. In: Gilbert LI (ed) Insect endocrinology. Academic, London

Sun Y, Fu D, Liu B, Wang L, Chen H (2022) Functional characterization of allatostatin C (PISCF/AST) and juvenile hormone acid *O*-methyltransferase in *Dendroctonus armandi*. Int J Mol Sci 23(5):2749. https://doi.org/10.3390/ijms23052749

Truman JW (2005) Hormonal control of insect ecdysis: endocrine cascades for coordinating behavior with physiology. Vitam Horm 73:1–30. https://doi.org/10.1016/S0083-6729(05)73001-6

Truman JW, Riddiford LM (2023) *Drosophila* postembryonic nervous system development: a model for the endocrine control of development. Genetics 223(3):iyac184. https://doi.org/10.1093/genetics/iyac184

Tsukamoto Y, Nagata S (2016) Newly identified allatostatin Bs and their receptor in the two-spotted cricket, *Gryllus bimaculatus*. Peptides 80:25–31. https://doi.org/10.1016/j.peptides.2016.03.015

Tu MP, Yin CM, Tatar M (2005) Mutations in insulin signaling pathway alter juvenile hormone synthesis in *Drosophila melanogaster*. Gen Comp Endocrinol 142(3):347–356. https://doi.org/10.1016/j.ygcen.2005.02.009

Uchibori-Asano M, Kayukawa T, Sezutusu H, Shinoda T, Daimon T (2017) Several developmental timing defects in the prothoracicotropic hormone (PTTH)-deficient silkworm, Bombyx mori. Insect Biochem Mol Biol 87:14–25. https://doi.org/10.1016/j.ibmb.2017.06.007

Vallejo DM, Juarez-Carreño S, Bolivar J, Morante J, Dominguez M (2015) A brain circuit that synchronizes growth and maturation revealed through Dilp8 binding to Lgr3. Science 350(6262):aac6767. https://doi.org/10.1126/science.aac6767

Verbakel L, Lenaerts C, Abou El Asrar R, Zandecki C, Bruyninckx E, Monjon E et al (2021) Prothoracicostatic activity of the ecdysis-regulating neuropeptide crustacean cardioactive peptide (CCAP) in the desert locust. Int J Mol Sci 22(24):13465. https://doi.org/10.3390/ijms222413465

Verlinden H, Gijbels M, Lismont E, Lenaerts C, Vanden Broeck J, Marchal E (2015) The pleiotropic allatoregulatory neuropeptides and their receptors: a mini-review. J Insect Physiol 80:2–14. https://doi.org/10.1016/j.jinsphys.2015.04.004

Wang C, Zhang J, Tobe SS, Bendena WG (2012) Defining the contribution of select neuropeptides and their receptors in regulating sesquiterpenoid biosynthesis by Drosophila melanogaster ring gland/corpus allatum through RNAi analysis. Gen Comp Endocrinol 176(3):347–353. https://doi.org/10.1016/j.ygcen.2011.12.039

Wegener C, Chen J (2022) Allatostatin A signalling: progress and new challenges from a paradigmatic pleiotropic invertebrate neuropeptide family. Front Physiol 13:920529. https://doi.org/10.3389/fphys.2022.920529

Wigglesworth VB (1936) Memoirs: the function of the corpus allatum in the growth and reproduction of *Rhodnius prolixus* (Hemiptera). J Cell Sci 2(313):91–121

Williams CM (1948) Physiology of insect diapause. III. The prothoracic glands in the Cecropia silkworm, with special reference to their significance in embryonic and postembryonic development. Biol Bull 94(1):60–65

Williams CM (1950) The metamorphosis of insects. Sci Am 182(4):24–29

Woodhead AP, Stay B, Seidel SL, Khan MA, Tobe SS (1989) Primary structure of four allatostatins: neuropeptide inhibitors of juvenile hormone synthesis. Proc Natl Acad Sci USA 86(15):5997–6001. https://doi.org/10.1073/pnas.86.15.5997

Yamada N, Kataoka H, Mizoguchi A (2017) Myosuppressin is involved in the regulation of pupal diapause in the cabbage army moth Mamestra brassiae. Sci Rep 7:41651. https://doi.org/10.1038/srep41651

Yamanaka N, Hua YJ, Mizoguchi A, Watanabe K, Niwa R, Tanaka Y et al (2005) Identification of a novel prothoracicostatic hormone and its receptor in the silkworm *Bombyx mori*. J Biol Chem 280(15):14684–14690. https://doi.org/10.1074/jbc.M413662200

Yamanaka N, Zitnan D, Kim YJ, Adams ME, Hua YJ, Suzuki Y et al (2006) Regulation of insect steroid hormone biosynthesis by innervating peptidergic neurons. Proc Natl Acad Sci USA 103(23):8622–8627. https://doi.org/10.1073/pnas.0511196103

Yamanaka N, Yamamoto S, Zitnan D, Watanabe K, Kawada T, Satake H et al (2008) Neuropeptide receptor transcriptome reveals unidentified neuroendocrine pathways. PLoS One 3(8):e3048. https://doi.org/10.1371/journal.pone.0003048

Yamanaka N, Hua YJ, Roller L, Spalovská-Valachová I, Mizoguchi A, Kataoka H et al (2010) *Bombyx* prothoracicostatic peptides activate the sex peptide receptor to regulate ecdysteroid biosynthesis. Proc Natl Acad Sci USA 107:2060–2065. https://doi.org/10.1073/pnas.0907471107

Yamanaka N, Rewitz KF, O'Connor MB (2013) Ecdysone control of developmental transitions: lessons from *Drosophila* research. Annu Rev Entomol 58:497–516. https://doi.org/10.1146/annurev-ento-120811-153608

Chapter 7
Growth Regulation Through Insulin-Like Peptides During Insect Development

Naoki Okamoto

Abstract Proper growth regulation in animals ensures species-specific body and organ sizes, with nutrition being the most significant environmental factor influencing growth. In insects, insulin-like peptides (ILPs), structurally similar to vertebrate insulin superfamily peptides, mediate nutrient-dependent growth. Most ILPs activate the insulin/insulin-like growth factor (IGF) signaling (IIS) pathway via insulin-like receptors (InR) to promote systemic and tissue growth. In insects, the primary source of ILPs is the medial neurosecretory cells (mNSCs) in the brain, where ILP expression and secretion are tightly controlled by nutritional status through multiple mechanisms. Various factors that regulate ILP expression and secretion have been identified using tissue- and cell-specific genetic tools in the fruit fly *Drosophila melanogaster*. ILP activity is further modulated by ILP-binding proteins, which control ILP availability and receptor binding, fine-tuning the IIS pathway. Organ-specific growth regulation is achieved through differences in IIS sensitivity. These differences in sensitivity allow precise control of organ size in response to nutritional and environmental fluctuations, emphasizing the evolutionary adaptability of the IIS pathway to adjust growth to environmental demands. In conclusion, ILPs play a central role in integrating nutritional signals to regulate systemic and organ-specific growth during insect development.

Keywords Insulin-like peptides (ILPs) · Bombyxin · DILPs · Insulin/insulin-like growth factor (IGF) signaling (IIS) · Insulin-like receptors (InR) · Medial neurosecretory cells (mNSCs) · ILP-binding proteins · Growth · Nutrition

---

N. Okamoto (✉)
Life Science Center for Survival Dynamics, Tsukuba Advanced Research Alliance (TARA), University of Tsukuba, Tsukuba, Japan
e-mail: naoki-okamoto@tara.tsukuba.ac.jp

A. Mizoguchi, N. Okamoto (eds.), *Insect Peptide Hormones and Neuropeptides*, Entomology Monographs, https://doi.org/10.1007/978-981-95-7510-7_7

7.1 Introduction

In multicellular organisms, proper growth regulation across the entire body and in specific organs during development is essential for achieving the correct size and proportion of the adult (Conlon and Raff 1999). Body and organ growth is influenced by both intrinsic developmental programs, which determine species-specific sizes, and various environmental factors. Among these, the most significant environmental factor impacting growth is nutritional status. The balance between the utilization of nutrients for storage and growth is finely regulated to ensure proper nutrition-dependent growth of the entire body and its organs.

In animals, the growth of the body and organs in response to nutritional status is primarily regulated by peptides of the insulin superfamily, represented by insulin/insulin-like growth factors (IGFs) in vertebrates and insulin-like peptides (ILPs) in invertebrates (Thissen et al. 1994; Okamoto and Yamanaka 2015). While insulin and IGFs share similar amino acid sequences in vertebrates, their structural features and physiological functions differ (Froesch and Zapf 1985). Both act on distinct receptor tyrosine kinases (RTKs), with insulin primarily acting on the insulin receptor (IR) and IGFs on the IGF type-I receptor (IGF-1R), although there is some crosstalk between each other (Nakae et al. 2001). The primary function of insulin is the regulation of carbohydrate and lipid metabolism (Saltiel and Kahn 2001), while the principal role of IGFs is to promote tissue and body growth (Jones and Clemmons 1995; Stewart and Rotwein 1996). Dietary nutrients and other hormones control insulin production and secretion in the pancreatic β-cells. The production of IGF-I in the liver and other peripheral tissues, including bone, is mainly regulated by growth hormone and nutritional status, playing a central role in properly regulating body and tissue size in vertebrates. When insulin or IGFs bind to their receptors, the activated receptors phosphorylate various substrates, including insulin receptor substrate (IRS) family (Fig. 7.1). Phosphorylated IRS proteins promote the activation of downstream cascades through the phosphoinositide 3-kinase (PI3K)/protein kinase B (AKT) axis (Nakae et al. 2001). This signaling pathway, collectively referred to as insulin/IGF signaling (IIS), constitutes a complex network that regulates cell proliferation, metabolism, and survival, playing an essential role in the growth and metabolism of the body and tissues.

In addition to insulin and IGF, other insulin superfamily peptides, such as relaxin (RLN) and RLN-like peptides (also known as insulin-like: INSL), function through leucine-rich repeat-containing G protein-coupled receptors (LGRs) and are mainly associated with reproductive functions (Bathgate et al. 2013). These peptides are often classified as the RLN family because they act through different receptor types than insulin and IGF. They are frequently not included in the insulin family peptides.

Since the 1980s, when bombyxin, the first insulin superfamily peptide identified in invertebrates, was discovered using the silkworm *Bombyx mori*, research on variety of insects, including the model organism fruit fly *Drosophila melanogaster*, has dramatically advanced our understanding of the function of evolutionarily conserved ILPs and their downstream IIS pathways (Fig. 7.1). This chapter reviews

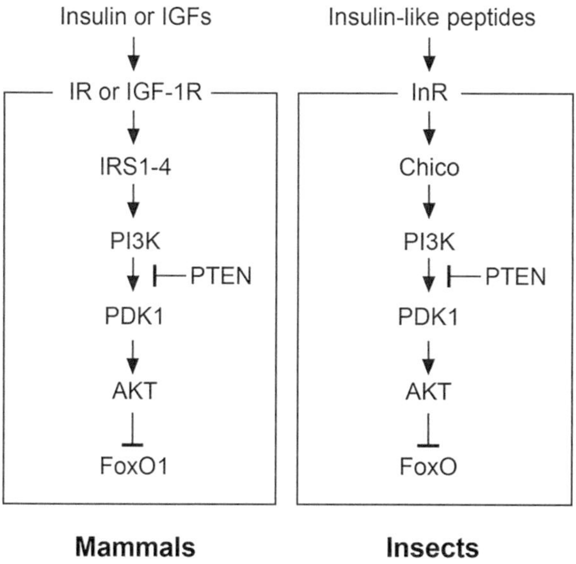

Fig. 7.1 Insulin/insulin-like growth factor (IGF) signaling (IIS) pathway in mammals and insects. Upon binding of the hormone (insulin/IGFs in mammals and insulin-like peptides in insects) to their receptors (insulin receptor (IR)/insulin-like growth factor type-1 receptor (IGF-1R) in mammals and insulin-like receptor (InR) in insects), evolutionarily conserved downstream signaling components are activated, triggering various responses involved in growth and metabolism. IRS Insulin receptor substrate, PDK1 phosphoinositide-dependent kinase 1, PTEN phosphatase and tensin Homolog, PI3K phosphoinositide 3-kinase, AKT protein kinase B, FoxO forkhead box O

studies related to the function of ILPs in regulating nutrition-dependent growth during insect development.

7.2 Insulin-Like Peptides (ILPs) in Insects

ILPs have been identified and characterized in various invertebrates, including insects (Antonova et al. 2012; Nässel and Vanden Broeck 2016; Murphy and Hu 2013). The primary structure of ILPs in invertebrates was first elucidated in the 1980s by a collaborative research group led by Hironori Ishizaki and Akinori Suzuki, who determined the primary structure of bombyxin purified from *B. mori* heads (Nagasawa et al. 1984, 1986; Ishizaki 2004; see Chap. 1 for details). Bombyxin is a heterodimeric peptide linked by two inter-chain disulfide bonds and one intra-chain disulfide bond like vertebrate insulin (Fig. 7.2). All six cysteine residues are conserved between human insulin and bombyxin. The bombyxin precursor has a typical insulin superfamily structure consisting of a signal peptide, B-chain, C-peptide, and A-chain in this order. Like vertebrate insulin, bombyxin undergoes post-translational processing where the signal peptide is cleaved, and then the C-peptide is removed, resulting in the secretion of a heterodimeric peptide

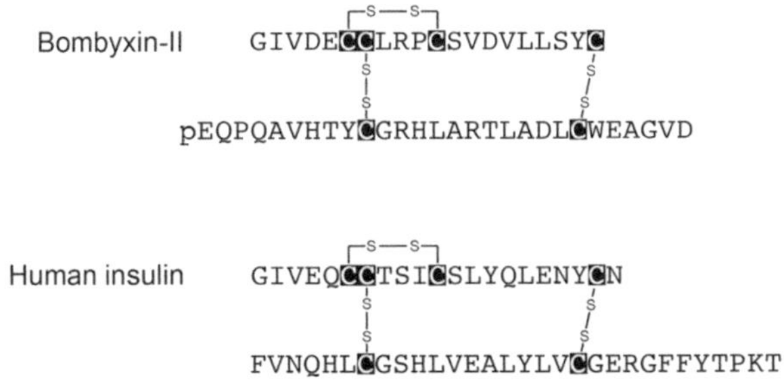

Fig. 7.2 Amino acid sequences of bombyxin II and human insulin. The identical residues between the two peptides are boxed. Disulfide bonds (S–S) are indicated by lines, and pyroglutamic acid is indicated by pE

composed of the A- and B-chains into the hemolymph (Ishizaki 2004). However, not all insect ILPs are secreted as heterodimeric peptides like insulin and bombyxin. Some insect ILPs, such as *Bommo*-IGF-like peptide (BIGFLP) in *B. mori*, are secreted as single-chain peptides without removing the C-peptide, similar to vertebrate IGF (Okamoto et al. 2009a).

Since the complete genome of *D. melanogaster* was decoded in 2000 (Adams et al. 2000; Myers et al. 2000), genome information from various insect species has become available, revealing the existence of numerous ILP-encoding genes (Antonova et al. 2012). ILP-coding genes have been identified in all insect species with available genomic data, with the number of ILP-encoding genes varying significantly between species. For example, in fully sequenced model insects, there are 8 ILPs in *D. melanogaster* (DILP1-8), 4 in the red flour beetle *Tribolium castaneum* (TcILP1–4), and 2 in the western honeybee *Apis mellifera* (AmILP1, 2) (Antonova et al. 2012; Nässel and Vanden Broeck 2016). Approximately 40 ILPs have been identified in *B. mori* (Mizoguchi and Okamoto 2013), although this large number may be related to the domestication of this insect.

Although insect ILPs have six conserved cysteine residues within the A- and B-chains, they exhibit substantial diversity beyond this common framework. Based on differences in primary sequence features, tissue- or cell-type-specific expression patterns, and receptor type, insect ILPs can be subdivided into several distinct categories. One major class comprises "typical" ILPs that closely resemble vertebrate insulin in structural features. In addition, insect ILPs include IGF-like peptides (IGFLPs) and RLN-related peptides (RLN-RPs), the latter of which can be further separated into DILP8-like/gonadulin and DILP7-like peptides. Because these classifications are primarily based on DILPs found in *D. melanogaster*, further subcategorization may be possible. In this chapter, we focus on the growth regulation mechanisms of insect ILPs, with particular emphasis on typical ILPs and IGFLPs, which are directly involved in growth control.

7.2.1 Typical ILP

Typical ILPs possess the canonical insulin-like structure and represent the largest ILP group in insects. In many species, including *D. melanogaster*, ILPs are predominantly produced in specific subsets of medial neurosecretory cells (mNSCs) located in the brain. In many studies on *D. melanogaster*, these neurons are often referred to as insulin-producing cells (IPCs). mNSCs that produce ILPs or IPCs will simply be referred to as mNSCs in this chapter.

A hallmark of typical ILPs is the presence of paired basic amino acid motifs (KK, RK, or RR) flanking the C-peptide region. These motifs serve as proteolytic cleavage sites, allowing the precursor to be processed into a mature heterodimeric hormone consisting of A- and B-chains, analogous to vertebrate insulin. In *D. melanogaster*, four of the eight ILPs (DILP1, DILP2, DILP3, and DILP5) fall into this category, based on both their amino acid sequences and their expression in mNSCs (Brogiolo et al. 2001; Ikeya et al. 2002; Rulifson et al. 2002; Liu et al. 2016). DILP4 shares the same overall structural features but is not produced in mNSCs; instead it is strongly expressed during embryogenesis (Brogiolo et al. 2001). Typical ILPs signal through the insulin-like receptor (InR), a RTK closely related to vertebrate IR and IGF-R1. Through this pathway, they exert central control systemic and tissue growth during development (Fernandez et al. 1995; Chen et al. 1996; Brogiolo et al. 2001).

While this chapter does not cover, these ILPs are also involved in a broad range of physiological processes, including metabolic homeostasis, reproduction, diapause, and lifespan determination (Tatar et al. 2003; Giannakou and Partridge 2007; Teleman 2010; Badisco et al. 2013; Sim and Denlinger 2013; Nässel and Vanden Broeck 2016; Chowański et al. 2021). For these functions of ILPs, please refer to other chapters of this book.

7.2.2 IGFLP

IGFLPs comprise a group of insect ILPs that resemble vertebrate IGFs in both structure and expression profile. In contrast to insulin, vertebrate IGFs are secreted as single-chain polypeptides without removal of the C-peptide and contain a characteristic C-terminal extension known as the D-domain. The first insect IGFLP to be structurally characterized was BIGFLP, purified from the hemolymph of *B. mori* pupae in 2009 (Okamoto et al. 2009a). Although BIGFLP contains dibasic residues (KR or RK) within the C-peptide region, motifs that could theoretically serve as cleavage sites, it is secreted intact as a single-chain peptide. A similar phenomenon is observed in vertebrate IGFs, which also harbor internal dibasic residues (RR) that are not processed (Rinderknecht and Humbel 1978). A further similarity between BIGFLP and vertebrate IGFs lies in its site of production. BIGFLP is primarily produced in the fat body (Okamoto et al. 2009a), an organ functionally analogous to the vertebrate liver. In vertebrates, the liver is the principal source of circulating IGFs, supporting the idea that insect IGFLPs may share functional parallels with IGFs. In *D. melanogaster*, DILP6 is highly expressed in the fat body during the pupal stage and is therefore classified as an IGFLP (Okamoto et al. 2009b; Slaidina et al. 2009). Notably, DILP6 contains a relatively short C-peptide (Brogiolo et al. 2001), a feature reminiscent of IGFs. Similarly, AeILP6 from the mosquito *Aedes aegypti* also displays a short C-peptide and a C-terminal extension and is predominantly expressed in the fat body (Riehle et al. 2006; Ling and Raikhel 2021). Despite these structural and anatomical similarities, insect IGFLPs are far more similar in amino acid sequence to other insect ILPs than to vertebrate IGF. This observation suggests that IGFs and IGFLPs arose independently in vertebrate and insect lineages.

Recent comparative analyses have uncovered additional IGF-like peptides in diverse insects, including a newly proposed group termed arthropod IGFs (aIGFs), which are characterized by short C-peptides and, in some cases, C-terminal extensions (Veenstra 2020a). However, the presence of two extra cysteine residues in aIGFs raises questions about whether they should be formally classified within the insulin superfamily.

IGFLPs are generally assumed to act through InR, the same receptor used by typical ILPs. In species such as *D. melanogaster* and *B. mori*, which possess only a single *InR* gene, ILPs and IGFLPs are presumed to share this receptor. Nevertheless, direct biochemical evidence demonstrating IGFLP–InR binding is still lacking. In contrast, some insects encode multiple *InR* genes (Kremer et al. 2018; Smýkal et al. 2020), raising the possibility that different ILP classes may preferentially engage distinct InRs or exhibit differing receptor affinities.

7.2.3 RLN-RP

(a) DILP8-like peptides/gonadulin: This subgroup consists of peptides homologous to DILP8, which signal through LGRs, analogous to vertebrate RLNs and INSLs. DILP8 was initially identified in 2012 as a hormone that delays puparium formation in response to tissue damage (Colombani et al. 2012; Garelli et al. 2012). Unlike typical ILPs, DILP8 does not activate InR; instead, its receptor is LGR3 (Gontijo and Garelli 2018). Homologs of DILP8 have since been detected in numerous arthropod genomes (Chandler et al. 2017; Veenstra 2016, 2020b), where they are often highly expressed in the gonads, leading to the alternative name "gonadulin." Although direct evidence for receptor usage is limited outside *D. melanogaster*, the presence of LGR3 orthologs in species that encode *gonadulins* strongly suggests that these peptides also signal via LGR3. Importantly, expression of DILP8-like genes is not restricted to gonadal tissues. In *D. melanogaster*, *DILP8* is also expressed in damaged imaginal discs and in the epidermis, where it contributes to the coordination of developmental timing, pupariation behavior, and bilateral wing size symmetry (Colombani et al. 2012; Garelli et al. 2012; Heredia et al. 2021; Blanco-Obregon et al. 2022).

(b) DILP7-like peptides: The second RLN-RP subgroup comprises DILP7 and its orthologs. In contrast to most ILPs, which exhibit limited sequence conservation beyond the six cysteine residues and processing motifs, DILP7-like peptides show remarkably high amino acid similarity across both the A- and B-chains (Veenstra 2010). Their receptor is predicted to be LGR4 (Veenstra et al. 2012). DILP7 orthologs are widely distributed among invertebrates, including mollusks, although they are absent from some lineages (Veenstra 2010). In *D. melanogaster*, DILP7 is produced by neurosecretory cells located in the abdominal ganglia rather than in mNSCs (Brogiolo et al. 2001; Miguel-Aliaga et al. 2008). Although the physiological roles of DILP7-like peptides remain incompletely understood, recent evidence suggests that DILP7 can act as a neuropeptide influencing photophobic behavior via LGR4 signaling (Imambocus et al. 2022).

7.3 Mechanisms Regulating the Function of ILPs during Development: Expression, Secretion, and Activity

In insects, the primary cells responsible for regulating systemic and tissue-specific growth in a nutrient-dependent manner are the mNSCs in the brain (Nijhout et al. 2013; Okamoto and Yamanaka 2015; Lin and Smagghe 2019; Texada et al. 2020). This section focuses on the mechanisms by which ILP expression, secretion, and activity are regulated in a nutrient-dependent manner during insect development. For details on the regulatory mechanisms of ILPs in the adult stage, readers may refer to recent reviews (Nässel and Vanden Broeck 2016; Chowański et al. 2021; Semaniuk et al. 2021).

7.3.1 Brain mNSCs as Primary ILP-Producing Cells

Insulin-like immunoreactivity in mNSCs in insect brains has been recognized in various species for approximately half a century. The direct evidence showing the production of insect ILPs by mNSCs was first demonstrated in 1987 by immunohistochemical analysis of *B. mori* brains using an anti-bombyxin antibody (Mizoguchi et al. 1987). In *B. mori* larvae, bombyxin produced by four pairs of large mNSCs is axonally transported to the corpora allata (CA) and released into the hemolymph (Mizoguchi et al. 1987; Ichikawa 1991). In *D. melanogaster* larvae, seven pairs of mNSCs extend their processes toward the dorsal vessel, and DILPs are directly released into the circulating hemolymph (Rulifson et al. 2002). Similar to *B. mori* and *D. melanogaster*, ILP production in brain mNSCs has also been observed in many other insect species (Nässel and Vanden Broeck 2016; Chowański et al. 2021), suggesting that ILP production by mNSCs fulfills an evolutionarily conserved function.

7.3.2 *Transcriptional Regulation of ILPs*

Although multiple ILPs are produced within the same population of brain medial neurosecretory cells (mNSCs), the transcriptional regulation of individual *ILP* genes is not uniform. Instead, each *ILP* expression is controlled by distinct regulatory programs. Comparative developmental analyses across several insect species have revealed that *ILP* genes display highly divergent temporal expression profiles (e.g., in *A. mellifera*, Wheeler et al. 2006 and de Azvedo and Hartfelder 2008; in *D. melanogaster*, Okamoto et al. 2009b, Slaidina et al. 2009, and Okamoto and Nishimura 2015; in the kissing bug *Rhodnius prolixus*, Defferrari et al. 2016; in the broad-horned flour beetle *Gnatocerus cornutus*, Okada et al. 2019; in the crickets *Modicogryllus siamensis*, Miki et al. 2020). In *D. melanogaster*, for instance, *DILP2* expression is already detectable in mNSCs during late embryogenesis (Wang et al. 2007). In contrast, *DILP5* and *DILP3* expression in mNSCs is upregulated later, peaking around the mid–first and mid–third larval instars, respectively (Ikeya et al. 2002; Okamoto and Nishimura 2015). *DILP1* expression in mNSCs rises during pupal development and diminishing rapidly after adult emergence (Liu et al. 2016). Despite these well-characterized expression dynamics, the molecular mechanisms responsible for establishing distinct developmental expression programs among *DILPs* in mNSCs remain poorly understood. In addition to developmental cues, each *DILP* is also independently regulated by nutritional status (Ikeya et al. 2002; Okamoto and Nishimura 2015).

During larval growth, transcriptional control of *DILP2*, *DILP3*, and *DILP5* in mNSCs involves multiple transcription factors (Winant et al. 2022). Notably, mNSCs express several transcription factors originally characterized for their roles in eye development (Wang et al. 2007; Winant et al. 2022). Functional analyses indicate that some of these factors influence *DILP* gene expression, although it is not yet clear whether they act directly on *DILP* promoters or through intermediate pathways (Winant et al. 2022). Among the *DILPs* expressed in mNSCs, *DILP5* exhibits the strongest sensitivity to nutritional conditions and is transcriptionally regulated by Eyeless (Ey) together with its partner Dachshund (Dac) (Clements et al. 2008; Okamoto et al. 2012). Interestingly, the mammalian homologs of these transcription factors, Pax6 and Dach1/Dach2, are highly expressed in pancreatic β-cells (Sander et al. 1997; Kalousova et al. 2010), suggesting that key elements of *insulin* gene regulation may be evolutionarily conserved between insects and mammals. In mammals, Forkhead box O (FoxO)1 is a central regulator of β-cell function (Glauser and Schlegel 2007). Consistently, the *D. melanogaster* ortholog FoxO plays a critical role in *DILP* gene expression (Okamoto and Nishimura 2015). Under starvation conditions, FoxO translocates into the nucleus of mNSCs, where it physically interacts with Ey and suppresses Ey-Dac-dependent transcription of *DILP5* (Okamoto and Nishimura 2015). The subcellular localization of FoxO in mNSCs is regulated by nutrient-dependent signals originating from glial cells that constitute the blood–brain barrier (BBB). When nutrients are abundant, DILP6 secreted from the BBB activates a neuronal relay involving cholinergic neurons, which subsequently release Jelly belly (Jeb) onto mNSCs. Binding of Jeb to its receptor, Anaplastic lymphoma kinase (Alk), triggers PI3K signaling in mNSCs, thereby modulating FoxO localization and promoting *DILP5* expression. This signaling cascade selectively controls *DILP5* transcription, while *DILP2*, despite being co-expressed in the same cells, remains largely insensitive to this pathway.

Additional extrinsic factors also modulate *DILP* expression in mNSCs during larval development. Short neuropeptide F (sNPF) has been shown to enhance larval growth by influencing *DILP1* and *DILP2* expression through MAPK signaling, particularly via ERK activation in mNSCs (Lee et al. 2008, 2009). However, because the sNPF receptor (sNPFR) is not expressed in larval mNSCs (Kapan et al. 2012; Carlsson et al. 2013), this regulation is thought to occur indirectly. Given that sNPF functions as a key feeding-related neuropeptide, analogous to vertebrate neuropeptide Y (NPY), this pathway likely provides a link between feeding behavior and systemic growth regulation. In contrast, in the adult brain, sNPFR is expressed in mNSCs (Kapan et al. 2012), and glucose-responsive sNPF neurons directly modulate mNSC activity (Oh et al. 2019).

Finally, developmental perturbations such as imaginal disc damage elicit systemic endocrine responses that affect *DILP* expression. In larval *D. melanogaster*, damaged discs secrete DILP8, which acts on two pairs of LGR3-expressing neurons in the brain to delay developmental progression (Gontijo and Garelli 2018). These neurons suppress both PTTH-producing cells, thereby reducing ecdysone synthesis, and mNSCs, leading to decreased *DILP* expression (Vallejo et al. 2015). While ubiquitous overexpression of *DILP8* does not alter DILP2 expression levels, it significantly reduces *DILP3* and *DILP5* expression (Vallejo et al. 2015). The precise neuronal mechanisms by which LGR3-positive neurons differentially regulate individual *DILP* genes remain to be elucidated.

7.3.3 Regulation of ILP Secretion

In insects, the secretion of ILPs from mNSCs is highly dependent on nutritional status. As in mammals, this response is mediated not only by direct nutrient sensing in mNSCs but also by various hormones and neurotransmitters whose expression and secretion are regulated by nutritional conditions.

Studies using *D. melanogaster* have established that nutrient status is primarily sensed by the fat body (Colombani et al. 2003), which subsequently communicates the nutritional state to brain mNSCs via humoral factors, called as fat body-derived signals (FDSs) (Géminard et al. 2009). It has become clear that ILP secretion from mNSCs is regulated by a complex interplay of multiple FDSs, including cytokines and peptide hormones produced in response to different nutrients (Fig. 7.3) (Meschi and Delanoue 2021). Since most of the knowledge regarding the regulation of ILP secretion during insect development comes from studies in *D. melanogaster*, the research discussed below uses *D. melanogaster* as a model organism.

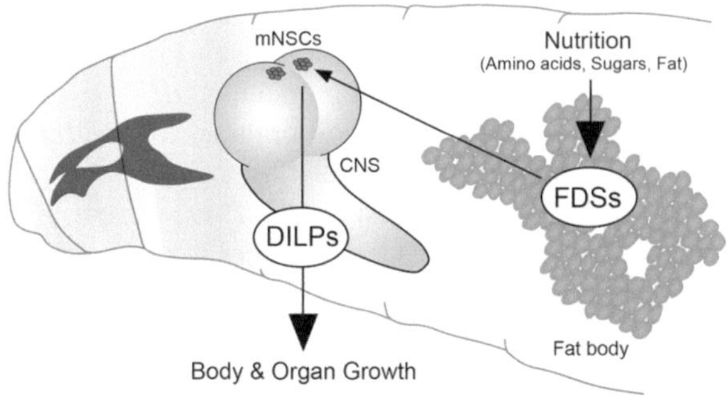

Fig. 7.3 DILP secretion from mNSCs is remotely regulated by fat body-derived signals (FDSs) in *D. melanogaster*. The availability of nutrition is mainly sensed by the fat body, a functional equivalent of the liver and adipocytes in vertebrates, which in turn remotely regulates DILP secretion from brain mNSCs through humoral signals called fat body-derived signals (FDSs). (Modified from Okamoto and Yamanaka 2015. CNS Central nervous system)

Larval growth is dependent on dietary nutrients, especially amino acids in *D. melanogaster* (Britton and Edgar 1998; Colombani et al. 2003). Amino acid sensing via the target of rapamycin (TOR) signaling pathway in the fat body remotely controls the release of various FDSs that regulate the activity of mNSCs in the brain (Géminard et al. 2009). Growth-blocking peptides (GBP1 and GBP2), which are TOR-dependent FDSs, are produced in an amino acid–dependent manner in the fat body (Koyama and Mirth 2016). GBPs stimulate DILP secretion from mNSCs via activating epidermal growth factor receptor (EGFR) expressed in a pair of intermediate inhibitory neurons, which directly synapse with mNSCs (Meschi et al. 2019). It was also reported that GBP functions as a ligand for GPCR called Methuselah-like 10 (Mthl10), which is expressed in adult brain mNSCs (Sung et al. 2017), suggesting that GBP may also directly activate mNSCs in the adults. Another TOR-dependent FDS, Stunted (Sun), acts on the GPCR Methuselah (Mth) expressed in mNSCs to directly stimulate DILP secretion from mNSCs (Delanoue et al. 2016). On the other hand, Eiger, a TNF-α homolog, is secreted into the hemolymph in response to decreased TOR signaling activity in the fat body and acts on Grindelwald, the Eiger receptor expressed in mNSCs, to directly inhibit ILP secretion (Agrawal et al. 2016). Recent studies have also shown that the Sun contributes to sex-dependent phenotypic plasticity in *D. melanogaster*. During the larval growth phase, the increase in IIS activity in females depends on the upregulation of *Sun* in the fat body (Millington et al. 2021). These findings suggest that, in addition to its amino acid–dependent role in growth regulation, Sun also contributes to differences in growth regulation between sexes during larval development.

FDS secretion is regulated not only by amino acids but also by other nutrients. Sugars and lipids regulate mNSC activity by triggering the release of the leptin-like cytokine Unpaired-2 (Upd2) from the fat body (Rajan and Perrimon 2012). Additionally, the fat body-derived peptide hormone CCHamide-2 (CCHa2) also regulates ILP secretion in a sugar-dependent manner (Sano et al. 2015).

Nutrient status information is also transmitted from the fat body to mNSCs via lipoprotein-mediated signaling. Lipid transfer protein (LTP) produced in the fat body is associated with dietary lipids, crosses the BBB, accumulates on neurons that directly contact mNSCs, and promotes DILP release (Brankatschk et al. 2014). Furthermore, dietary cholesterol regulates body growth by promoting DILP release (Texada et al. 2022). This nutrient input information is mediated by the nutrient sensor TOR and the cholesterol transporter Niemann-Pick disease type C1 (Npc1) in both the BBB and fat body.

In addition to nutritional status, DILP secretion from mNSCs also depends on oxygen sensing in the fat body (Texada et al. 2019a). The fat body senses oxygen levels via the TOR and conserved transcription factor hypoxia inducible factor (HIF)-1a and modulates mNSC activity through FDS(s). Interestingly, oxygen-sensing information is conveyed to mNSC through an unidentified FDS(s) that is distinct from the nutrient-dependent signals.

ILP secretion from mNSCs during larval development is also influenced by temperature. When *D. melanogaster* larvae are reared at lower temperatures (18 °C) compared to the standard 25 °C, they achieve larger body sizes (Ghosh et al. 2013). Around ten pairs of cold-sensing neurons located in the anterior tip of larval head directly interact with mNSC and promote DILP secretion (Li and Gong 2015). Although the significance of the increase of body size at lower temperatures is unclear, similar phenomena are known in a wide range of organisms known as the temperature-size rule (Atkinson 1994). DILP secretion from mNSCs is also regulated by internal sensory neurons located in the larval body surface (Gr28a-expressing v'td neurons) (Ohhara and Yamanaka 2022). The axons of Gr28a-expressing v'td neurons project directly to mNSCs. Silencing of Gr28a-expressing neurons inhibits the release of DILPs from the mNSCs and suppresses larval growth, although the nature of the sensory input these neurons receive is unknown.

In addition to the fat body and central nervous system, other organs regulate ILP secretion from mNSCs. Recent studies have reported that growth via DILP secretion is regulated by a salivary gland-derived (Sgsf) peptide (Li et al. 2022). However, the mechanism, including the receptor for Sgsf peptide, remains completely unknown. Enterocytes (ECs), which are involved in nutrient absorption in the gut, are also critical tissues for the regulation of larval growth. In response to nutrient deprivation, larval ECs release Hedgehog (Hh), which acts on the fat body to inhibit larval growth (Rodenfels et al. 2014). It remains unclear whether this fat body–mediated growth inhibition is DILP dependent. Recent studies have identified zinc-sensing ECs that support larval feeding and growth (Redhai et al. 2020). Zinc-gated chloride channel Hodor activates TOR signaling in ECs, leading to increased food intake and DILP secretion from mNSCs via an unidentified systemic humoral signal.

Additionally, the gut microbiota also promotes systemic larval growth via DILP secretion. The growth-promoting effect of the microbiota is linked to the positive effects on systemic IIS activity, partly through promoting DILP secretion via the fat body (Shin et al. 2011; Storelli et al. 2011). These findings reveal that the gut functions as a tissue that regulates systemic growth in response to dietary nutrients and microbial-derived metabolites. Although recent studies in adult flies have shown that several peptide hormones secreted by enteroendocrine cells in the gut regulate DILP secretion from mNSCs (Yoshinari et al. 2021; Chen et al. 2024), it remains unclear whether a similar mechanism operates in larvae.

Recent research has revealed that under a high-sugar diet, which induces an obesity-like phenotype including insulin resistance, complex inter-organ communication among muscle tissue, neurons, and the fat body regulates DILP secretion from the mNSCs (Kubrak et al. 2024). Under such dietary conditions, muscle-derived bone morphogenetic protein (BMP) signals to neurons expressing the peptide hormone bursicon. Bursicon directly stimulates DILP secretion through the LGR-type B receptor LGR2 and further enhances insulin signaling indirectly by increasing the insulin sensitivity of the fat body.

It has been believed that the secretion of all DILPs produced by mNSCs is simultaneously induced by mNSC depolarization. However, the secretion of specific types of DILPs is selectively regulated depending on the nutrients. During larval development, DILP2 and DILP5 secretion responds to amino acids, while sugar stimulates the corpora cardiaca (CC) to release the adipokinetic hormone (AKH), which acts directly on mNSCs to promote the secretion of DILP3 (Kim and Neufeld 2015). Thus, despite being produced in the same mNSCs, amino acids selectively induce the secretion of DILP2 and DILP5, whereas sugars selectively induce DILP3 secretion. The mechanism underlying this selective regulation of the secretion of different DILPs within the same cells remains unknown.

7.3.4 Regulation of ILP by Binding Proteins

The biological actions of insect ILPs are regulated not only by their expression and secretion but also by extracellular factors that associate with ILPs after secretion. ILP-binding proteins fine-tune IIS by stabilizing circulating ILPs, restricting their access to target tissues, or blocking receptor engagement. This mode of regulation is conceptually analogous to the modulation of vertebrate IGFs by IGF-binding proteins (IGFBPs). Although insects do not appear to possess bona-fide IGFBP orthologs containing both the IGFBP-like N-terminal domain and the thyroglobulin-like C-terminal domain, several secreted proteins have been identified that directly interact with ILPs, including neuroparsins, imaginal morphogenesis protein–late 2 (Imp-L2), and the secreted decoy of the insulin receptor (SDR).

Neuroparsins constitute a group of cysteine-rich secreted proteins widely distributed among insects and crustaceans and display marked sequence similarity to the N-terminal region of vertebrate IGFBPs (Badisco et al. 2007; see also Sect. 2.5.2). Functional evidence for a direct interaction with ILPs was obtained in the desert locust *Schistocerca gregaria*, where neuroparsin binds insulin-related peptides (Scg-IRP) in vitro (Badisco et al. 2008), supporting the idea that neuroparsins can modulate ILP activity. However, not all neuroparsin family members appear to act as ILP-binding proteins. In mosquitoes, the neuroparsin-like hormone ovarian ecdysteroidogenic hormone (OEH) stimulates vitellogenesis in *A. aegypti* by activating a RTK, the venus kinase receptor, functioning in parallel with ILPs (Vogal et al. 2015). These observations suggest that neuroparsins comprise a functionally diverse protein family with both ILP-dependent and ILP-independent roles. Additional details regarding the endocrine functions of neuroparsins are discussed in Sect. 2.5.2.

Imp-L2 is another secreted modulator of IIS and is characterized by the presence of immunoglobulin (Ig)-like domains. It was originally identified as an ecdysone-inducible gene, and its Ig-like domain resembles that of the mammalian tumor suppressor Mac25, previously referred to as IGFBP7 but no longer classified as a true IGFBP. In *D. melanogaster*, Imp-L2 physically associates with multiple DILPs in the hemolymph and interferes with their binding to the InR (Honegger et al. 2008; Alic et al. 2011; Okamoto et al. 2013). Genetic analyses have shown that loss of *Imp-L2* results in increased body size, whereas overexpression of *Imp-L2* leads to growth suppression (Honegger et al. 2008). Consistent with a role in nutritional adaptation, *Imp-L2* expression in the fat body is upregulated under starvation conditions (Honegger et al. 2008), implying that Imp-L2 sequesters circulating DILPs to dampen IIS when nutrients are limited. Moreover, Imp-L2 can associate with the acid-labile subunit (ALS), forming a trimeric complex with DILPs, reminiscent of IGF-IGFBP-ALS complexes in mammals (Arquier et al. 2008), thereby further influencing DILP stability and signaling capacity. Beyond its systemic function, Imp-L2 also exerts local effects within the nervous system. In *D. melanogaster*, DILPs secreted from mNSCs are selectively taken up by certain neurons in an Imp-L2–dependent manner (Bader et al. 2013). In addition, axonally transported DILPs are captured and stored by the CC, paired neurohemal organs within the ring gland (Ghosh et al. 2022). The release of DILPs from the CC is regulated independently of mNSCs (Ghosh et al. 2022), suggesting that DILPs may be used for different physiological roles through transport to neurons and organs that are regulated separately from mNSCs.

SDR represents a third class of ILP-binding proteins and is structurally related to the extracellular domain of the InR (Okamoto et al. 2013). This protein is mainly produced by glial cells and released into the hemolymph, where it directly associates with circulating DILPs. Functional studies demonstrate that knockdown or loss of *SDR* leads to increased systemic growth, consistent with a role as a negative regulator of IIS. In contrast to Imp-L2, *SDR* expression shows relatively stable in response to nutritional status, suggesting that SDR act as a consistent modulator of IIS that is largely independent of environmental nutrient conditions.

7.4 Regulation of Growth During Development by ILPs

The regulation of systemic growth during development by ILPs ensures that various tissues grow in a coordinated manner, maintaining proper proportions of the body structure. At the same time, tissue-specific growth regulation ensures that individual organs achieve their appropriate sizes. This section focuses on the mechanisms by which insect ILPs regulate growth.

7.4.1 ILPs Function as Growth Factors

The direct growth-promoting effects of insect ILPs were first demonstrated using synthetic bombyxin-II from *B. mori* on ex vivo tissue culture system (Nijhout and Grunert 2002). Wing imaginal discs from final-instar larvae of the butterfly *Precis coenia* exhibited a dose-dependent increase in mitotic rate when bombyxin-II was added to the culture medium along with ecdysone (Nijhout and Grunert 2002). Furthermore, when hemolymph from *P. coenia* was pretreated with a bombyxin antibody-bound column to remove bombyxin, the growth-promoting effects of the hemolymph on the wing imaginal discs were abolished. Based on these findings, the authors concluded that bombyxin functions as a growth factor for wing imaginal discs (Nijhout and Grunert 2002). Similarly, the growth-promoting effects of bombyxin-II on ex vivo cultured wing disc have also been reported in the hawkmoth *Manduca sexta* (Nijhout et al. 2007). Bombyxin-II also promotes cell proliferation in the hematopoietic organs (HPOs) of *B. mori*. When larval HPOs were cultured ex vivo with bombyxin-II, the number of released hemocytes increased in a dose-dependent manner (Nakahara et al. 2006). In contrast to the studies on *P. coenia* wing imaginal discs, bombyxin-II was effective for *Bombyx* HPOs even in the absence of ecdysone.

The growth-promoting effects of BIGFLP, which is produced in the fat body of *B. mori* during pupa-adult development, on various tissues have also been analyzed using ex vivo tissue culture systems (Okamoto et al. 2009a; Fujinaga et al. 2017). Purified BIGFLP exhibited dose-dependent growth-promoting effects on cultured genital discs, sperm ducts, and muscle anlagen (Okamoto et al. 2009a). Moreover, using ex vivo cultured genital discs, it was demonstrated that the growth-promoting effects of BIGFLP are mediated via the IIS pathway (Fujinaga et al. 2017). Interestingly, ecdysone, which is secreted during the same developmental stage as BIGFLP secretion, was found to activate the MAPK pathway in ex vivo cultured genital discs (Fujinaga et al. 2017), suggesting that the growth and development of the adult-specific tissues are coordinately regulated by ILPs and ecdysone during pupa-adult development.

7.4.2 Systemic Growth Regulation by ILPs During Development

In insects, the larval or nymphal stage is specialized for nutrient intake and systemic growth. After the embryonic stage, holometabolous insects develop through several larval instars before pupation, during which they can increase in size by several hundred times (Robertson 1966). After completion of metamorphosis, the adult does not grow in body size. This means that the final size of an insect body size largely depends on regulating its growth during the larval stage, particularly regulating two key parameters: "growth rate" and "growth duration" (Tennessen and Thummel

2011; Nijhout et al. 2013; Texada et al. 2020). The hormone that primarily regulates the "growth rate" during insect development is ILP, while the hormone responsible for controlling "growth duration" is the insect steroid hormone ecdysone. Although the effects of insect ILPs on ecdysteroidogenesis in the prothoracic glands (PGs) are demonstrated in many insects, including *B. mori* and *D. melanogaster* (Kannangara et al. 2021; Gu et al. 2022), this chapter focuses on how insect ILPs regulate the "growth rate" during development. For the effects of bombyxin on the PGs, please refer to Chap. 1.

Previous research has demonstrated that mutations or knockdown of ILPs, InRs, and IIS components in various insect species affects body size regulation (e.g., in the cricket *Gryllus bimaculatus*, Dabour et al. 2011; in *R. prolixus*, Defferrari et al. 2016; in the moth *Maruca vitrata*, Al Baki et al. 2019, 2021; in the Asian tiger mosquito *Aedes albopictus*, Dai et al. 2023). Particularly in *D. melanogaster*, advances in tissue-specific genetic manipulation techniques since the 2000s have led to significant progress in understanding the growth-regulating role of DILPs (Edgar 2006; Texada et al. 2022). In *D. melanogaster*, genetic ablation of mNSCs results in severe growth delay and significantly reduced pupal and adult body size (Fig. 7.4) (Rulifson et al. 2002). This phenotype is similar to that observed in wild-type flies raised under poor nutrition or hypomorphic mutants of *InR* (Chen et al.

Fig. 7.4 mNSC-ablated flies show severe growth defect. Pupal and adult body size of mNSC-ablated flies. DILP-producing mNSCs in the brain were genetically ablated using a *DILP2* promoter to express the pro-apoptotic gene, *reaper*

1996; Tatar et al. 2001; Brogiolo et al. 2001; Banzai and Nishimura 2023). The similarity in phenotypes across these conditions demonstrates that ILPs secreted from mNSCs function as systemic growth regulators during larval stages, tightly linking nutrient intake and growth.

Interestingly, knockout of each of the *DILP* genes (*DILP2*, *DILP3*, or *DILP5*), which are known to be produced in mNSCs during the larval stage, does not result in severe phenotypes. The strongest growth defect is observed in *DILP2* mutants, where adult body weight is reduced by only 5–10% (Grönke et al. 2010). Under normal rearing conditions, clear growth defects are not observed in *DILP3* or *DILP5* single mutants. However, *DILP2*, *DILP3*, and *DILP5* triple mutants show significantly delayed larval growth (Okamoto and Nishimura 2015), with adult body size reduced by over 40% (Grönke et al. 2010). This indicates that DILPs produced by mNSCs function redundantly in regulating systemic body growth. Moreover, a fat body-derived IGFLP, DILP6, also plays a key role in compensating for the loss of mNSC-derived DILPs. In *DILP2*, *DILP3*, and *DILP5* triple mutants, expression of *DILP6* in the fat body is dramatically upregulated, compensating for the lack of DILP production from mNSCs (Grönke et al. 2010). This upregulation is directly induced by transcription factor FoxO (Slaidina et al. 2009), which is activated in response to decreased IIS. In fact, the quadruple mutant of *DILP2*, *DILP3*, *DILP5*, and *DILP6* is lethal during development, further highlighting the compensatory role of fat body–derived DILP6 in the absence of mNSC-derived DILPs (Grönke et al. 2010).

Although ecdysone primarily plays a role for regulating "growth duration," it also affects systemic growth through IIS. Before the large ecdysone pulse that induces pupation, low basal levels of ecdysone negatively regulate IIS through the ecdysone receptor (EcR) expressed in the fat body, controlling systemic growth during the larval stage (Colombani et al. 2005; Boulan et al. 2013; Moeller et al. 2017; Texada et al. 2019b). In the fat body, ecdysone represses the function of transcription factor Myc through EcR, which influences the release of FDS(s) that regulates DILP secretion from mNSCs, thereby controlling the larval growth rate (Delanoue et al. 2010). Another mechanism by which ecdysone negatively regulates DILP-mediated systemic growth involves microRNA *miR-8* (Hyun et al. 2009; Jin et al. 2012). *miR-8* directly targets and suppresses *u-shaped* (*ush*), an ortholog of mammalian *FOG2*, which leads to the upregulation of PI3 kinase, a component of insulin signaling in the fat body, promoting systemic growth (Hyun et al. 2009; Jin et al. 2012). As ecdysone levels rise in late-stage larvae prior to pupation, *miR-8* expression decreases, leading to reduced systemic growth. This ecdysone-IIS crosstalk explains how the increase in ecdysone levels at the end of the final larval instar suppresses tissue growth and reduces systemic body growth.

On the other hand, ecdysone also promotes systemic growth through fat body–derived DILP6, which is produced in response to ecdysone during the wandering and pupal stages after larval feeding has ceased. Among all the *DILP* single mutants, *DILP6* mutants show the smallest adult body size (Grönke et al. 2010). Unlike the *DILP2*, *DILP3*, and *DILP5* triple mutants, which show severe growth defects during larval feeding stage (Okamoto and Nishimura 2015), *DILP6* mutants do not exhibit obvious growth defects during this stage, but their adult body weight is reduced by 10–20% compared to wild-type flies (Okamoto et al. 2009a, b; Slaidina et al. 2009; Grönke et al. 2010). This suggests that DILP6, produced by the fat body during post-feeding stages, facilitates the utilization of stored nutrients accumulated during the feeding period to support additional systemic growth. This fat body–derived IGFLP-mediated post-feeding growth is also linked to the differential growth rates of various organs, which will be discussed in the next section.

7.4.3 Regulation of Organ Growth by ILPs During Development

The systemic growth regulation mechanisms discussed above ensure appropriate final proportions by maintaining a proper balance between overall body size and organ size. However, these proportions can vary depending on environmental factors, such as nutrition, affecting different organs (Emlen and Nijhout 2000; Shingleton et al. 2007; Koyama et al. 2013; Lavine et al. 2015). For example, under low-nutrient conditions during development, a reduction in overall body size is generally accompanied by a proportional decrease in the size of most organs. However, some organs may exhibit minimal changes in size or, in extreme cases, may shrink disproportionately. This organ-specific sensitivity to nutritional cues is a phenomenon observed in many insects. In some beetles, for instance, the availability of nutrients during development strongly influences organ scaling (Emlen and Nijhout 2000). In these species, the "horns," which males use as weapons in combat, exhibit exceptionally high sensitivity to nutrition, unlike the rest of the body and other organs. In contrast, the size of genitalia essential for mating remains largely unaffected by changes in nutritional conditions.

Recent studies have revealed that the differential sensitivity of specific organs to nutrition can be explained by their responsiveness to insect ILPs and the IIS pathway. In the beetle *Trypoxylus dichotomus*, knockdown of *InR* during the early prepupal stage results in a 16% reduction in adult horn length, whereas wing length decreases by only 2%, and reproductive organs remain unaffected (Emlen et al. 2012). These results clearly indicate that during the post-feeding period, the horn primordia exhibit higher sensitivity to IIS compared to other organs. Similarly, in *G. cornutus*, knockdown of *InR* exclusively reduces the size of the weapon-like mandibles (Okada et al. 2019). Furthermore, when each of five ILPs in *G. cornutus* is knocked down, only knockdown of *GcorILP2*, which is highly expressed in the fat body, specifically reduces mandible size (Okada et al. 2019). These findings suggest that GcorILP2 acts as a messenger linking nutritional status with weapon development. The fact that the growth of the mandibles in *G. cornutus* is regulated by only one of the ILPs is a significant example of functional specialization among ILPs for specific organ growth. Interestingly, although GcorILP2 is structurally classified as a typical ILP, its expression is very low in the brain and extremely high in the fat body

(Okada et al. 2019). This expression pattern resembles that of IGFLP secreted from the fat body as an ecdysone dependent-growth factor, but GcorILP2 appears to be regulated by nutritional status. It remains an intriguing question whether fat body–derived ILPs or IGFLPs function in the development of weapons in other insect species as well.

What mechanisms explain the differential sensitivity of organs to ILPs? In both *T. dichotomus* and *G. cornutus*, no clear organ-specific differences in *InR* expression have been observed (Emlen et al. 2012; Okada et al. 2019), and therefore the varying nutritional responsiveness of different organs cannot be explained by differences in receptor expression levels. However, studies using *D. melanogaster* and the horned dung beetle *Onthophagus taurus* suggest that differences in the expression levels of *FoxO*, a negative transcription factor downstream of IIS, across organs may explain organ-specific differences in IIS sensitivity (Tang et al. 2011; Casasa and Moczek 2018). Organs with higher *FoxO* expression are likely to be more sensitive to changes in IIS, while organs like the genital primordia, where FoxO expression is low, may be less affected by changes in IIS, making them less dependent on nutritional status. In fact, overexpression of *FoxO* in the genital disc (or primordia) restores nutritional dependence, resulting in reduced genital size in adult males in *D. melanogaster* (Tang et al. 2011). In lepidopteran insects, such as *P. coenia* and *M. sexta*, the wing disc change their sensitivity to nutrition or ILPs depending on the developmental stage. This seems to be explained by changes in ecdysone concentrations during the final larval stage, as ILPs and ecdysone act synergistically to regulate wing growth (Nijhout and Grunert 2002; Nijhout et al. 2007).

Research using *D. melanogaster* has also revealed that the development of the central nervous system undergoes a transition from high to low nutritional sensitivity depending on the developmental stage. In early larval stage, dietary amino acids are essential for neuroblasts (NBs) to return to the cell cycle after quiescence (Britton and Edgar 1998). This nutritional reactivation of NBs involves an interorgan relay, where amino acid sensing in the fat body remotely stimulates the production of DILP6 in the glial cell niche (Chell and Brand 2010; Sousa-Nunes et al. 2011). As a result, glial DILP6 acts on InR of nearby NBs, activating IIS in a nutrient-dependent manner and promoting cell cycle re-entry (Chell and Brand 2010; Sousa-Nunes et al. 2011). Later in larval development, another secreted factor Jeb is constitutively secreted from the glial cell niche. This factor binds to Alk receptor, an RTK expressed in NBs, to activate the PI3K pathway, allowing NB proliferation to continue even under nutrient-poor conditions by bypassing InR-mediated nutritional responses (Cheng et al. 2011). The low nutritional sensitivity mechanisms seen in genital organs and the central nervous system in many insects are thought to be important adaptive mechanisms that ensure the proper development of these critical systems under poor nutritional conditions.

In some insect species, not only nutritional cues but also environmental factors, such as population density, temperature, and photoperiod, affect the growth of specific organs during development. For instance, in planthoppers, wing length significantly changes in response to population density and the quality of host plants, and this mechanism involves IIS (Zhang et al. 2019). Interestingly, in stressful conditions such as high population density or poor food quality, the insects develop longer wings, allowing them to fly to more favorable environments. In the brown planthopper *Nilaparvata lugens*, two InRs (Nl-InR1 and Nl-InR2) act as switches that control wing morphology by regulating FoxO activity in response to environmental cues (Xu et al. 2015). Activation of Nl-InR1, which functions as a typical InR, leads to the development of long-winged morphs via IIS. However, Nl-InR2 physically interacts with Nl-InR1 and functions as a negative regulator. This inhibition shuts down IIS in developing wings, resulting in short-winged adult morphs. Thus, changes in Nl-InR2 activity in response to environmental changes ultimately regulate wing morph switching in *N. lugens*. Since many insects possess multiple InRs (Kremer et al. 2018; Smýkal et al. 2020) and some InRs lack intracellular or membrane domains (decoy InRs) (Smýkal et al. 2020), similar mechanisms may be conserved in other species. Further research on various organs and insect species is needed to fully understand how organ-specific environmental responses are achieved.

7.5 Future Perspectives

In conclusion, ILPs are central in integrating nutritional signals to regulate systemic and organ-specific growth during insect development. The intricate regulation of ILP expression, secretion, and activity, along with their interactions with binding proteins and receptors, underscores their crucial role in maintaining systemic and tissue growth and adapting to diverse environmental conditions.

Since the 2000s, advances in cell-type and organ-specific genetic tools, particularly in the model organism *D. melanogaster*, have greatly enhanced our understanding of the evolutionarily conserved functions of ILPs and their downstream IIS pathways. As highlighted in this chapter, significant progress has been made in elucidating how various factors influence ILP-mediated growth regulation in a nutrient-dependent manner. However, much of this understanding is still limited to *D. melanogaster*, and many questions remain for other insects with diverse evolutionary and ecological backgrounds. With the recent development of powerful techniques such as RNAi and CRISPR/Cas9 genome editing, it is hoped that future research will uncover additional roles of insect ILPs in growth regulation that could not be revealed by studies using *D. melanogaster*.

References

Agrawal N, Delanoue R, Mauri A, Basco D, Pasco M, Thorens B et al (2016) The *Drosophila* TNF eiger is an adipokine that acts on insulin-producing cells to mediate nutrient response. Cell Metab 23:675–684. https://doi.org/10.1016/j.cmet.2016.03.003

Al Baki MA, Lee DW, Jung JK, Kim Y (2019) Insulin-like peptides of the legume pod borer, *Maruca vitrata*, and their mediation effects on hemolymph trehalose level, larval development, and adult reproduction. Arch Insect Biochem Physiol 100:e21524. https://doi.org/10.1002/arch.21524

Al Baki MA, Kyo Jung J, Kim Y (2021) Physiological alterations in deletion mutants of two insulin-like peptides encoded in maruca vitrata using Crispr/Cas9. Front Physiol 12:701616. https://doi.org/10.3389/fphys.2021.701616

Alic N, Hoddinott MP, Vinti G, Partridge L (2011) Lifespan extension by increased expression of the *Drosophila* homologue of the IGFBP7 tumour suppressor. Aging Cell 10:137–147. https://doi.org/10.1111/j.1474-9726.2010.00653.x

Antonova Y, Arik AJ, Moore W, Riehle MR, Brown MR (2012) Insulin-like peptides: structure, signaling, and function. In: Gilbert LI (ed) Insect endocrinology. Elsevier/Academic, pp 63–92. https://doi.org/10.1016/b978-0-12-384749-2.10002-0

Arquier N, Géminard C, Bourouis M, Jarretou G, Honegger B, Paix A et al (2008) *Drosophila* ALS regulates growth and metabolism through functional interaction with insulin-like peptides. Cell Metab 7:333–338. https://doi.org/10.1016/j.cmet.2008.02.003

Atkinson D (1994) Temperature and organism size: a biological law for ectotherms? Adv Ecol Res 25:1–58. https://doi.org/10.1016/S0065-2504(08)60212-3

Bader R, Sarraf-Zadeh L, Peters M, Moderau N, Stocker H, Köhler K et al (2013) The IGFBP7 homolog Imp-L2 promotes insulin signaling in distinct neurons of the *Drosophila* brain. J Cell Sci 126:2571–2576. https://doi.org/10.1242/jcs.120261

Badisco L, Claeys I, Van Loy T, Van Hiel M, Franssens V, Simonet G et al (2007) Neuroparsins, a family of conserved arthropod neuropeptides. Gen Comp Endocrinol 153:64–71. https://doi.org/10.1016/j.ygcen.2007.03.008

Badisco L, Claeys I, Van Hiel M, Clynen E, Huybrechts J, Vandersmissen T et al (2008) Purification and characterization of an insulin-related peptide in the desert locust, *Schistocerca gregaria*: immunolocalization, cDNA cloning, transcript profiling and interaction with neuroparsin. J Mol Endocrinol 40:137–150. https://doi.org/10.1677/jme-07-0161

Badisco L, Van Wielendaele P, Vanden Broeck J (2013) Eat to reproduce: a key role for the insulin signaling pathway in adult insects. Front Physiol 4:202. https://doi.org/10.3389/fphys.2013.00202

Banzai K, Nishimura T (2023) Isolation of a novel missense mutation in insulin receptor as a spontaneous revertant in *ImpL2* mutants in *Drosophila*. Development 150:dev201248. https://doi.org/10.1242/dev.201248

Bathgate RAD, Halls ML, van der Westhuizen ET, Callander GE, Kocan M, Summers RJ (2013) Relaxin family peptides and their receptors. Physiol Rev 93:405–480. https://doi.org/10.1152/physrev.00001.2012

Blanco-Obregon D, El Marzkioui K, Brutscher F, Kapoor V, Valzania L, Andersen DS et al (2022) A Dilp8-dependent time window ensures tissue size adjustment in *Drosophila*. Nat Commun 13:5629. https://doi.org/10.1038/s41467-022-33387-6

Boulan L, Martín D, Milán M (2013) *bantam* miRNA promotes systemic growth by connecting insulin signaling and ecdysone production. Curr Biol 23:473–478. https://doi.org/10.1016/j.cub.2013.01.072

Brankatschk M, Dunst S, Nemetschke L, Eaton S (2014) Delivery of circulating lipoproteins to specific neurons in the *Drosophila* brain regulates systemic insulin signaling. eLife 3:e02862. https://doi.org/10.7554/eLife.02862

Britton JS, Edgar BA (1998) Environmental control of the cell cycle in *Drosophila*: nutrition activates mitotic and endoreplicative cells by distinct mechanisms. Development 125:2149–2158. https://doi.org/10.1242/dev.125.11.2149

Brogiolo W, Stocker H, Ikeya T, Rintelen F, Fernandez R, Hafen E (2001) An evolutionarily conserved function of the *Drosophila* insulin receptor and insulin-like peptides in growth control. Curr Biol 11:213–221. https://doi.org/10.1016/s0960-9822(01)00068-9

Carlsson MA, Enell LE, Nässel DR (2013) Distribution of short neuropeptide F and its receptor in neuronal circuits related to feeding in larval *Drosophila*. Cell Tissue Res 353:511–523. https://doi.org/10.1007/s00441-013-1660-4

Casasa S, Moczek AP (2018) Insulin signalling's role in mediating tissue-specific nutritional plasticity and robustness in the horn-polyphenic beetle *Onthophagus taurus*. Proc R Soc B 285:20181631. https://doi.org/10.1098/rspb.2018.1631

Chandler J, Gandhi N, Mancera R, Smith G, Elizur A, Ventura T (2017) Understanding insulin endocrinology in decapod Crustacea: molecular modelling characterization of an insulin-binding protein and insulin-like peptides in the Eastern spiny lobster, *Sagmariasus verreauxi*. Int J Mol Sci 18(9):1832. https://doi.org/10.3390/ijms18091832

Chell JM, Brand AH (2010) Nutrition-responsive glia control exit of neural stem cells from quiescence. Cell 143:1161–1173. https://doi.org/10.1016/j.cell.2010.12.007

Chen C, Jack J, Garofalo RS (1996) The *Drosophila* insulin receptor is required for normal growth. Endocrinology 137:846–856. https://doi.org/10.1210/en.137.3.846

Chen J, Nouzová M, Noriega FG, Tatar M (2024) Gut-to-brain regulation of *Drosophila* aging through neuropeptide F, insulin, and juvenile hormone. Proc Natl Acad Sci USA 121: e2411987121. https://doi.org/10.1073/pnas.2411987121

Cheng LY, Bailey AP, Leevers SJ, Ragan TJ, Driscoll PC, Gould AP (2011) Anaplastic lymphoma kinase spare organ growth during nutrient restriction in *Drosophila*. Cell 146:435–447. https://doi.org/10.1016/j.cell.2011.06.040

Chowański S, Walkowiak-Nowicka K, Winkiel M, Marciniak P, Urbański A, Pacholska-Bogalska J (2021) Insulin-Like peptides and cross-talk with other factors in the regulation of insect metabolism. Front Physiol 12:973. https://doi.org/10.3389/fphys.2021.701203

Clements J, Hens K, Francis C, Schellens A, Callaerts P (2008) Conserved role for the Drosophila Pax6 homolog eyeless in differentiation and function of insulin-producing neurons. Proc Natl Acad Sci USA 105:16183–16188. https://doi.org/10.1073/pnas.0708330105

Colombani J, Raisin S, Pantalacci S, Radimerski T, Montagne J, Léopold P (2003) A nutrient sensor mechanism controls *Drosophila* growth. Cell 114:739–749. https://doi.org/10.1016/s0092-8674(03)00713-x

Colombani J, Bianchini L, Layalle S, Pondeville E, Dauphin-Villemant C, Antoniewski C et al (2005) Antagonistic actions of ecdysone and insulins determine final size in *Drosophila*. Science 310:667–670. https://doi.org/10.1126/science.1119432

Colombani J, Andersen DS, Léopold P (2012) Secreted peptide Dilp8 coordinates *Drosophila* tissue growth with developmental timing. Science 336:582–585. https://doi.org/10.1126/science.1216689

Conlon I, Raff M (1999) Size control in animal development. Cell 96:235–244. https://doi.org/10.1016/s0092-8674(00)80563-2

Dabour N, Bando T, Nakamura T, Miyawaki K, Mito T, Ohuchi H et al (2011) Cricket body size is altered by systemic RNAi against insulin signaling components and epidermal growth factor receptor. Develop Growth Differ 53:857–869. https://doi.org/10.1111/j.1440-169x.2011.01291.x

Dai Y, Li X, Ding J, Liang Z, Guo R, Yi T et al (2023) Molecular and expression characterization of insulin-like signaling in development and metabolism of *Aedes albopictus*. Parasit Vectors 16: 134. https://doi.org/10.1186/s13071-023-05747-8

Defferrari MS, Orchard I, Lange AB (2016) An insulin-like growth factor in *Rhodnius prolixus* is involved in post-feeding nutrient balance and growth. Front Neurosci 10:566. https://doi.org/10.3389/fendo.2018.00745

Delanoue R, Slaidina M, Léopold P (2010) The steroid hormone ecdysone controls systemic growth by repressing dMyc function in *Drosophila* fat cells. Dev Cell 18:1012–1021. https://doi.org/10.1016/j.devcel.2010.05.007

Delanoue R, Meschi E, Agrawal N, Mauri A, Tsatskis Y, McNeill H et al (2016) *Drosophila* insulin release is triggered by adipose Stunted ligand to brain Methuselah receptor. Science 353:1553–1556. https://doi.org/10.1126/science.aaf8430

Edgar BA (2006) How flies get their size: genctics meets physiology. Nat Rev Genet 7:907–916. https://doi.org/10.1038/nrg1989

Emlen DJ, Nijhout HF (2000) The development and evolution of exaggerated morphologies in insects. Annu Rev Entomol 45:661–708. https://doi.org/10.1146/annurev.ento.45.1.661

Emlen DJ, Warren IA, Johns A, Dworkin I, Lavine LC (2012) A mechanism of extreme growth and reliable signaling in sexually selected ornaments and weapons. Science 337:860–864. https://doi.org/10.1126/science.1224286

Fernandez R, Tabarini D, Azpiazu N, Frasch M, Schlessinger J (1995) The *Drosophila* insulin receptor homolog: a gene essential for embryonic development encodes two receptor isoforms with different signaling potential. EMBO J 14:3373–3384. https://doi.org/10.1002/j.1460-2075.1995.tb07343.x

Froesch ER, Zapf J (1985) Insulin-like growth factors and insulin: comparative aspects. Diabetologia 28:485–493. https://doi.org/10.1007/BF00281982

Fujinaga D, Kohmura Y, Okamoto N, Kataoka H, Mizoguchi A (2017) Insulin-like growth factor (IGF)-like peptide and 20-hydroxyecdysone regulate the growth and development of the male genital disk through different mechanisms in the silkmoth, *Bombyx mori*. Insect Biochem Mol Biol 87:35–44. https://doi.org/10.1016/j.ibmb.2017.06.003

Garelli A, Gontijo AM, Miguela V, Caparros E, Dominguez M (2012) Imaginal discs secrete insulin-like peptide 8 to mediate plasticity of growth and maturation. Science 336:579–582. https://doi.org/10.1126/science.1216735

Géminard C, Rulifson EJ, Léopold P (2009) Remote control of insulin secretion by fat cells in *Drosophila*. Cell Metab 10:199–207. https://doi.org/10.1016/j.cmet.2009.08.002

Ghosh SM, Testa ND, Shingleton AW (2013) Temperature-size rule is mediated by thermal plasticity of critical size in *Drosophila melanogaster*. Proc R Soc Lond B Biol Sci 280: 20130174. https://doi.org/10.1098/rspb.2013.0174

Ghosh S, Leng W, Wilsch-Bräuninger M, Barrera-Velázquez M, Léopold P, Eaton S (2022) A local insulin reservoir in *Drosophila* alpha cell homologs ensures developmental progression under nutrient shortage. Curr Biol 32:1788–1797.e5. https://doi.org/10.1016/j.cub.2022.02.068

Giannakou ME, Partridge L (2007) Role of insulin-like signalling in *Drosophila* lifespan. Trends Biochem Sci 32:180–188. https://doi.org/10.1016/j.tibs.2007.02.007

Glauser DA, Schlegel W (2007) The emerging role of FOXO transcription factors in pancreatic beta cells. J Endocrinol 193:195–207. https://doi.org/10.1677/JOE-06-0191

Gontijo AM, Garelli A (2018) The biology and evolution of the Dilp8-Lgr3 pathway: a relaxin-like pathway coupling tissue growth and developmental timing control. Mech Dev 154:44–50. https://doi.org/10.1016/j.mod.2018.04.005

Grönke S, Clarke DF, Broughton S, Andrews TD, Partridge L (2010) Molecular evolution and functional characterization of *Drosophila* insulin-like peptides. PLoS Genet 6:e1000857. https://doi.org/10.1371/journal.pgen.1000857

Gu S-H, Chang C-H, Lin P-L (2022) Bombyxin-stimulated ecdysteroidogenesis in relation to sugar transporter/trehalase expressions in *Bombyx* prothoracic glands. Insect Biochem Mol Biol 151: 103864. https://doi.org/10.1016/j.ibmb.2022.103864

de Azevedo SV, Hartfelder K (2008) The insulin signaling pathway in honey bee (*Apis mellifera*) caste development—differential expression of insulin-like peptides and insulin receptors in queen and worker larvae. J Insect Physiol 54:1064–1071. https://doi.org/10.1016/j.jinsphys.2008.04.009

Heredia F, Volonté Y, Pereirinha J, Fernandez-Acosta M, Casimiro AP, Belém CG et al (2021) The steroid-hormone ecdysone coordinates parallel pupariation neuromotor and morphogenetic subprograms via epidermis-to-neuron Dilp8-Lgr3 signal induction. Nat Commun 12:1–20. https://doi.org/10.1038/s41467-021-23218-5

Honegger B, Galic M, Köhler K, Wittwer F, Brogiolo W, Hafen E et al (2008) Imp-L2, a putative homolog of vertebrate IGF-binding protein 7, counteracts insulin signaling in *Drosophila* and is essential for starvation resistance. J Biol 7:10. https://doi.org/10.1186/jbiol72

Hyun S, Lee JH, Jin H, Nam JW, Namkoong B, Lee G et al (2009) Conserved microRNA miR-8/miR-200 and its target USH/FOG2 control growth by regulating PI3K. Cell 139:1096–1108. https://doi.org/10.1016/j.cell.2009.11.020

Ichikawa T (1991) Architecture of cerebral neurosecretory cell systems in the silkworm *Bombyx mori*. J Exp Biol 161:217–237. https://doi.org/10.1242/jeb.161.1.217

Ikeya T, Galic M, Belawat P, Nairz K, Hafen E (2002) Nutrient-dependent expression of insulin-like peptides from neuroendocrine cells in the CNS contributes to growth regulation in *Drosophila*. Curr Biol 12:1293–1300. https://doi.org/10.1016/s0960-9822(02)01043-6

Imambocus BN, Zhou F, Formozov A, Wittich A, Tenedini FM, Hu C et al (2022) A neuropeptidergic circuit gates selective escape behavior of *Drosophila* larvae. Curr Biol 32: 149–63.e8. https://doi.org/10.1016/j.cub.2021.10.069

Ishizaki H (2004) Molecular characterization of the brain secretory peptides, prothoracicotropic hormone (PTTH) and bombyxin, of the silkmoth *Bombyx mori*. Proc Jpn Acad Ser B Phys Biol Sci 80:195–203. https://doi.org/10.2183/pjab.80.195

Jin H, Narry Kim V, Hyun S (2012) Conserved microRNA miR-8 controls body size in response to steroid signaling in *Drosophila*. Genes Dev 26:1427–1432. https://doi.org/10.1101/gad.192872.112

Jones JI, Clemmons DR (1995) Insulin-like growth factors and their binding proteins: biological actions. Endocr Rev 16:3–34. https://doi.org/10.1210/edrv-16-1-3

Kalousova A, Mavropoulos A, Adams BA, Nekrep N, Li Z, Krauss S et al (2010) Dachshund homologues play a conserved role in islet cell development. Dev Biol 348:143–152. https://doi.org/10.1016/j.ydbio.2010.09.007

Kannangara JR, Mirth CK, Warr CG (2021) Regulation of ecdysone production in *Drosophila* by neuropeptides and peptide hormones. Open Biol 11:200373. https://doi.org/10.1098/rsob.200373

Kapan N, Lushchak OV, Luo J, Nässel DR (2012) Identified peptidergic neurons in the *Drosophila* brain regulate insulin-producing cells, stress responses and metabolism by coexpressed short neuropeptide F and corazonin. Cell Mol Life Sci 69:4051–4066. https://doi.org/10.1007/s00018-012-1097-z

Kim J, Neufeld TP (2015) Dietary sugar promotes systemic TOR activation in *Drosophila* through AKH-dependent selective secretion of Dilp3. Nat Commun 6:6846. https://doi.org/10.1038/ncomms7846

Koyama T, Mirth CK (2016) Growth-blocking peptides as nutrition-sensitive signals for insulin secretion and body size regulation. PLoS Biol 14:e1002392. https://doi.org/10.1371/journal.pbio.1002392

Koyama T, Mendes CC, Mirth CK (2013) Mechanisms regulating nutrition-dependent developmental plasticity through organ-specific effects in insects. Front Physiol 4:263. https://doi.org/10.3389/fphys.2013.00263

Kremer LPM, Korb J, Bornberg-Bauer E (2018) Reconstructed evolution of insulin receptors in insects reveals duplications in early insects and cockroaches. J Exp Zool B Mol Dev Evol 330: 305–311. https://doi.org/10.1002/jez.b.22809

Kubrak O, Jørgensen AF, Koyama T, Lassen M, Nagy S, Hald J et al (2024) LGR signaling mediates muscle-adipose tissue crosstalk and protects against diet-induced insulin resistance. Nat Commun 15:6126. https://doi.org/10.1038/s41467-024-50468-w

Lavine L, Gotoh H, Brent CS, Dworkin I, Emlen DJ (2015) Exaggerated trait growth in insects. Annu Rev Entomol 60:453–472. https://doi.org/10.1146/annurev-ento-010814-021045

Lee K-S, Kwon OY, Lee JH, Kwon K, Min K-J, Jung S-A et al (2008) *Drosophila* short neuropeptide F signalling regulates growth by ERK-mediated insulin signalling. Nat Cell Biol 10:468–475. https://doi.org/10.1038/ncb1710

Lee KS, Hong SH, Kim AK, Ju SK, Kwon OY, Yu K (2009) Processed short neuropeptide F peptides regulate growth through the ERK-insulin pathway in *Drosophila melanogaster*. FEBS Lett 583:2573–2577. https://doi.org/10.1016/j.febslet.2009.07.024

Li Q, Gong Z (2015) Cold-sensing regulates *Drosophila* growth through insulin-producing cells. Nat Commun 6:10083. https://doi.org/10.1038/ncomms10083

Li Z, Qian W, Song W, Zhao T, Yang Y, Wang W et al (2022) A salivary gland-secreted peptide regulates insect systemic growth. Cell Rep 38:110397. https://doi.org/10.1016/j.celrep.2022.110397

Lin XY, Smagghe G (2019) Roles of the insulin signaling pathway in insect development and organ growth. Peptides 122:169923. https://doi.org/10.1016/j.peptides.2018.02.001

Ling L, Raikhel AS (2021) Cross-talk of insulin-like peptides, juvenile hormone, and 20-hydroxyecdysone in regulation of metabolism in the mosquito *Aedes aegypti*. Proc Natl Acad Sci USA 118:e2023470118. https://doi.org/10.1073/pnas.2023470118

Liu Y, Liao S, Veenstra JA, Nässel DR (2016) *Drosophila* insulin-like peptide 1 (DILP1) is transiently expressed during non-feeding stages and reproductive dormancy. Sci Rep 6:26620. https://doi.org/10.1038/srep26620

Meschi E, Delanoue R (2021) Adipokine and fat body in flies: connecting organs. Mol Cell Endocrinol 533:111339. https://doi.org/10.1016/j.mce.2021.111339

Meschi E, Léopold P, Delanoue R (2019) An EGF-responsive neural circuit couples insulin secretion with nutrition in *Drosophila*. Dev Cell 48:76–86.e5. https://doi.org/10.1016/j.devcel.2018.11.029

Miguel-Aliaga I, Thor S, Gould AP (2008) Postmitotic specification of *Drosophila* insulinergic neurons from pioneer neurons. PLoS Biol 6:e58. https://doi.org/10.1371/journal.pbio.0060058

Miki T, Shinohara T, Chafino S, Tomioka K (2020) Photoperiod and temperature separately regulate nymphal development through JH and insulin/TOR signaling pathways in an insect. Proc Natl Acad Sci USA 117:5525–5531. https://doi.org/10.1073/pnas.1922747117

Millington JW, Brownrigg GP, Chao C, Sun Z, Basner-Collins PJ, Wat LW et al (2021) Female-biased upregulation of insulin pathway activity mediates the sex difference in *Drosophila* body size plasticity. eLife 10:e58341. https://doi.org/10.7554/eLife.58341

Mizoguchi A, Okamoto N (2013) Insulin-like and IGF-like peptides in the silkmoth *Bombyx mori*: discovery, structure, secretion, and function. Front Physiol 4:217. https://doi.org/10.3389/fphys.2013.00217

Mizoguchi A, Ishizaki H, Nagasawa H, Kataoka H, Isogai A, Tamura S et al (1987) A monoclonal antibody against a synthetic fragment of bombyxin (4K-prothoracicotropic hormone) from the silkmoth, *Bombyx mori*: characterization and immunohistochemistry. Mol Cell Endocrinol 51:227–235. https://doi.org/10.1016/0303-7207(87)90032-3

Moeller ME, Nagy S, Gerlach SU, Soegaard KC, Danielsen ET, Texada MJ et al (2017) Warts signaling controls organ and body growth through regulation of ecdysone. Curr Biol 27:1652–1659.e4. https://doi.org/10.1016/j.cub.2017.04.048

Murphy CT, Hu PJ (2013) Insulin/insulin-like growth factor signaling in *C. elegans*. WormBook 26:1–43. https://doi.org/10.1895/wormbook.1.164.1

Myers EW, Sutton GG, Delcher AL, Dew IM, Fasulo DP, Flanigan MJ et al (2000) A whole-genome assembly of *Drosophila*. Science 287:2196–2204. https://doi.org/10.1126/science.287.5461.2196

Nagasawa H, Kataoka H, Isogai A, Tamura S, Suzuki A, Ishizaki H et al (1984) Amino-terminal amino acid sequence of the silkworm prothoracicotropic hormone: homology with insulin. Science 226:1344–1346. https://doi.org/10.1126/science.226.4680.1344

Nagasawa H, Kataoka H, Isogai A, Tamura S, Suzuki A, Mizoguchi A et al (1986) Amino acid sequence of a prothoracicotropic hormone of the silkworm *Bombyx mori*. Proc Natl Acad Sci USA 83:5840–5843. https://doi.org/10.1073/pnas.83.16.5840

Nakae J, Kido Y, Accili D (2001) Distinct and overlapping functions of insulin and IGF-I receptors. Endocr Rev 22:818–835. https://doi.org/10.1210/edrv.22.6.0452

Nakahara Y, Matsumoto H, Kanamori Y, Kataoka H, Mizoguchi A, Kiuchi M et al (2006) Insulin signaling is involved in hematopoietic regulation in an insect hematopoietic organ. J Insect Physiol 52:105–111. https://doi.org/10.1016/j.jinsphys.2005.09.009

Nässel DR, Vanden Broeck J (2016) Insulin/IGF signaling in *Drosophila* and other insects: factors that regulate production, release and post-release action of the insulin-like peptides. Cell Mol Life Sci 73:271–290. https://doi.org/10.1007/s00018-015-2063-3

Nijhout HF, Grunert LW (2002) Bombyxin is a growth factor for wing imaginal disks in Lepidoptera. Proc Natl Acad Sci USA 99:15446–15450. https://doi.org/10.1073/pnas.242548399

Nijhout HF, Smith WA, Schachar I, Subramanian S, Tobler A, Grunert LW (2007) The control of growth and differentiation of the wing imaginal disks of *Manduca sexta*. Dev Biol 302:569–576. https://doi.org/10.1016/j.ydbio.2006.10.023

Nijhout HF, Riddiford LM, Mirth C, Shingleton AW, Suzuki Y, Callier V (2013) The developmental control of size in insects. WIRE Dev Biol 3:113–134. https://doi.org/10.1002/wdev.124

Oh Y, Lai JS-Y, Mills HJ, Erdjument-Bromage H, Giammarinaro B, Saadipour K et al (2019) A glucose-sensing neuron pair regulates insulin and glucagon in *Drosophila*. Nature 574:559–564. https://doi.org/10.1038/s41586-019-1675-4

Ohhara Y, Yamanaka N (2022) Internal sensory neurons regulate stage-specific growth in *Drosophila*. Development 149:dev200440. https://doi.org/10.1242/dev.200440

Okada Y, Katsuki M, Okamoto N, Fujioka H, Okada K (2019) A specific type of insulin-like peptide regulates the conditional growth of a beetle weapon. PLoS Biol 17:e3000541. https://doi.org/10.1371/journal.pbio.3000541

Okamoto N, Nishimura T (2015) Signaling from glia and cholinergic neurons controls nutrient-dependent production of an insulin-like peptide for *Drosophila* body growth. Dev Cell 35:295–310. https://doi.org/10.1016/j.devcel.2015.10.003

Okamoto N, Yamanaka N (2015) Nutrition-dependent control of insect development by insulin-like peptides. Curr Opin Insect Sci 11:21–30. https://doi.org/10.1016/j.cois.2015.08.001

Okamoto N, Yamanaka N, Satake H, Saegusa H, Kataoka H, Mizoguchi A (2009a) An ecdysteroid-inducible insulin-like growth factor-like peptide regulates adult development of the silkmoth *Bombyx mori*. FEBS J 276:1221–1232. https://doi.org/10.1111/j.1742-4658.2008.06859.x

Okamoto N, Yamanaka N, Yagi Y, Nishida Y, Kataoka H, O'Connor MB et al (2009b) A fat body-derived IGF-like peptide regulates postfeeding growth in *Drosophila*. Dev Cell 17:885–891. https://doi.org/10.1016/j.devcel.2009.10.008

Okamoto N, Nishimori Y, Nishimura T (2012) Conserved role for the Dachshund protein with *Drosophila* Pax6 homolog eyeless in insulin expression. Proc Natl Acad Sci USA 109:2406–2411. https://doi.org/10.1073/pnas.1116050109

Okamoto N, Nakamori R, Murai T, Yamauchi Y, Masuda A, Nishimura T (2013) A secreted decoy of InR antagonizes insulin/IGF signaling to restrict body growth in *Drosophila*. Genes Dev 27: 87–97. https://doi.org/10.1101/gad.204479.112

Rajan A, Perrimon N (2012) *Drosophila* cytokine unpaired 2 regulates physiological homeostasis by remotely controlling insulin secretion. Cell 151:123–137. https://doi.org/10.1016/j.cell.2012.08.019

Redhai S, Pilgrim C, Gaspar P, Giesen LV, Lopes T, Riabinina O et al (2020) An intestinal zinc sensor regulates food intake and developmental growth. Nature 580:263–268. https://doi.org/10.1038/s41586-020-2111-5

Riehle MA, Fan Y, Cao C, Brown MR (2006) Molecular characterization of insulin-like peptides in the yellow fever mosquito, *Aedes aegypti*: expression, cellular localization, and phylogeny. Peptides 27:2547–2560. https://doi.org/10.1016/j.peptides.2006.07.016

Rinderknecht E, Humbel RE (1978) The amino acid sequence of human insulin-like growth factor I and its structural homology with proinsulin. J Biol Chem 253:2769–2776. https://doi.org/10.1016/S0021-9258(17)40889-1

Robertson FW (1966) The ecological genetics of growth in *Drosophila*. 8. Adaptation to a new diet. Genet Res 8:165–179. https://doi.org/10.1017/S0016672300010028

Rodenfels J, Lavrynenko O, Ayciriex S, Sampaio JL, Carvalho M, Shevchenko A et al (2014) Production of systemically circulating Hedgehog by the intestine couples nutrition to growth and development. Genes Dev 28:2636–2651. https://doi.org/10.1101/gad.249763.114

Rulifson EJ, Kim SK, Nusse R (2002) Ablation of insulin-producing neurons in flies: growth and diabetic phenotypes. Science 296:1118–1120. https://doi.org/10.1126/science.1070058

Saltiel AR, Kahn CR (2001) Insulin signaling and the regulation of glucose and lipid metabolism. Nature 414:799–806. https://doi.org/10.1038/414799a

Sander M, Neubüser A, Kalamaras J, Ee HC, Martin GR, German MS (1997) Genetic analysis reveals that PAX6 is required for normal transcription of pancreatic hormone genes and islet development. Genes Dev 11:1662–1673. https://doi.org/10.1101/gad.11.13.1662

Sano H, Nakamura A, Texada MJ, Truman JW, Ishimoto H, Kamikouchi A et al (2015) The nutrient-responsive hormone CCHamide-2 controls growth by regulating insulin-like peptides in the brain of *Drosophila melanogaster*. PLoS Genet 11:e1005209. https://doi.org/10.1371/journal.pgen.1005209

Semaniuk U, Strilbytska O, Malinovska K, Storey KB, Vaiserman A, Lushchak V et al (2021) Factors that regulate expression patterns of insulin-like peptides and their association with physiological and metabolic traits in *Drosophila*. Insect Biochem Mol Biol 135:103609. https://doi.org/10.1016/j.ibmb.2021.103609

Shin SC, Kim SH, You H, Kim B, Kim AC, Lee KA et al (2011) *Drosophila* microbiome modulates host developmental and metabolic homeostasis via insulin signaling. Science 334:670–674. https://doi.org/10.1126/science.1212782

Shingleton AW, Frankino WA, Flatt T, Nijhout HF, Emlen DJ (2007) Size and shape: the developmental regulation of static allometry in insects. BioEssays 29:536–548. https://doi.org/10.1002/bies.20584

Sim C, Denlinger DL (2013) Insulin signaling and the regulation of insect diapause. Front Physiol 4:189. https://doi.org/10.3389/fphys.2013.00189

Slaidina M, Delanoue R, Gronke S, Partridge L, Léopold P (2009) A *Drosophila* insulin-like peptide promotes growth during nonfeeding states. Dev Cell 17:874–884. https://doi.org/10.1016/j.devcel.2009.10.009

Smýkal V, Pivarči M, Provazník J, Bazalová O, Jedlička P, Lukšan O et al (2020) Complex evolution of insect insulin receptors and homologous decoy receptors, and functional significance of their multiplicity. Mol Biol Evol 37:1775–1789. https://doi.org/10.1093/molbev/msaa048

Sousa-Nunes R, Yee LL, Gould AP (2011) Fat cells reactivate quiescent neuroblasts via TOR and glial insulin relays in *Drosophila*. Nature 471:508–513. https://doi.org/10.1038/nature09867

Stewart CE, Rotwein P (1996) Growth, differentiation, and survival: multiple physiological functions for insulin-like growth factors. Physiol Rev 76:1005–1026. https://doi.org/10.1152/physrev.1996.76.4.1005

Storelli G, Defaye A, Erkosar B, Hols P, Royet J, Leulier F (2011) *Lactobacillus plantarum* promotes *Drosophila* systemic growth by modulating hormonal signals through TOR-dependent nutrient sensing. Cell Metab 14:403–414. https://doi.org/10.1016/j.cmet.2011.07.012

Sung EJ, Ryuda M, Matsumoto H, Uryu O, Ochiai M, Cook ME et al (2017) Cytokine signaling through *Drosophila* Mthl10 ties lifespan to environmental stress. Proc Natl Acad Sci USA 114:13786–13791. https://doi.org/10.1073/pnas.1712453115

Tang HY, Smith-Caldas MSB, Driscoll MV, Salhadar S, Shingleton AW (2011) FOXO regulates organ-specific phenotypic plasticity in *Drosophila*. PLoS Genet 7:e1002373. https://doi.org/10.1371/journal.pgen.1002373

Tatar M, Kopelman A, Epstein D, Tu M-P, Yin C-M, Garofalo R (2001) A mutant *Drosophila* insulin receptor homolog that extends life-span and impairs neuroendocrine function. Science 292:107–110. https://doi.org/10.1126/science.1057987

Tatar M, Bartke A, Antebi A (2003) The endocrine regulation of aging by insulin-like signals. Science 299:1346–1351. https://doi.org/10.1126/science.1081447

Teleman AA (2010) Molecular mechanisms of metabolic regulation by insulin in *Drosophila*. Biochem J 425:13–26. https://doi.org/10.1042/BJ20091181

Tennessen JM, Thummel CS (2011) Coordinating growth and maturation—insights from *Drosophila*. Curr Biol 21:R750–R757. https://doi.org/10.1016/j.cub.2011.06.033

Texada MJ, Jørgensen AF, Christensen CF, Koyama T, Malita A, Smith DK et al (2019a) A fat-tissue sensor couples growth to oxygen availability by remotely controlling insulin secretion. Nat Commun 10:1–16. https://doi.org/10.1038/s41467-019-09943-y

Texada MJ, Malita A, Christensen CF, Dall KB, Faergeman NJ, Nagy S et al (2019b) Autophagy-mediated cholesterol trafficking controls steroid production. Dev Cell 48:659–671.e4. https://doi.org/10.1016/j.devcel.2019.01.007

Texada MJ, Koyama T, Rewitz K (2020) Regulation of body size and growth control. Genetics 216: 269–313. https://doi.org/10.1534/genetics.120.303095

Texada MJ, Lassen M, Pedersen LH, Koyama T, Malita A, Rewitz K (2022) Insulin signaling couples growth and early maturation to cholesterol intake in *Drosophila*. Curr Biol 32:1548–1562.e6. https://doi.org/10.1016/j.cub.2022.02.021

Thissen JP, Ketelslegers JM, Underwood LE (1994) Nutritional regulation of the insulin-like growth factors. Endocr Rev 15:80–101. https://doi.org/10.1210/edrv-15-1-80

Vallejo DM, Juarez-Carreño S, Bolivar J, Morante J, Dominguez M (2015) A brain circuit that synchronizes growth and maturation revealed through Dilp8 binding to Lgr3. Science 350: aac6767. https://doi.org/10.1126/science.aac6767

Veenstra JA (2010) Neurohormones and neuropeptides encoded by the genome of *Lottia gigantea*, with reference to other mollusks and insects. Gen Comp Endocrinol 167:86–103. https://doi.org/10.1016/j.ygcen.2010.02.010

Veenstra JA (2016) Similarities between decapod and insect neuropeptidomes. PeerJ 4:e2043. https://doi.org/10.7717/peerj.2043

Veenstra JA (2020a) Arthropod IGF, relaxin and gonadulin, putative orthologs of *Drosophila* insulin-like peptides 6, 7 and 8, likely originated from an ancient gene triplication. PeerJ 8: e9534. https://doi.org/10.7717/peerj.9534

Veenstra JA (2020b) Gonadulins, the fourth type of insulin-related peptides in decapods. Gen Comp Endocrinol 296:113528. https://doi.org/10.1016/j.ygcen.2020.113528

Veenstra JA, Rombauts S, Grbic M (2012) In silico cloning of genes encoding neuropeptides, neurohormones and their putative G-protein coupled receptors in a spider mite. Insect Biochem Mol Biol 42:277–295. https://doi.org/10.1016/j.ibmb.2011.12.009

Vogal KJ, Brown MR, Strand MR (2015) Ovary ecdysteroidogenic hormone requires a receptor tyrosine kinase to activate egg formation in the mosquito *Aedes aegypti*. Proc Natl Acad Sci USA 112:5057–5062. https://doi.org/10.1073/pnas.1501814112

Wang S, Tulina N, Carlin DL, Rulifson EJ (2007) The origin of islet-like cells in *Drosophila* identifies parallels to the vertebrate endocrine axis. Proc Natl Acad Sci USA 104:19873–19878. https://doi.org/10.1073/pnas.0707465104

Wheeler DE, Buck N, Evans JD (2006) Expression of insulin pathway genes during the period of caste determination in the honey bee. *Apis mellifera*. Insect Mol Biol 15:597–602. https://doi.org/10.1111/imb.12065

Winant M, Buhler K, Clements J, De Groef S, Hens K, Vulsteke V, Callaerts P (2022) Genome-wide analysis identifies Homothorax and Extradenticle as regulators of insulin in *Drosophila* insulin-producing cells. PLoS Genet 18:e1010380. https://doi.org/10.1371/journal.pgen.1010380

Xu HJ, Xue J, Lu B, Zhang XC, Zhuo JC, He SF et al (2015) Two insulin receptors determine alternative wing morphs in planthoppers. Nature 519:464–467. https://doi.org/10.1038/nature14286

Yoshinari Y, Kosakamoto H, Kamiyama T, Hoshino R, Matsuoka R, Kondo S et al (2021) The sugar-responsive enteroendocrine neuropeptide F regulates lipid metabolism through glucagon-like and insulin-like hormones in *Drosophila melanogaster*. Nat Commun 12:4818. https://doi.org/10.1038/s41467-021-25146-w

Zhang CX, Brisson JA, Xu HJ (2019) Molecular mechanisms of wing polymorphism in insects. Annu Rev Entomol 64:297–314. https://doi.org/10.1146/annurev-ento-011118-112448

Chapter 8
Regulation of Diapause

Kunihiro Shiomi

Abstract Many insect species enter diapause to ensure survival under unfavorable environmental conditions. Recent molecular biology approaches have revealed details of the molecular mechanisms of diapause processes. The initiation and termination of diapause are mediated by ecdysteroids, juvenile hormone (JH), and diapause hormone (DH). Various factors, including neuropeptides involved in the secretion of these hormones in upstream and downstream pathways, have been identified. These findings indicate that diapause involves hierarchical signaling pathways. This chapter provides an overview of the hormones and neuropeptides involved in diapause at each developmental stage, including the egg (embryo), larva, pupa, and adult.

Keywords Diapause hormone (DH) · Prothoracicotropic hormone (PTTH) · Diapause · Seasonal adaptation · *Bombyx mori*

8.1 Introduction

To ensure survival under unfavorable environmental conditions and to synchronize the developmental stage of populations, many insect species enter diapause, which is a seasonal adaptation that evolves as a specific subtype of dormancy (Denlinger 2022; Kostal 2011). In addition, there may be sufficient variation within natural populations to allow selection to modify the diverse features of the diapause response. Thus, diapause can be seen as a bet-hedging strategy in addition to the synchronization of reproductive opportunities as a seasonal adaptation (Masaki 1980).

Diapause is not a passive response to changing conditions, but rather an actively induced state that precedes adverse natural conditions, and the end of diapause does not need to coincide with the end of adversity (Tauber et al. 1986; Danks 1987).

K. Shiomi (✉)
Division of Applied Biology, Faculty of Textile Science & Technology, Shinshu University, Nagano, Japan
e-mail: shiomi@shinshu-u.ac.jp

A. Mizoguchi, N. Okamoto (eds.), *Insect Peptide Hormones and Neuropeptides*,
Entomology Monographs, https://doi.org/10.1007/978-981-95-7510-7_8

Therefore, diapause is different from simple quiescence, which is an immediate response to a decline in any limiting environmental factor(s) below physiological thresholds with immediate resumption of the processes if the factor(s) rise above them (Kostal 2006). Diapause is primarily divided into the pre-diapause, diapause, and post-diapause processes. The events in between are far from static but instead reflect a dynamic progression from the establishment of developmental arrest (induction and preparation phases) to the maintenance of the arrested state (initiation, maintenance, and termination phases) and finally to the transition that ends diapause (postdiapause quiescence phase).

The diapause phenotypes in morphological, physiological, behavioral, and reproductive responses, also known as diapause syndrome, improve cold/stress tolerance in later life stages or progeny (Tauber et al. 1986). Diapausing insects arrest normal development and activate a unique metabolic system called "diapause development," which enables them to survive harsh environments such as extreme cold or heat, with changes in energy metabolism, storage, and responsiveness to the environment (Hahn and Denlinger 2011; Kostal 2006).

Each insect species has an inherent developmental stage for diapause induction. Depending on the species, egg (embryo), larval, pupal, and adult (reproductive) diapauses occur at specific developmental stages. Furthermore, whether or not a species enters diapause can be determined genetically or by environmental conditions, referred to as obligatory or facultative diapause, respectively (Denlinger et al. 2012). In facultative diapause, various environmental conditions, including day length (photoperiod), temperature, humidity, and nutritional conditions, determine diapause, and changes in these conditions determine the subsequent developmental fate (diapause or non-diapause) of the individual or its progeny. In addition, diapause can be induced by plant-derived volatile substances, pheromones, and population density (Denlinger et al. 2012). Furthermore, diapause is not determined by a single environmental condition but is modified (overwritten) and induced by complex environmental factors, such as day length, temperature, and nutritional conditions (Denlinger et al. 2012; Tauber et al. 1986). Seasonal environmental changes involve fluctuations in multiple environmental factors. Seasonal polymorphisms, predetermined by a single environmental factor, are subsequently overwritten and induced by multiple environmental factors that change daily. This is thought to have evolved as an ingenious and robust life-history strategy for organisms via the neuroendocrine system.

A conceptual model for the induction of diapause has been proposed (Tauber et al. 1986), which suggests that environmental information is first received by sensor molecules and then stored for a relatively long period, integrated, and translated into the neuroendocrine system, which then induces plastic changes in the phenotypes of diapause syndrome by regulating hormone secretion. The brain is central to both the reception of photoperiodic cues and the translation of this information into a meaningful developmental program that prompts the initiation and termination of diapause (Denlinger 2022).

The endocrine system finely orchestrates the progression of development from the embryonic stage through larval and pupal development to adulthood and finally to reproductive maturation (see Chap. 6). The halt in the developmental characteristics due to diapause implies a blockade in the endocrine signaling pathway that, in time, is countered to reinitiate development. Development is regulated by the progressive increase and decrease in the titers of ecdysteroids and juvenile hormone (JH), and the synthesis and release of these growth-regulating hormones are regulated by upstream neuropeptides. Both the rise and decline in hormone titers are critical for developmental progression. Development can be halted by either failing to elevate a specific hormone or maintaining a high hormone titer, that is, by preventing its decline. Both mechanisms have been observed in certain diapauses (Denlinger 2022).

8.2 Embryonic and Pharate Larval Diapause

Embryonic diapause occurs at diverse stages of embryogenesis in various insects. This suggests that diverse mechanisms control embryonic diapause. Diapause hormone (DH), a neuropeptide, has been extensively studied as an embryonic diapause-inducing hormone.

8.2.1 Diapause Hormone

The domestic silkworm *Bombyx mori* is a typical insect that ceases its normal development during early embryogenesis and undergoes metabolic changes during diapause (Yamashita and Hasegawa 1985; Shiomi et al. 2015). The development of diapause-destined embryos is arrested during the G2 cell cycle stage immediately after the formation of the cephalic lobe and telson and sequential segmentation of the mesoderm and ectoderm (Nakagaki et al. 1991). Bivoltine strains of *B. mori* that have two generations annually show a facultative diapause phenotype, which is determined by environmental conditions such as photoperiod, temperature, and nutrient conditions during embryonic and larval development in the mother's generation (Yamashita and Hasegawa 1985). In general, the progeny diapause of bivoltine strains is predominantly determined by environmental conditions during maternal embryonic development, with long days and high temperatures inducing embryonic diapause (Yamashita and Hasegawa 1985; Kogure 1933; Watanabe 1924). For the bivoltine Kosetsu strain, the temperature signal during the mother's embryonic development is the predominant factor affecting diapause determination, regardless of the photoperiod during embryonic and larval development, when reared on a certain artificial diet (Shiomi et al. 2015; Tsuchiya et al. 2021). When eggs are incubated at 25 °C under continuous darkness, the resultant female moths lay nearly 100% diapause eggs. In contrast, incubation of eggs at 15 °C in the dark conditions results in moths laying nearly 100% non-diapause eggs. Thus, Kosetsu is considered to be a typical strain that is susceptible to temperature-dependent diapause induction.

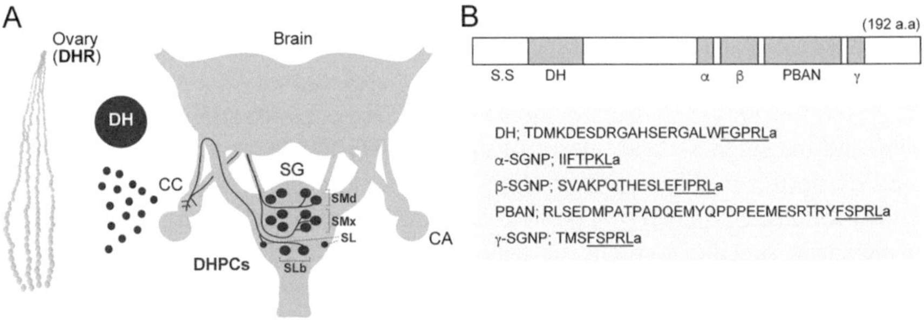

Fig. 8.1 Neuropeptide hormone and neurosecretory cells involved in diapause induction of the silkworm, *Bombyx mori*. (**a**) *DH-PBAN* encodes a polyprotein precursor containing diapause hormone (DH) and is exclusively expressed in seven pairs of neurosecretory cells, the DH-PBAN-producing neurosecretory cells (DHPCs), in the subesophageal ganglion (SG). DHPCs consist of three neuromeres: four mandibular cells [SMd], six maxillary cells [SMx], two labial cells [SLb]) along the ventral midline, and the lateral cells (two cells of lateral [SL]). Axons of DHPCs are represented in the hemisphere. (**b**) Schematic drawing of the biosynthesis of DH; α-, ß-, γ-SGNP; and PBAN neuropeptides in *Bombyx* SG. DH-PBAN cDNA encoding preprohormone undergoes post-translational processing via a series of enzymatic steps that cleave and further modify intermediate peptide substrates to yield the signal sequence (SS); DH; α-, ß-, and γ-SGNP; and PBAN. DH; α-, ß-, and γ-SGNP; and PBAN neuropeptides belong to the FXPRLa peptide family that contains Phe-X-Pro-Arg (Lys)-Leu-NH2 at the C-terminus (underlined)

The molecular mechanisms underlying the induction of embryonic diapause in *B. mori* have been investigated extensively (Yamashita 1996; Denlinger et al. 2012). Embryonic diapause is induced by the diapause hormone (DH) signaling pathway, which consists of a highly sensitive and specific interaction between DH and the DH receptor (DHR) during pupal-adult development (Shiomi et al. 2015) (Fig. 8.1a). DH is a 24-amino-acid peptide that is post-translationally processed from the precursor polypeptide DH-PBAN. DH belongs to the FXPRLamide peptide family (see Sect. 2.3.13), which also includes pheromone biosynthesis-activating neuropeptides (PBAN), myotropins, and pyrokinins (Holman et al. 1986; Imai et al. 1991; Kitamura et al. 1989; Raina et al. 1989; Schoofs et al. 1990). *DH-PBAN* encodes a polyprotein precursor containing DH, PBAN, and α-, ß-, and γ-SGNP, which are exclusively expressed in seven pairs of neurosecretory cells, known as the DH-PBAN-producing neurosecretory cells (DHPCs), in the subesophageal ganglion (SG) (Sato et al. 1993, 1994, 1998; Shiomi et al. 2003) (Fig. 8.1b). DH is released into the hemolymph from the corpus cardiacum (CC) (Sato et al. 1998), the major release site for neuropeptide hormones, and acts on DHR in the ovary (Homma et al. 2006). The cerebral γ-aminobutyric acid (GABA)ergic and corazonin (CRZ) pathways modulate DH release during temperature-dependent expression of the plasma membrane GABA transporter (GAT) (Tsuchiya et al. 2021) by controlling clock genes (Fig. 8.2) (Homma et al. 2022). Furthermore, *B. mori* transient receptor potential ankyrin 1 (BommoTRPA1) acts as a thermosensitive channel activated at temperatures above ~21 °C and affects diapause induction through DH release. Thus, BommoTRPA1 acts as a molecular switch for temperature-dependent diapause induction (Sato et al. 2014).

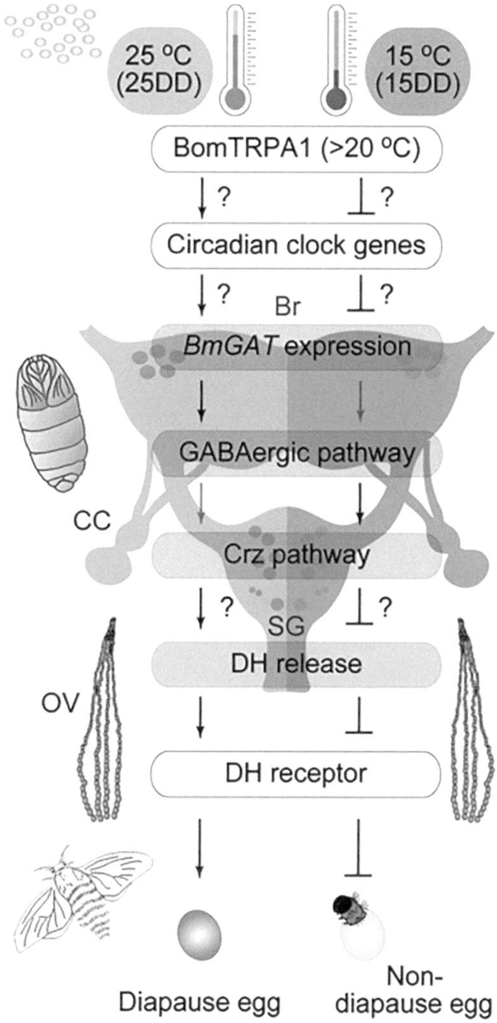

Fig. 8.2 Schematic representation of putative hierarchical pathway in diapause induction. BomTRPA1 acts as a molecular switch for temperature-dependent diapause induction in Kosetsu strain of *B. mori*. GABAergic and corazonin (CRZ) pathways modulate DH release during temperature-dependent expression of the plasma membrane GABA transporter (GAT) via the control of clock genes. TRPA1 Thermosensitive channel A1, Br brain, CC corpus cardiacum, SG subesophageal ganglion, OV ovary

B. mori was domesticated from the wild mulberry silkworm *B. mandarina* approximately 5000–10,000 years ago in China (Goldsmith et al. 2005; Sasaki 1898; Yoshitake 1968; Yukuhiro et al. 2002). *B. mori* and *B. mandarina* are morphologically, cytologically, and physiologically similar, copulation between them is possible, and the resultant progeny are never sterile (Miyata 1970; Murakami and Imai 1974; Yokoyama 2019), although both species are genetically differentiated (Xia et al. 2009). In addition, *B. mandarina* enters embryonic diapause at a developmental stage similar to that of *B. mori* (Umeya 1946; Yamashita and Yaginuma 1991). Embryonic diapause in *B. mandarina* is photoperiod dependent, although the DH signaling pathway and thermal sensitivity of TRPA1 are conserved in both species. Interestingly, in the *BommoTRPA1* knockout mutant of the Kosetsu strain, temperature-dependent induction of diapause was replaced by photoperiodic induction (Yokoyama et al. 2021). Based on these findings, it was hypothesized that BommoTRPA1-activated signals are strongly linked to the signaling pathway participating in diapause induction in Kosetsu to selectively utilize temperature information as a cue. Given that thermoregulation is easier for humans than the photoperiodic control, it is likely that the temperature-dependent diapause induction phenotype was artificially selected and spread throughout *B. mori* populations during ancient domestication.

The white-spotted tussock moth *Orgyia thyellina* (family Lymantriidae) also appears to use DH to induce embryonic diapause. Interestingly, in this species, all four FXPRLa neuropeptides encoded in the same reading frame (Fig. 8.1b) are capable of eliciting the diapause response (Uehara et al. 2011). It was revealed that not only does *Orgyia* DH induce embryonic diapause and seasonal polyphenism but also participate in the hypertrophy of follicles and ovaries. DH may also contribute to embryonic diapause induction in insects other than Lepidoptera, as noted in the migratory locust *Locusta migratoria* (Hao et al. 2019).

8.2.2 Ecdysteroids and Juvenile Hormone

Ecdysteroids are involved in the maintenance and termination of embryonic diapause in silkworms. The amount of 20-hydroxyecdysone (20E) in the diapause eggs of *B. mori* is lower than that in non-diapause eggs (Sonobe and Yamada 2004). Diapause can be avoided by injecting 20E into diapause eggs (Makka et al. 2002), and the addition of 20E to culture systems promotes the development of diapausing embryos (Fujiwara et al. 2006a). Furthermore, activation of extracellular signal-regulated kinase (ERK), a member of the mitogen-activated protein kinase (MAPK) family, induces the expression of various genes, leading to the termination of diapause in silkworm eggs (Fujiwara et al. 2006a, b). In the eggs of *L. migratoria*, diapause is thought to be maintained by a lack of ecdysteroids, as in *B. mori* (Tawfik et al. 2002). The concentrations of ecdysteroids are lower in locust diapause eggs than those in non-diapause eggs. Locusts enter diapause during mid-embryonic development, and diapause is broken not only by ecdysteroids but also by JH treatment (Kidokoro et al. 2006). It has been suggested that JH treatment breaks diapause indirectly by promoting ecdysteroid secretion from the prothoracic glands (PG). The diapause fate of *L. migratoria* embryos also appears to be influenced by reactive oxygen species (ROS) and FoxO phosphorylation downstream of the insulin signaling pathway (Chen et al. 2020).

On the other hand, excessive secretion of ecdysteroids has been suggested to induce diapause in the gypsy moth *Lymantria dispar*. *L. dispar* enters diapause as pharate first-instar larvae, which are fully developed but remain encased within the chorion of the egg. KK-42, an imidazole derivative that is thought to block ecdysteroid synthesis, averts diapause in *L. dispar* (Suzuki et al. 1993; Lee and Denlinger 1996), and diapause can be reinstated by ecdysteroid administration (Lee and Denlinger 1997). These results suggest that diapause in *L. dispar* is induced and maintained by high titers of ecdysteroids. It is well known that a drop in the ecdysteroid titer is an essential component of the signal for permitting the progression of late embryonic development (Sláma 1980; Žďárek and Denlinger 1987). The pharate first-instar larval diapause of the mosquito *Aedes albopictus* appears to be induced by the absence of juvenile hormones (Batz et al. 2019), because although the titers of ecdysteroids present in diapausing and non-diapausing eggs do not differ, the abundance of JH I is considerably lower at diapause initiation.

8.2.3 Diapause-Promoting Factor

The wild silkmoth *Antheraea yamamai* enters diapause during the pharate first instar larval stage (Suzuki et al. 1990). A factor has been identified as a candidate that may function in maintaining diapause. This is a small amidated peptide with the sequence DILRG-NH_2 and is called Yamamarin (Yang et al. 2007). However, the mechanisms by which Yamamarin regulates diapause remains unknown.

8.3 Larval and Pupal Diapause

Locking of a larva or pupa in developmental arrest implies that the hormonal signal for advancement to the next stage has not yet been received. This suggests that activation of the PG to produce ecdysteroids is required to transition to the next developmental stage. Indeed, a shutdown in the brain-PG axis is typically a conspicuous feature of larval and pupal diapause. Whether prothoracicotropic hormone (PTTH) synthesis or release is regulated varies between species. However, the effect of the shutdown of PTTH secretion on ecdysteroid synthesis is consistent. Numerous reports have documented low or undetectable ecdysteroid titers in larvae or pupae during diapause and their elevation at diapause termination. JH can also play an important role, particularly in larvae that undergo stationary ecdysis.

8.3.1 PTTH and Ecdysteroid in Pupal Diapause

The most widely studied form of insect diapause is pupal diapause, and research on this subject has a long history since Williams' extensive works with diapausing pupae of the giant silk moth *Hyalophora cecropia* (Williams 1946, 1952). Based on his works, it was hypothesized that diapause is induced by the inactivation of the PG due to the cessation of brain hormone secretion and the resulting lack of the secretion of molting hormone. At that time, the chemical nature of both the brain hormone and the molting hormone was unknown, but this hypothesis was derived through ingenious organ removal and transplantation experiments. This hypothesis is widely supported in other insects. The brain hormone was renamed PTTH, and the molting hormone was renamed ecdysteroid (see Chap. 1).

With recent molecular advances in insect hormone research, it is possible to examine the role of PTTH with greater precision, as has been elegantly demonstrated for pupal diapause (winter diapause) in the armyworm *Mamestra brassicae* (Mizoguchi et al. 2013). When *M. brassicae* larvae are reared under long-day conditions, they do not enter diapause, whereas diapause is induced under short-day conditions. Under long-day conditions, a large increase in PTTH concentration in the hemolymph was observed after pupation, followed by a major surge in ecdysteroids (Fig. 8.3). In contrast, little PTTH or ecdysteroids were detected in the hemolymph of individuals destined for pupal diapause. Furthermore, when PTTH was injected into pupae reared under short-day conditions immediately after pupation, the pupae did not enter diapause and continued to develop into adults. Therefore, the cessation of PTTH secretion immediately after pupation is a necessary and sufficient condition for the initiation of diapause, consistent with expectations from earlier models for pupal diapause. In this species, the PTTH content in the brains of diapausing pupae was higher than that in non-diapausing pupae, indicating that the diapausing pupae have sufficient PTTH, but it is not released.

Diapause in *M. brassicae* pupae lasts for several months at approximately 25 °C, but if they experience low temperatures (approximately 4 °C) for 7 weeks, diapause is broken, and if the temperature is high enough, they begin to develop into adults. The development from pupae to adults also requires PTTH secretion. At this time, the hemolymph PTTH concentration is only approximately one-third of that during pupal-adult development of non-diapausing pupae. However, the PTTH responsiveness of the PG is significantly higher, suggesting that even a small amount of PTTH is sufficient to activate the PG after diapause (Yamada and Mizoguchi 2017).

Interestingly, *M. brassicae* also enters summer diapause in some strains, in which winter diapause is induced by photoperiods with a light period shorter than 13 h, and adult emergence is observed uniformly several months later. In contrast, summer diapause is induced by photoperiods with a light period longer than 14.5 h, and adult emergence occurs asynchronously. That is, non-diapause occurs only under photoperiods with a light period of a narrow range of 13–14.5 h (Masaki 1980; Grüner and Masaki 1994). One of the most intriguing endocrinological problems is the mechanism that distinguishes summer from winter diapause. As inferred from their seasonal incidence, the two types of diapause differ in their environmental requirements for induction, maintenance, and termination, and there may be endocrinological differences. It has been reported that the ecdysteroid titers in summer diapause pupae are intermediate between the high levels in developing pupae and the low levels in winter diapause pupae (Maslennikova et al. 1976).

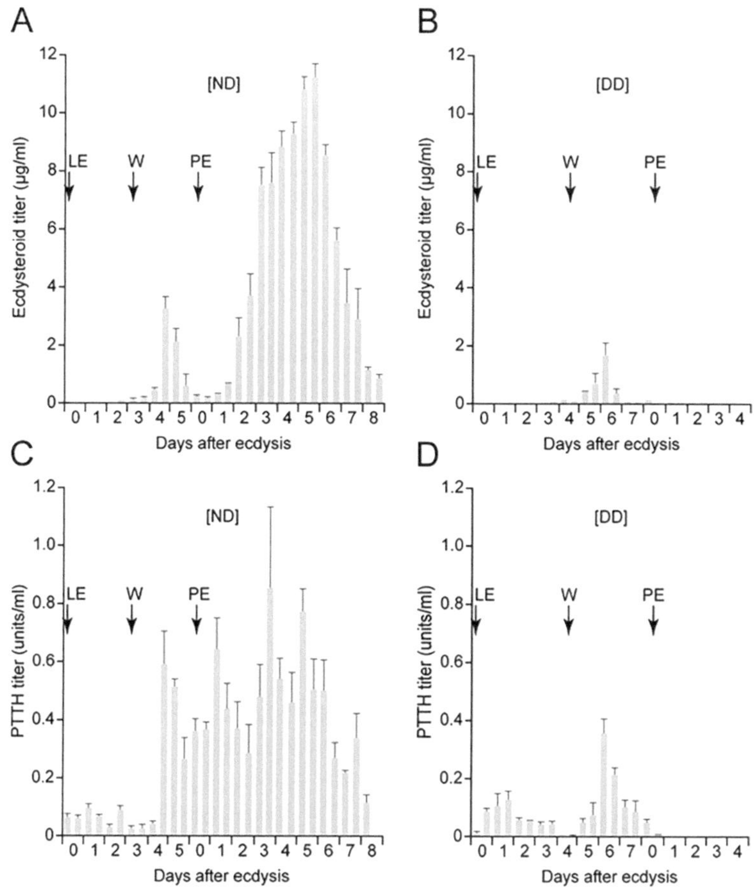

Fig. 8.3 Developmental changes in hemolymph ecdysteroid and PTTH titers in *M. brassicae*. (**a**, **b**) Ecdysteroid titers in the hemolymph of non-diapause-distined (ND) (**a**) and diapause-distined (DD) (**b**) animals. (**c**, **d**) PTTH titers in the hemolymph of ND (**c**) and DD (**d**) animals. The arrows indicate the times of the final larval ecdysis (LE), the onset of wandering (W), and pupal ecdysis (PE). The values shown are the mean (+SEM) of four independent determinations. https://doi.org/10.1371/journal.pone.0060824.g002

8.3.2 Regulation of PTTH and Ecdysteroid Secretion in Pupal Diapause

The molecular mechanisms involved in the regulation of PTTH and ecdysteroid secretion during pupal diapause have been investigated in various insects. The regulatory mechanisms of PTTH and ecdysteroids vary among species, and the mechanisms of cessation of their synthesis and release are not the same in all species. In some lepidopterans, *PTTH* expression is maintained high in larvae and pupae that are not yet ready to enter diapause, but it decreases before the initiation of diapause and remains low during diapause (Xu and Denlinger 2003; Wei et al. 2005; Zhang and Denlinger 2012). These results suggest that PTTH is regulated at the transcriptional level in these species.

Aminergic neurotransmitters, such as dopamine and serotonin, are involved in pupal diapause. High dopamine levels in the brain are important for diapause induction. Serotonin receptors present in PTTH-producing cells of the Chinese oak silkmoth *Antheraea pernyi* may facilitate the synthesis and release of PTTH (Wang et al. 2013).

8.3.3 JH and Ecdysteroid in Larval Diapause

Numerous insect species undergo larval diapause. However, the developmental stage at which diapause begins varies widely. Among lepidopteran insects, many species undergo diapause in their final instars. However, some species undergo diapause during their pharate larval stage before hatching (see Sect. 8.2). In addition, among the species that undergo diapause in the final instar, some species do so during the feeding period, while others do so after the cessation of feeding or in the prepupal stage just before metamorphosis. Therefore, there is a large variation in the timing of diapause initiation between species.

Larval diapause can be broadly divided into two types: one in which the larvae do not molt during diapause and extend a specific instar, such as the European corn borer *Ostrinia nubilalis*, and one in which the larvae molt repeatedly and the number of larval instars increases, such as the rice stem borer *Cydia pomonella* and southwestern corn borer (*Diatraea grandiosella*) (Yin and Chippendale 1973; Chippendale 1977). In the latter, there is no or only a slight increase in body size associated with molting. This type of molting is called "stationary ecdysis."

A stationary ecdysis suggests the presence of JH, because the larvae would likely pupate if JH is absent. In fact, JH hemolymph titers remained high, and the corpora allata (CA) were active during diapause (Yin and Chippendale 1973), suggesting the involvement of JH in stationary ecdysis. At the termination of diapause, the JH titer decreases, which is an important signal that permits pupation (Yin and Chippendale 1973). Knocking down the transcription factor FoxO in diapausing nymphs of the brown planthopper *Laodelphax striatellus* resulted in a reduced JH titer and a concomitant reduction in the duration of nymphal diapause (Yin et al. 2018), indicating that JH may also maintain this diapause. Conversely, there are known cases in which JH treatment breaks diapause. The rice field moth *Omphisa fuscidentalis* enters diapause as a fully grown final instar larvae, and the treatment of diapausing larvae with JH activates the PG without the involvement of the brain to promote the secretion of ecdysteroids, resulting in the termination of diapause (Singtripop et al. 2000).

Stationary ecdysis during diapause suggests periodic secretion of ecdysteroids, which induces ecdysis. Therefore, the PGs are not inactivated even during diapause. It is unclear whether the periodic changes in PG activity are due to autonomous feedback regulation of the PGs, periodic stimulation by PTTH, or other factors. Furthermore, the mechanisms through which high JH concentrations are maintained in diapausing larvae remain largely unknown. It is also unknown whether the significant decrease in food intake during diapause is due to JH.

In cases where larval diapause is not accompanied by stationary ecdysis, ecdysteroids are deeply involved in the control mechanism of larval diapause, similar to pupal diapause (Richard and Saunders 1987; Gelman et al. 1992). This is because the concentration of ecdysteroids in the larval hemolymph during diapause remained low in most cases examined. Since ecdysteroids are secreted from the PGs, it is thought that inactivation of the PGs is deeply involved in the induction and maintenance of larval diapause. Although the mechanism of PG inactivation is not fully understood, several studies have suggested that PTTH secretion is halted or the PGs become less responsive to PTTH (Richard and Saunders 1987; Gelman et al. 1992).

These cases clearly show that there may be multiple hormone regulatory mechanisms for the same larval diapause, and it is interesting to consider the diversity of endocrine mechanisms involved in adaptation.

8.3.4 Regulation of JH and Ecdysteroid Secretion in Larval Diapause

Non-diapausing larvae of the longicorn beetle *Psacothea hilaris* are distinguished from diapausing larvae by the high activity of JH esterase (JHE), which is involved in the degradation of JH (Munyiri and Ishikawa 2004). However, no such difference has been observed in the Mediterranean corn borer *Sesamia nonagrioides* (Schafellner et al. 2008). This suggests that JH titer is regulated by multiple synthesis and degradation mechanisms and the importance of each may vary among species, as reported in several studies. In the cricket *Modicogryllus siamensis*, short day lengths appear to elicit extra nymphal molts characteristic of diapause through photoperiodic upregulation of juvenile hormone acid O-methyltransferase (JHAMT), an enzyme that converts inactive JH precursors to active JH via the TGF-ß family signaling pathway (Miki et al. 2020).

8.3.5 Other Hormones

Although DH induces embryonic diapause in some insects, it also plays a role in pupal diapause. Rather than promoting diapause induction, DH breaks pupal diapause in *Heliothis*/*Helicoverpa* species (Zhang et al. 2015). DH titers are high in larvae and pupae not destined for diapause. However, in diapause-destined individuals, the titers drop at the initiation of diapause and remain low until diapause ends, suggesting that low levels of DH are essential for the manifestation of pupal diapause. DH stimulates ecdysteroidogenesis in the PGs both in vivo and in vitro, and labeled DH binds to the plasma membrane of PG cells, suggesting that DH breaks diapause by activating the PGs to synthesize ecdysone (Zhang et al. 2004).

PTTH retains its role as a major player in the regulation of larval and pupal diapause through its action on the PGs. However, supplementary pathways also contribute to PG regulation. Myosuppressin (MS), a PG-inhibiting peptide secreted from the brain, is thought to stabilize diapause in *M. brassicae*. MS appears to play an important role in inactivating the PGs during the initiation of diapause, at least in *M. brassicae* (Yamada et al. 2017). The PGs become highly refractory to stimulation by PTTH within a day after pupation, a time point that coincides with an increase in the hemolymph titer of MS. PTTH-stimulated activation of the PGs in vitro can be effectively countered by MS (Yamada et al. 2017). Thus, during diapause, not only is PTTH synthesis and release suspended, but its targets, the PGs, also become refractory to PTTH. This double guarantee ensures that the synthesis of ecdysteroids, which promote development, is suppressed throughout diapause.

Insulin signaling is a conserved mechanism that underlies diverse forms of animal dormancy (Hand et al. 2016). The link between insulin-like peptides (ILPs; see Sect. 2.2.1) and insect diapause has been widely examined in adult diapause. However, some information is also available for pupal diapause, with the most widely studied being pupal diapause in the cotton bollworm *Helicoverpa armigera* (Zhang et al. 2017). In *H. armigera*, under long-day conditions, high levels of ILPs relieve FoxO-mediated repression of PTTH gene expression through activation of insulin/IGF signaling (IIS; see Chap. 7), thereby allowing adult development. In contrast, low levels of ILPs and high levels of ROS under short-day conditions inhibit FoxO phosphorylation, promote FoxO nuclear localization, and repress PTTH gene expression, thereby inducing diapause (Palli 2017; Zhang et al. 2017).

8.4 Adult Diapause

Adult diapause refers to cessation of reproductive function. In females, the most prominent feature is the arrest of oocyte development, whereas in males, the most prominent feature is a failure to mate with receptive females. Some species that migrate long distances before diapause have flight muscles that degenerate during diapause and regenerate after diapause termination (Pener 1992). Most early studies on adult diapause focused on the role of JH. Although JH remains the central focus of adult diapause, the roles of ecdysteroids, ILPs, and other neuropeptides have recently gained attention.

8.4.1 Juvenile Hormone

Hormonal control of adult diapause has primarily been studied in the Colorado potato beetle *Leptinotarsa decemlineata*, the common house mosquito *Culex pipiens*, and the firebug *Pyrrhocoris apterus*. Pioneering work by de Wilde et al. in the Netherlands in the 1950s revealed that JH plays a central role in diapause control in *L. decemlineata* (de Kort 1990; Jindra et al. 2013). The internal concentration of JH decreased rapidly after adult eclosion and increased after diapause ended. The CA, which synthesizes JH, becomes smaller during diapause, and JH synthesis activity also decreases. In addition, when the CA were removed from non-diapausing adults, behavioral changes similar to those of diapausing individuals were observed, such as ceasing feeding and burrowing into the soil. In contrast, JH injection into diapausing individuals promoted female egg-laying. These results indicate that a decrease in JH concentration induces and maintains diapause, whereas an increase in JH concentration breaks diapause. As the destruction of the pars intercerebralis in non-diapausing adults has the same effect as the removal of the CA, it is thought that the changes in CA activity associated with diapause are regulated by a neurosecretory hormone(s) secreted from the brain.

8.4.2 Regulation of JH Secretion

Studies in other insects have revealed several variations in adult diapause and its hormonal regulation, although the central role of JH in diapause regulation appears to be common. In *C. pipiens*, the amount of mRNA encoding allatotropin (see Sect. 2.3.4) is low during diapause in the entire body, supporting the role of allatotropin in the regulation of the CA. Knockdown of allatotropin mRNA using RNAi in females programmed for non-diapause resulted in the cessation of ovarian development akin to diapause. In contrast, no difference in the expression of allatostatin C (AST-C; see Sect. 2.3.3), a neuropeptide that inhibits JH biosynthesis, has been observed between diapausal and non-diapausal females (Kang et al. 2014). This suggests that the inhibition of allatotropin secretion plays an important role in controlling the cessation of CA function during diapause. To investigate whether the JH signaling pathway is involved in diapause breaking in *P. apterus*, RNAi-knockdown experiments were performed (Smykal et al. 2014). Inhibition of Methoprene-tolerant (Met) and Taiman (Tai) functions by RNAi was consistent with the effect of CA ablation, suggesting that the JH receptor (Met-Tai complex) may be required for vitellogenesis during diapause breaking.

8.4.3 Ecdysteroids

A relationship between ecdysteroids and adult diapause has also been suggested in several species. In some cases, it is related to diapause induction, whereas in others, it plays a role in diapause termination. In adults, ecdysteroids are not produced in the PGs, which degenerate during adult development, but are secreted primarily by the ovaries. In *D. melanogaster*, developmental arrest occurs during the previtellogenic stage of ovarian follicular development (Saunders et al. 1989), and oocytes cease to incorporate yolk proteins (YPs) (Saunders 1990). Diapausing ovaries produce low levels of ecdysteroids; however, as diapause ends, ecdysteroid levels increase, activating endocytosis and allowing vitellogenesis initiation (Richard et al. 1998). How these results on ecdysteroids relate to JH and the precise molecular mechanisms involved remain unclear.

8.4.4 Insulin Signaling

Several studies have highlighted the importance of ILPs in adult diapause. Genetic analysis of the insulin signaling pathway in *D. melanogaster* has shown that ILPs control diapause (Kubrak et al. 2014; Schiesari et al. 2016). The removal of insulin-producing cells (IPCs) in the brain, which produce *Drosophila* ILP (DILP)-2, DILP-3, and DILP-5, suppression of their excitability, or suppression of insulin signaling using chico mutants significantly increased diapause rates. In contrast, overexpression of *DILP-2* and *DILP-5* and increased excitability of IPCs inhibited diapause. These findings support the hypothesis that diapause is caused by decreased insulin signaling.

Short days and low temperatures also cause developmental arrest in males of *D. melanogaster*. Development of the testes and accessory glands and spermatogenesis are halted under these conditions (Kubrak et al. 2016). As in females, lipid reserves increase at the initiation of diapause and then decrease as they are utilized, expression of genes related to stress response and immunity increases, and males stop mating. Insulin signaling is suppressed in diapausing males as in females.

Furthermore, aminergic signaling has been speculated to be involved in the control of adult diapause in *D. melanogaster* (Andreatta et al. 2018). High concentrations of serotonin or dopamine induce diapause, and inhibition of these signaling pathways leads to diapause inhibition. Serotonin and dopamine inhibit the production and release of ILP in IPCs, thereby inducing diapause. In contrast, suppression of dopamine receptor (DopR1) expression reduces the rate of diapause, suggesting that DopR1 signaling contributes to the control of adult diapause.

In adult diapause in the mosquito *C. pipiens*, insulin signaling is involved in the pathway that promotes JH synthesis (Sim and Denlinger 2008). JH is known to terminate diapause in this species. When RNAi experiment was performed on the insulin receptor in non-diapausing female mosquitoes under long-day conditions, the primary follicles were arrested at a stage comparable to diapause. These mosquitoes are rescued from this developmental arrest with the application of JH. Furthermore, activation of the insulin signaling pathway suppresses *FoxO* transcription. Females with suppressed *FoxO* expression are unable to accumulate large amounts of fat, which is a characteristic of non-diapause.

8.4.5 Other Neuropeptides

It has been reported that a subset of dorsal neurons (DN3s) that express the neuropeptide AST-C facilitates recovery from cold-induced reproductive dormancy in *D. melanogaster*. Neural activity of AST-C-expressing DN3s and systemic AST-C peptide levels was suppressed by cold exposure. Cold temperatures also affect AST-C levels in other *Drosophila* species and mosquitoes. The stimulatory effect of AST-C on egg production is mediated by cholinergic AST-C-R2 neurons (Meiselman et al. 2022).

It has also been demonstrated that the neuropeptide diuretic hormone 31 (DH31), also called CT-like diuretic hormone (CT-DH; see Sect. 2.3.8), produced by brain neurons that project to the CA plays an essential role in regulating reproductive dormancy by suppressing JH biosynthesis in adult *D. melanogaster* (Kurogi et al. 2023). The CA expresses a gene encoding the DH31 receptor, which is required for DH31-triggered elevation of intracellular cAMP in the CA. Knockdown of DH31 in these CA-projecting neurons or the DH31 receptor in the CA suppresses the decrease in JH titers, which is normally observed under dormancy-inducing conditions, leading to abnormal yolk accumulation in the ovaries. These findings suggest that CA-projecting peptidergic neurons play an essential role in the regulation of reproductive dormancy by suppressing JH biosynthesis.

Pigment-dispersing factor (PDF; see Sect. 2.4.6) is a known brain neuropeptide involved in circadian control in various insects. RNAi-mediated knockdown of *PDF* inhibited oviposition arrest induced by a transfer from long- to short-day conditions in females of the brown-winged green bug *Plautia stali*. Additionally, knockdown of *pdf* delayed oviposition onset after a change from short- to long-day conditions, suggesting that PDF is involved in the photoperiodic control of reproduction in *P. stali* females (Hasebe et al. 2022).

Adipokinetic hormone (AKH; see Sect. 2.3.1) has been proposed to be involved in diapause (Socha and Kodrik 1999; Kubrak et al. 2014). In *P. apterus*, both diapause and wing polymorphism are controlled by photoperiod and temperature. Under constant short-day conditions, the bugs from the temperate population enter diapause and virtually all individuals become brachypters (short winged). In contrast, under longer days, the bugs do not enter diapause, and a fraction of the population becomes macropterous (long-winged). Diapausing females of *P. apterus* mobilize approximately twice the amount of lipid as non-diapausing females when injected with AKH. This indicates that diapausing females are highly sensitive to AKH. In *D. melanogaster*, *AKH* expression increased more than fourfold during the first 3 weeks of diapause and remained approximately twofold higher for the remainder of the diapause period.

8.5 Prospection

Recent molecular biology approaches have revealed details of the molecular mechanisms of diapause. Although ecdysteroids, JH, and DH mediate the initiation and termination of diapause, it has been found that various factors, including neuropeptides, are involved upstream or downstream of the secretion of these hormones. These findings indicate that diapause involves hierarchical signaling pathways. Diapause is also considered a phenotype that each species has acquired uniquely during evolution. Therefore, it is assumed that hierarchical signaling pathways also have various diversities depending on the species, which can be inferred from previous findings. In contrast, the connection between the circadian clock and the endocrine system is thought to be conserved as a common denominator. As pointed out by Denlinger (2022), the evolutionary story of diapause simply requires a connection between the circadian clock, specific genes, and the neuroendocrine system regulating development (Denlinger 2022). Recent studies using genome-editing technology have shown that clock genes regulate diapause not only in *D. melanogaster* but also in other species, including *B. mori* (Homma et al. 2022; Ikeda et al. 2021). In the future, it is expected that the molecular mechanisms involved in the reception of environmental information that determine diapause and the function of clock factors will be elucidated, and the differences in the diapause mechanisms that each species has acquired during the evolutionary process will become clear.

References

Andreatta G, Kyriacou CP, Flatt T, Costa R (2018) Aminergic signaling controls ovarian dormancy in *Drosophila*. Sci Rep 8:2030. https://doi.org/10.1038/s41598-018-20407-z

Batz ZA, Brent CS, Marias MR, Sugijanto J, Armbruster PA (2019) Juvenile hormone III but not 20-hydroxyecdysone regulates the embryonic diapause of *Aedes albopictus*. Front Physiol 10: 1352. https://doi.org/10.3389/fphys.2019.01352

Chen J, Cui DN, Ullah H, Li S, Pan F, Xu CM et al (2020) The function of LmPrx6 in diapause regulation in *Locusta migratoria* through the insulin signaling pathway. Insects 11. https://doi.org/10.3390/insects11110763

Chippendale GM (1977) Hormonal regulation of larval diapause. Ann Rev Ent 22:121–138

Danks HV (1987) Insect dormancy: an ecological perspective. Biological Survey of Canada

de Kort CAD (1990) Thirty-five years of diapause research with the colorado potato beetle. Entomol Exp Appl 56:1–13

Denlinger DL (2022) Insect diapause. Cambridge University Press

Denlinger DL, Yocum GD, Rinehart JP (2012) Hormonal control of diapause. In: Gilbert LI (ed) Insect endocrinology. San Diego: Academic Press, pp 430–463

Fujiwara Y, Tanaka Y, Iwata K, Rubio RO, Yaginuma T, Yamashita O et al (2006a) ERK/MAPK regulates ecdysteroid and sorbitol metabolism for embryonic diapause termination in the silkworm, *Bombyx mori*. J Insect Physiol 52:569–575. https://doi.org/10.1016/j.jinsphys.2006.02.004

Fujiwara Y, Shindome C, Takeda M, Shiomi K (2006b) The roles of ERK and P38 MAPK signaling cascades on embryonic diapause initiation and termination of the silkworm, *Bombyx mori*. Insect Biochem Mol Biol 36:47–53. https://doi.org/10.1016/j.ibmb.2005.10.005

Gelman DB, Thyagaraja BS, Kelly TJ, Masler EP, Bell RA, Borkovec AB (1992) Prothoracicotropic hormone levels in brains of the European corn borer, *Ostrinia nubilalis*: diapause vs the non-diapause state. J Insect Physiol 38:383–395

Goldsmith MR, Shimada T, Abe H (2005) The genetics and genomics of the silkworm, *Bombyx mori*. Annu Rev Entomol 50:71–100. https://doi.org/10.1146/annurev.ento.50.071803.130456

Grüner C, Masaki S (1994) Summer diapause in the polymorphic life cycle of the noctuid moth *Mamestra brassicae*. In: Danks HV (ed) Insect life-cycle polymorphism. Kluwer Academic, pp 191–204

Hahn DA, Denlinger DL (2011) Energetics of insect diapause. Annu Rev Entomol 56:103–121. https://doi.org/10.1146/annurev-ento-112408-085436

Hand SC, Denlinger DL, Podrabsky JE, Roy R (2016) Mechanisms of animal diapause: recent developments from nematodes, crustaceans, insects, and fish. Am J Physiol Regul Integr Comp Physiol 310:R1193–R1211. https://doi.org/10.1152/ajpregu.00250.2015

Hao K, Tu X, Ullah H, McNeill MR, Zhang Z (2019) Novel *Lom-dh* genes play potential role in promoting egg diapause of *Locusta migratoria* L. Front Physiol 10:767. https://doi.org/10.3389/fphys.2019.00767

Hasebe M, Kotaki T, Shiga S (2022) Pigment-dispersing factor is involved in photoperiodic control of reproduction in the brown-winged green bug, *Plautia stali*. J Insect Physiol 137:104359. https://doi.org/10.1016/j.jinsphys.2022.104359

Holman GM, Cook BJ, Nachman RJ (1986) Primary structure and synthesis of a blocked myotropic neuropeptide isolated from the cockroach, *Leucophaea madera*e. Comp Biochem Physiol C Comp Pharmacol Toxicol 85:219–224. https://doi.org/10.1016/0742-8413(86)90077-0

Homma T, Watanabe K, Tsurumaru S, Kataoka H, Imai K, Kamba M, Niimi T, Yamashita O, Yaginuma T (2006) G protein-coupled receptor for diapause hormone, an inducer of *Bombyx* embryonic diapause. Biochem Biophys Res Commun 344:386–393. https://doi.org/10.1016/j.bbrc.2006.03.085

Homma S, Murata A, Ikegami M, Kobayashi M, Yamazaki M, Ikeda K et al (2022) Circadian clock genes regulate temperature-dependent diapause induction in silkworm *Bombyx mori*. Front Physiol 13:863380. https://doi.org/10.3389/fphys.2022.863380

Ikeda K, Daimon T, Shiomi K, Udaka H, Numata H (2021) Involvement of the clock gene *Period* in the Photoperiodism of the silkmoth *Bombyx mori*. Zool Sci 38:523–530. https://doi.org/10.2108/zs210081

Imai K, Konno T, Nakazawa Y, Komiya T, Isobe M, Koga K et al (1991) Isolation and structure of diapause hormone of the silkworm, *Bombyx mori*. Proc Jpn Acad Ser B67:98–101

Jindra M, Palli SR, Riddiford LM (2013) The juvenile hormone signaling pathway in insect development. Annu Rev Entomol 58:181–204. https://doi.org/10.1146/annurev-ento-120811-153700

Kang DS, Denlinger DL, Sim C (2014) Suppression of allatotropin simulates reproductive diapause in the mosquito *Culex pipiens*. J Insect Physiol 64:48–53. https://doi.org/10.1016/j.jinsphys.2014.03.005

Kidokoro K, Iwata K, Fujiwara Y, Takeda M (2006) Effects of juvenile hormone analogs and 20-hydroxyecdysone on diapause termination in eggs of *Locusta migratoria* and *Oxya yezoensis*. J Insect Physiol 52:473–479. https://doi.org/10.1016/j.jinsphys.2006.01.001

Kitamura A, Nagasawa H, Kataoka H, Inoue T, Matsumoto S, Ando T et al (1989) Amino acid sequence of pheromone-biosynthesis-activating neuropeptide (PBAN) of the silkworm, *Bombyx mori*. Biochem Biophys Res Commun 163:520–526

Kogure M (1933) The influence of light and temperature on certain characters of the silkworm, *Bombyx mori*. J Agric Kyushu Imp Univ 4:1–93

Kostal V (2006) Eco-physiological phases of insect diapause. J Insect Physiol 52:113–127. https://doi.org/10.1016/j.jinsphys.2005.09.008

Kostal V (2011) Insect photoperiodic calendar and circadian clock: independence, cooperation, or unity? J Insect Physiol 57:538–556. https://doi.org/10.1016/j.jinsphys.2010.10.006
Kubrak OI, Kucerova L, Theopold U, Nassel DR (2014) The sleeping beauty: how reproductive diapause affects hormone signaling, metabolism, immune response and somatic maintenance in *Drosophila melanogaster*. PLoS One 9:e113051. https://doi.org/10.1371/journal.pone.0113051
Kubrak OI, Kucerova L, Theopold U, Nylin S, Nassel DR (2016) Characterization of reproductive dormancy in male *Drosophila melanogaster*. Front Physiol 7:572. https://doi.org/10.3389/fphys.2016.00572
Kurogi Y, Imura E, Mizuno Y, Hoshino R, Nouzova M, Matsuyama S et al (2023) Female reproductive dormancy in *Drosophila* is regulated by DH31-producing neurons projecting into the corpus allatum. Development 150. https://doi.org/10.1242/dev.201186
Lee K-Y, Denlinger DL (1996) Diapause-regulated proteins in the gut of pharate first instar larvae of the gypsy moth, *Lymantria dispar*, and the effect of KK-42 and neck ligation on expression. J Insect Physiol 42:423–431
Lee K-Y, Denlinger DL (1997) A role for ecdysteroids in the induction and maintenance of the pharate first instar diapause of the gypsy moth, *Lymantria dispar*. J Insect Physiol 43:289–296
Makka T, Seino A, Tomita S, Fujiwara H, Sonobe H (2002) A possible role of 20-hydroxyecdysone in embryonic development of the silkworm *Bombyx mori*. Arch Insect Biochem Physiol 51: 111–120. https://doi.org/10.1002/arch.10055
Masaki S (1980) Summer diapause. Annu Rev Entomol 25:1–25
Maslennikova VA, Chernysh JI, Abel Nabi AA (1976) Ecdysone titre during induction of winter and summer diapause in *Barathra brassicae* L. Entomol Obozr 55:768–776
Meiselman MR, Alpert MH, Cui X, Shea J, Gregg I, Gallio M et al (2022) Recovery from cold-induced reproductive dormancy is regulated by temperature-dependent AstC signaling. Curr Biol 32:1362–1375 e1368. https://doi.org/10.1016/j.cub.2022.01.061
Miki T, Shinohara T, Chafino S, Noji S, Tomioka K (2020) Photoperiod and temperature separately regulate nymphal development through JH and insulin/TOR signaling pathways in an insect. Proc Natl Acad Sci USA 117:5525–5531. https://doi.org/10.1073/pnas.1922747117
Miyata T (1970) A generic revision of the Japanese Bombycidae, with description of a new genus (Lepidoptera). Tinea 8:190–199
Mizoguchi A, Ohsumi S, Kobayashi K, Okamoto N, Yamada N, Tateishi K et al (2013) Prothoracicotropic hormone acts as a neuroendocrine switch between pupal diapause and adult development. PLoS One 8:e60824. https://doi.org/10.1371/journal.pone.0060824
Munyiri FN, Ishikawa Y (2004) Endocrine changes associated with metamorphosis and diapause induction in the yellow-spotted longicorn beetle, *Psacothea hilaris*. J Insect Physiol 50:1075–1081. https://doi.org/10.1016/j.jinsphys.2004.09.009
Murakami A, Imai HT (1974) Cytological evidence for holocentric chromosomes of the silkworms, *Bombyx mori* and *B. mandarina*, (Bombycidae, Lepidoptera). Chromosoma 47:167–178
Nakagaki M, Takei R, Nagashima E, Yaginuma T (1991) Cell cycles in embryos of the silkworm, *Bombyx mori*: G_2-arrest at diapause stage. Roux Arch Dev Biol 200:223–229. https://doi.org/10.1007/Bf00361341
Palli SR (2017) New roles for old actors, ROS and PRMT1. Proc Natl Acad Sci USA 114:10810–10812. https://doi.org/10.1073/pnas.1715062114
Pener MP (1992) Environmental cues, endocrine factors, and reproductive diapause in male insects. Chronobiol Int 9:102–113. https://doi.org/10.3109/07420529209064521
Raina AK, Jaffe H, Kempe TG, Keim P, Blacher RW, Fales HM et al (1989) Identification of a neuropeptide hormone that regulates sex pheromone production in female moths. Science 244: 796–798
Richard DS, Saunders DS (1987) Prothoracic gland function in diapause and non-diapause *Sarcophaga argyrostoma* and *Calliphora vicina*. J Insect Physiol 33:385–392
Richard DS, Watkins NL, Serafin RB, Gilbert LI (1998) Ecdysteroids regulate yolk protein uptake by *Drosophila melanogaster* oocytes. J Insect Physiol 44:637–644. https://doi.org/10.1016/s0022-1910(98)00020-1

Sasaki C (1898) On the affinity of our wild and domestic silkworms. Annot Zool Jap 2:33–41
Sato Y, Oguchi M, Menjo N, Imai K, Saito H, Ikeda M et al (1993) Precursor polyprotein for multiple neuropeptides secreted from the subesophageal ganglion of the silkworm *Bombyx mori*: characterization of the cDNA encoding the diapause hormone precursor and identification of additional peptides. Proc Natl Acad Sci USA 90:3251–3255. https://doi.org/10.1073/pnas.90.8.3251
Sato Y, Ikeda M, Yamashita O (1994) Neurosecretory cells expressing the gene for common precursor for diapause hormone and pheromone biosynthesis-activating neuropeptide in the suboesophageal ganglion of the silkworm, *Bombyx mori*. Gen Comp Endocrinol 96:27–36
Sato Y, Shiomi K, Saito H, Imai K, Yamashita O (1998) Phe-X-Pro-Arg-Leu-NH_2 peptide producing cells in the central nervous system of the silkworm, *Bombyx mori*. J Insect Physiol 44:333–342
Sato A, Sokabe T, Kashio M, Yasukochi Y, Tominaga M, Shiomi K (2014) Embryonic thermosensitive TRPA1 determines transgenerational diapause phenotype of the silkworm, *Bombyx mori*. Proc Natl Acad Sci USA 111:E1249–E1255. https://doi.org/10.1073/pnas.1322134111
Saunders DS (1990) The circadian basis of ovarian diapause regulation in *Drosophila melanogaster*: is the period gene causally involved in photoperiodic time measurement? J Biol Rhythm 5:315–331. https://doi.org/10.1177/074873049000500404
Saunders DS, Henrich VC, Gilbert LI (1989) Induction of diapause in *Drosophila melanogaster*: photoperiodic regulation and the impact of arrhythmic clock mutations on time measurement. Proc Natl Acad Sci USA 86:3748–3752. https://doi.org/10.1073/pnas.86.10.3748
Schafellner C, Eizaguirre M, Lopez C, Sehnal F (2008) Juvenile hormone esterase activity in the pupating and diapausing larvae of *Sesamia nonagrioides*. J Insect Physiol 54:916–921
Schiesari L, Andreatta G, Kyriacou CP, O'Connor MB, Costa R (2016) The Insulin-like proteins dILPs-2/5 determine diapause inducibility in *Drosophila*. PLoS One 11:e0163680. https://doi.org/10.1371/journal.pone.0163680
Schoofs L, Holman GM, Hayes TK, Tips A, Nachman RJ, Vandesande F, De Loof A (1990) Isolation and identification of locusta-myotropin (Lom-MT), a novel biologically active insect neuropeptide. Peptides 11:427–433
Shiomi K, Kajiura Z, Nakagaki M, Yamashita O (2003) Baculovirus-mediated efficient gene transfer into the central nervous system of the silkworm, *Bombyx mori*. J Insect Biotechnol Sericol 72:149–155
Shiomi K, Takasu Y, Kunii M, Tsuchiya R, Mukaida M, Kobayashi M et al (2015) Disruption of diapause induction by TALEN-based gene mutagenesis in relation to a unique neuropeptide signaling pathway in *Bombyx*. Sci Rep 5:15566. https://doi.org/10.1038/srep15566
Sim C, Denlinger DL (2008) Insulin signaling and FOXO regulate the overwintering diapause of the mosquito *Culex pipiens*. Proc Natl Acad Sci USA 105:6777–6781. https://doi.org/10.1073/pnas.0802067105
Singtripop T, Wanichacheewa S, Sakurai S (2000) Juvenile hormone-mediated termination of larval diapause in the bamboo borer, *Omphisa fuscidentalis*. Insect Biochem Mol Biol 30:847–854. https://doi.org/10.1016/s0965-1748(00)00057-6
Sláma K (1980) Homeostatic functions of ecdysteroids in ecdysis and oviposition. Acta Entomol Bohem 77:145–168
Smykal V, Bajgar A, Provaznik J, Fexova S, Buricova M, Takaki K et al (2014) Juvenile hormone signaling during reproduction and development of the linden bug, *Pyrrhocoris apterus*. Insect Biochem Mol Biol 45:69–76. https://doi.org/10.1016/j.ibmb.2013.12.003
Socha R, Kodrik D (1999) Differences in adipokinetic response of *Pyrrhocoris apterus* (Heteroptera) in relation to wing dimorphism and diapause. Physiol Entomol 24:278–284
Sonobe H, Yamada R (2004) Ecdysteroids during early embryonic development in silkworm *Bombyx mori*: metabolism and functions. Zool Sci 21:503–516. https://doi.org/10.2108/zsj.21.503

Suzuki K, Minagawa T, Kumagai T, Naya S, Endo Y, Osanai M et al (1990) Control mechanism of diapause of the pharate first-instar larvae of the silkmoth *Antheraea yamamai*. J Insect Physiol 36:855–860

Suzuki K, Nakamura T, Yanbe T, Kurihara M, Kuwano E (1993) Termination of diapause in pharate first-instar larvae of the gypsy moth *Lymantria dispar japonica* by an imidazole derivative KK-42. J Insect Physiol 39:107–110

Tauber MJ, Tauber CA, Masaki S (1986) Seasonal adaptations of insects. Oxford University press

Tawfik AI, Tanaka Y, Tanaka S (2002) Possible involvement of ecdysteroids in photoperiodically induced suppression of ovarian development in a Japanese strain of the migratory locust, *Locusta migratoria*. J Insect Physiol 48:411–418. https://doi.org/10.1016/s0022-1910(02)00058-6

Tsuchiya R, Kaneshima A, Kobayashi M, Yamazaki M, Takasu Y, Sezutsu H et al (2021) Maternal GABAergic and GnRH/corazonin pathway modulates egg diapause phenotype of the silkworm *Bombyx mori*. Proc Natl Acad Sci USA 118:e2020028118. https://doi.org/10.1073/pnas.2020028118

Uehara H, Senoh Y, Yoneda K, Kato Y, Shiomi K (2011) An FXPRLamide neuropeptide induces seasonal reproductive polyphenism underlying a life-history tradeoff in the tussock moth. PLoS One 6:e24213. https://doi.org/10.1371/journal.pone.0024213

Umeya Y (1946) Embryonic hibernation and diapause in insects from the viewpoint of the hibernating-eggs of the silkworm. Bull Sericul Exp Sta 12:393–480

Wang Q, Mohamed AA, Takeda M (2013) Serotonin receptor B may lock the gate of PTTH release/synthesis in the Chinese silk moth, *Antheraea pernyi*; a diapause initiation/maintenance mechanism? PLoS One 8:e79381. https://doi.org/10.1371/journal.pone.0079381

Watanabe K (1924) Studies on the voltinism of the silkworm, *Bombyx mori*. Bull Sericult Exp Sta (Tokyo) 6:411–455

Wei Z-J, Zhang Q-R, Kang L, Xu W-H, Denlinger DL (2005) Molecular characterization and expression of prothoracicotropic hormone during development and pupal diapause in the cotton bollworm, *Helicoverpa armigera*. J Insect Physiol 51:691–700

Williams CN (1946) Physiology of insect diapause; the role of the brain in the production and termination of pupal dormancy in the giant silkworm, *Platysamia cecropia*. Biol Bull 90:234–243

Williams CN (1952) Physiology of insect diapause. IV. The brain and prothoracic glands as an endocrine system in the cecropia silkworm. Biol Bull 103:120–138

Xia QY, Guo YR, Zhang Z, Li D, Xuan ZL, Li Z et al (2009) Complete resequencing of 40 genomes reveals domestication events and genes in silkworm (*Bombyx*). Science 326:433–436. https://doi.org/10.1126/science.1176620

Xu WH, Denlinger DL (2003) Molecular characterization of prothoracicotropic hormone and diapause hormone in *Heliothis virescens* during diapause, and a new role for diapause hormone. Insect Mol Biol 12:509–516. https://doi.org/10.1046/j.1365-2583.2003.00437.x

Yamada N, Mizoguchi A (2017) Endocrine changes during diapause development in the cabbage army moth *Mamestra brassicae*. Physiol Entomol 42:239–245

Yamada N, Kataoka H, Mizoguchi A (2017) Myosuppressin is involved in the regulation of pupal diapause in the cabbage army moth *Mamestra brassicae*. Sci Rep 7:41651. https://doi.org/10.1038/srep41651

Yamashita O (1996) Diapause hormone of the silkworm, *Bombyx mori*: structure, gene expression and function. J Insect Physiol 42:669–679

Yamashita O, Hasegawa K (1985) Embryonic diapause. Pergamon Press, Oxford

Yamashita O, Yaginuma T (1991) Silkworm eggs at low temperatures: implication for sericulture. In: Lee JRE, Denlinger DL (eds) Insects at low temperature. Chapman and Hall, New York, pp 424–445

Yang P, Abe S, Sato Y, Yamashita T, Matsuda F, Hamayasu T et al (2007) A palmitonyl conjugate of an insect pentapeptide causes growth arrest in mammalian cells and mimics the action of diapause hormone. J Insect Biotechnol Sericol 76:63–69

Yin CM, Chippendale GM (1973) Juvenile hormone regulation of the larval diapause of the southwestern corn borer, *Diatraea grandiosella*. J Insect Physiol 19:2403–2420
Yin ZJ, Dong XL, Kang K, Chen H, Dai XY, Wu GA et al (2018) FoxO transcription factor regulate hormone mediated signaling on nymphal diapause. Front Physiol 9:1654. https://doi.org/10.3389/fphys.2018.01654
Yokoyama T (2019) Reproduction and natural enemies of the wild murberry silkmoth (*Bombyx mandarina*). Sanshi-Konchu Biotec 88:25–38
Yokoyama T, Saito S, Shimoda M, Kobayashi M, Takasu Y, Sezutsu H et al (2021) Comparisons in temperature and photoperiodic-dependent diapause induction between domestic and wild mulberry silkworms. Sci Rep 11:8052. https://doi.org/10.1038/s41598-021-87590-4
Yoshitake N (1968) Phylogenetic aspects on the origin of Japanese race of the silkworm, *Bombyx mori* L. J Seric Sci Jpn 37:83–87
Yukuhiro K, Sezutsu H, Itoh M, Shimizu K, Banno Y (2002) Significant levels of sequence divergence and gene rearrangements have occurred between the mitochondrial genomes of the wild mulberry silkmoth, *Bombyx mandarina*, and its close relative, the domesticated silkmoth, *Bombyx mori*. Mol Biol Evol 19:1385–1389. https://doi.org/10.1093/oxfordjournals.molbev.a004200
Žďárek J, Denlinger DL (1987) Pupal ecdysis in flies: the role of ecdysteroids in its regulation. J Insect Physiol 33:123–128
Zhang Q, Denlinger DL (2012) Dynamics of diapause hormone and prothoracicotropic hormone transcript expression at diapause termination in pupae of the corn earworm, *Helicoverpa zea*. Peptides 34:120–126. https://doi.org/10.1016/j.peptides.2011.05.019
Zhang T-Y, Sun J-S, Zhang Q-R, Xu R-J, Xu W-H (2004) The diapause hormone-pheromone biosynthesis activating neuropeptide gene of *Helicoverpa armigera* encodes multiple peptides that break, rather than induce, diapause. J Insect Physiol 50:547–554
Zhang Q, Nachman RJ, Denlinger DL (2015) Diapause hormone in the *Helicoverpa/Heliothis* complex: a review of gene expression, peptide structure and activity, analog and antagonist development, and the receptor. Peptides 72:196–201. https://doi.org/10.1016/j.peptides.2015.05.005
Zhang XS, Wang T, Lin XW, Denlinger DL, Xu WH (2017) Reactive oxygen species extend insect life span using components of the insulin-signaling pathway. Proc Natl Acad Sci USA 114: E7832–E7840. https://doi.org/10.1073/pnas.1711042114

Chapter 9
Hormonal Circuits for Homeostatic Regulation of Energy and Water Balances in Insects

Takashi Koyama

Abstract Organisms adjust their growth, development, physiology, metabolism, and behavior in response to internal and environmental fluctuations to enhance their chances of survival and reproduction. In animals, specialized cells and/or organs detect these fluctuations and release endocrine factors that regulate growth and homeostatic responses. One of the major functions of these endocrine factors is to maintain energy and water balance. In insects, energy (metabolic) homeostasis, which is crucial for facilitating adult physiology, is primarily regulated by insulin signaling for energy uptake and glucagon-like adipokinetic hormone for energy mobilization. Water (osmotic) homeostasis is achieved through a balance between diuretic hormones (DHs) and antidiuretic hormones (ADHs), which optimize osmoregulatory responses. Notably, osmoregulatory circuits interact with metabolic homeostasis networks, highlighting that these hormonal pathways integrate osmotic and metabolic regulation. Conventional DH and ADH pathways play a pivotal role in regulating metabolic homeostasis and influencing behaviors that indirectly affect metabolic outcomes. This interplay suggests that these hormonal circuits coordinate both water balance and metabolic regulation. This chapter overviews the action of these hormones as systemic regulators, providing a framework for understanding how metabolic and osmotic homeostasis are orchestrated to achieve optimal systemic regulation.

Keywords Water (osmotic) homeostasis · Energy (metabolic) homeostasis · Insulin-like peptides (ILPs) · Adipokinetic hormone (AKH) · Diuretic hormones (DHs) · Antidiuretic hormones (ADHs) · Inter-organ communication

T. Koyama (✉)
Department of Biology, University of Copenhagen, Copenhagen, Denmark
e-mail: takashi.koyama@bio.ku.dk

A. Mizoguchi, N. Okamoto (eds.), *Insect Peptide Hormones and Neuropeptides*, Entomology Monographs, https://doi.org/10.1007/978-981-95-7510-7_9

9.1 Introduction: Hormonal Regulation of Metabolic and Osmotic Homeostasis in Insects

Organisms experience changes in external environmental conditions and their internal states. Therefore, they adjust their development, growth, physiology, metabolism, and behavior in response to these changes to maximize their fitness, thereby optimizing survival and reproductive success. These adaptive responses rely on sensing fluctuations in both environmental conditions and internal states, communicating them to the rest of the body, and initiating systemic actions to counteract these deviations. Developing juvenile insects sense various extrinsic conditions, such as nutrition, oxygen concentrations, daylight length, and humidity, and intrinsic conditions, such as nutritional stores, ion balance, and osmotic pressure, to adjust their growth and development, maximizing their body size and reproductive success while minimizing developing time (Koyama and Mirth 2018; Koyama et al. 2020; Texada et al. 2020). In contrast, once they undergo metamorphosis and become sexually mature adults, insects do not grow or develop further. However, like developing juveniles, they continue to sense changes in intrinsic and extrinsic conditions via specialized cells/organs and adjust their physiology, metabolism, and behavior to maintain systemic homeostasis, ensuring survival and reproductive success. This adaptive mechanism often relies on hormones and neuropeptides that ultimately modulate genetic programs to adjust physiology, metabolism, and behavior (Koyama et al. 2020, 2023). This systemic homeostasis in insects includes metabolic (energy) homeostasis and osmotic (water) homeostasis.

9.1.1 Hormonal Regulation of Metabolic Homeostasis

Metabolic homeostasis primarily relies on extracting energy from the environment (usually through food) and making it available for organismal and cellular activities. When insects consume excess energy from the environment, sugar levels in their circulation increase. This increased circulating sugar is stored to prepare for potential energy scarcity in the future. Insects store excess energy as triacylglycerides (TAGs) and diacylglycerides (DAGs) mainly in the fat body, which functions analogously to the mammalian liver and adipocytes, and glycogen mainly in the muscle and fat body (Wigglesworth 1949; Ruaud et al. 2011; Yamada et al. 2018; Koyama et al. 2020, 2021b). This process of energy uptake in insects is primarily regulated by insulin-like peptides (ILPs), which are homologous to mammalian insulin and insulin-like growth factors (IGFs) (Fig. 9.1) (Brogiolo et al. 2001; Ikeya et al. 2002).

Under unfavorable environmental conditions, insects face energy scarcity, resulting in a negative energy balance. In response, they mobilize stored energy reserves to maintain organismal and cellular functions. Under such conditions,

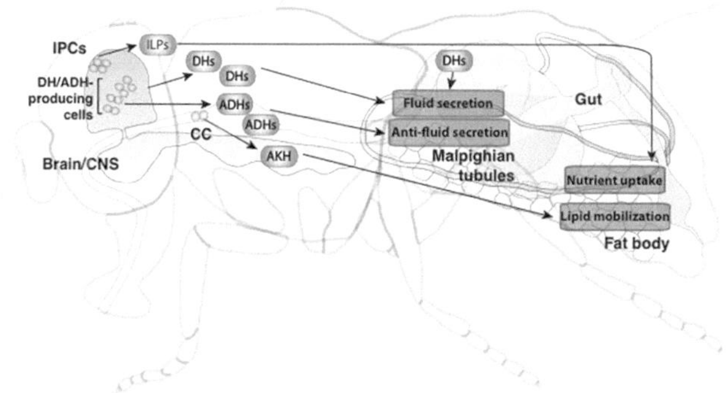

Fig. 9.1 Schematic diagram for hormonal regulation of metabolic homeostasis and water homeostasis. *ILPs* Insulin-like peptides, *AKH* adipokinetic hormone, *IPCs* insulin-producing cells, *CC* corpora cardiaca, *CNS* central nervous system, *DHs* diuretic hormones, *ADHs* antidiuretic hormones (see Sect. 9.1 for details)

several hormones are released to increase circulating sugar and lipid levels. In insects, the most well-studied energy mobilization hormone is adipokinetic hormone (AKH), which is functionally analogous to mammalian glucagon (Van der Horst et al. 2001; Van der Horst 2003). However, the AKH and glucagon pathways are not evolutionarily homologous, as AKH/AKH receptor (AKH-R) signaling likely evolved from an ancient gonadotropin-releasing hormone lineage (see Sect. 2.3.1). AKH is produced and secreted by a paired neurosecretory organ, corpora cardiaca (CC; singular, corpus cardiacum; Fig. 9.1) and primarily functions to mobilize stored lipids (Van der Horst et al. 2001, 2003; Gáliková et al. 2015).

When nutritional conditions are insufficient, insects fine-tune energy consumption, uptake, and mobilization to maintain circulating sugar and lipid levels by regulating the production and release of ILPs and AKH. In addition, recent studies have revealed that various neuropeptides are secreted in response to different internal and environmental conditions to regulate production and secretion of ILPs and AKH. Furthermore, the insulin and AKH pathways are also involved in controlling feeding and food preference behaviors, which indirectly contribute to metabolic homeostasis. Finally, both insulin and AKH signaling pathways appear to affect each other either directly or indirectly to fine-tune feedback homeostatic regulations. Taken together, the homeostatic regulation of metabolism appears to be a complex yet crucial process for maximizing fitness and optimizing energy homeostasis, especially under fluctuating nutritional conditions (Nässel and Zandawala 2019).

9.1.2 Hormonal Regulation of Osmotic Homeostasis

Osmotic homeostasis is regulated by a number of diuretic hormones (DHs) and antidiuretic hormones (ADHs) (Fig. 9.1). Systemic regulation of salt and water balance is among the most tightly set homeostatic mechanisms in animals, because even small osmotic perturbations can have serious consequences for organisms. In insects, osmotic homeostasis is primarily regulated by the balanced actions of fluid secretion from the Malpighian (renal) tubules and water reabsorption in the hindgut (Benoit et al. 2023; Halberg and Denholm 2024). When internal osmolarity is low (excess water and insufficient salts), various DHs are released to excrete excess water from the body. Conversely, under high osmolarity conditions (insufficient water and excess salts), ADHs are released to reduce fluid secretion. Therefore, sensing internal osmotic fluctuations and releasing appropriate cocktails of DHs and ADHs are key for fine-tuning fluid secretion and water reabsorption (Koyama et al. 2023; Halberg and Denholm 2024).

Various osmoregulatory networks are sensitive to changes in water and salt balance, and osmotic disturbances are detected by specific cells and organs, which regulate the release of DHs and ADHs. In insects, osmosensory mechanisms detect changes in both volume and osmotic pressure of the circulating hemolymph (insect blood). The kissing bug *Rhodnius prolixus* is capable of taking in a bloodmeal equivalent to approximately 10 times its own body weight. In this species, a bloodmeal stimulates a pair of mechanosensory cells located in the ventral abdominal body wall, which relay information on abdominal distention. Once the mechanosensory cells are activated, this signal then activates the brain neurosecretory cells via afferent neurons projecting from the abdomen (Chiang and Davey 1988; Chiang et al. 1990). Although the molecular mechanism underlying this osmosensing remains unclear, transient receptor potential (TRP) cation channel family members are likely involved, as TRPs mediate the detection of a wide range of stimuli in both mammals and insects (Rosenbaum and Islas 2023). In fact, in the fruit fly *Drosophila melanogaster*, four neurons expressing the mechanosensitive TRP vanilloid (TRPV) channel called Nanchung sense changes in extracellular osmolality to control water consumption (Jourjine et al. 2016). Interestingly, stimulation of Nanchung-positive neurons reduces the activation of several downstream peptidergic neurons, including the insulin-producing cells (IPCs), suggesting that the osmoregulation circuit involving Nanchung-positive neurons also influences metabolic homeostasis (González Segarra et al. 2023). In *D. melanogaster*, in addition to Nanchung-positive neurons, allatostatin A (AST-A)-producing neurons are also known as osmosensors, which suppress the neurons producing neuropeptide F (NPF), the insect homolog of neuropeptide Y (NPY) in mammals. In turn, this promotes water-seeking behavior and inhibits feeding behavior (Landayan et al. 2021). These results reveal that osmoregulation is also under control of other neurosecretory hormones. Interestingly, gut-derived NPF appears to control sugar satiety and protein consumption by regulating AKH signaling

(Yoshinari et al. 2021; Malita et al. 2022), suggesting that a number of neurosecretory hormones can control homeostasis in both water balance and metabolism, thereby facilitating optimal systemic homeostatic regulation.

As described above, neuronal osmoreceptors play a major role in controlling osmotic homeostasis in insects. In addition to these neuronal osmoreceptors, peripheral organs also sense osmotic perturbations, similar to those observed in mammals (Carlson et al. 1997; Kuramochi and Kobayashi 2000; Lechner et al. 2011). Indeed, water intake induces satiety in desiccated flies before hemolymph osmotic pressure is fully restored (Folk and Bradley 2004; Lin et al. 2014; Jourjine et al. 2016), suggesting that insects can predict near-future osmotic disturbances and orchestrate appropriate osmoregulation in advance of the osmotic disturbance. Taken together, both homeostatic regulations in metabolism and water balance are highly complex and tightly controlled by an extensive network of feedback circuits.

The following sections explore endocrine mechanisms that regulate systemic homeostasis—specifically metabolic homeostasis and osmotic homeostasis—in *D. melanogaster* and a broad range of other species. Although homeostatic regulations of metabolism and water balance are well defined and generally considered separate processes, they are highly interconnected. Both stored energy sources and body water are primarily ingested from the environment (food) through the digestive system, so controlling one process can significantly impact the other. Additionally, a wide range of neurosecretory hormones controls multidimensional processes. For instance, the release of certain neuropeptides is sensitive to osmotic pressures as well as particular macronutrients, while other hormones regulate water homeostasis as well as gut contractions for effective food digestion. Therefore, the following sections discuss first how different neurosecretory hormones control homeostasis in metabolism and water balance, then how other neuropeptides influence secretion of ILPs, AKH, DHs, and ADHs that are essential for homeostasis in metabolism and water balance, and finally how various neuroendocrine factors simultaneously influence these two processes as part of a comprehensive program for optimal systemic homeostasis.

9.2 Hormonal Regulation of Sugar and Lipid Metabolism

9.2.1 Nutrient Uptake and Insulin-Like Peptides (ILPs)

In mammals, pancreatic β-cells produce insulin, and its secretion is sensitive to high blood sugar levels. Secreted insulin promotes the cellular uptake and storage of glucose to avoid hyperglycemia. The insulin signaling is evolutionarily highly conserved. In insects, several ILPs are encoded in their genomes and produced in several different cells/organs (Brogiolo et al. 2001; Ikeya et al. 2002). The primary source of ILP production is the IPCs in the brain, which are functionally analogous to pancreatic β-cells (Wang et al. 2007; Clements et al. 2008). In *D. melanogaster*, ILP1, ILP2, ILP3, and ILP5 are produced in the IPCs during juvenile stages. In addition to these cells, larval corticotropin-releasing factor (CRF)–like diuretic hormone (CRF-DH, also known as DH44 in *D. melanogaster*)–producing cells exhibit ILP2-producing function, in addition to CRF-DH production, during the adult stage (Ohhara et al. 2018). These ILPs are known to be major hormones regulating larval growth (see Chap. 7). In the fat body, ILP6 is produced during the post-feeding pre-pupal and pupal stages for metamorphic growth (Okamoto et al. 2009; Slaidina et al. 2009). ILP6 production in the fat body is also enhanced under starvation (Slaidina et al. 2009; Bai et al. 2012).

Regulations of growth and metabolic homeostasis by ILPs are both mediated by a single insulin-like receptor (InR) in *D. melanogaster*, although in the case of mammals, growth is regulated by IGF signaling, and metabolic homeostasis is controlled by insulin signaling. Consistent with this notion, although there are multiple InRs in several insect species, no clear division of task between growth control and homeostatic regulation by different receptors has been observed. Therefore, a "single" InR signaling pathway controls both growth and metabolism in insects. To achieve this, different ILPs show different responses to changes in external or internal conditions. Different ILPs are under independent regulation of transcription and secretion (see Sect. 7.3.3). Indeed, their expression patterns during development differ from each other (Okamoto and Nishimura 2015). Interestingly, the expression of each *ILP* gene is sensitive to different nutritional cues in *D. melanogaster* (Ikeya et al. 2002; Kim and Neufeld 2015; Post and Tatar 2016). This difference enables animals to adapt their metabolisms to different nutritional inputs. In addition, several hemolymph proteins are known to modulate ILP-InR interaction, such as *Drosophila* Acid-labile subunit (dALS), Ecdysone-inducible gene L2 (Imp-L2), and Secreted decoy of InR (SDR) (see Sect. 7.3.4).

In *D. melanogaster*, dALS is necessary for effective signaling activity for ILP2 and ILP5, whereas it cannot bind ILP3 (Arquier et al. 2008). Under unfavorable nutritional conditions, Imp-L2 is released into the circulation and binds strongly to ILP1, ILP2, ILP5, and ILP6 and weakly to ILP3 and ILP4, sequestering circulating ILPs to reduce insulin signaling activity in target cells (Sarraf-Zadeh et al. 2013). At the same time, Imp-L2 promotes ILP2 actions locally in specific neurons (Bader et al. 2013). Finally, SDR binds strongly to ILP3; moderately to ILP1, ILP2, and ILP7; and weakly to ILP5 and ILP6, thereby reducing insulin signaling activity by sequestering ILPs in the circulation (Okamoto et al. 2013). These studies reveal that, although all ILPs act through the single InR, the ILPs function flexibly in response to different dietary inputs. Furthermore, ILPs themselves have complex feedback regulatory networks that affect *ILP* expression. ILP2, ILP5, and ILP6 are known to act as negative regulators for *ILP* genes, whereas ILP3 acts as a positive transcriptional regulator (Grönke et al. 2010; Bai et al. 2012). Moreover, different ILPs are known to stimulate downstream components differently. In *Drosophila* S2 cells, ILP5 induces sustained Akt phosphorylation, whereas ILP2 induces acute and transient AKT phosphorylation (Post et al. 2018). Taken together, through the single receptor, insulin signaling controls both growth and metabolism in response to complex environmental fluctuations.

Given the small body size and high energy demands of locomotion, particularly flight, metabolic regulation in insects must be tightly regulated and sensitive to internal nutritional conditions (Antonova et al. 2012). Once insects feed, nutrients are allocated between immediate energy consumption and storage for growth and reproduction. ILPs are responsible for a number of key steps in energy uptake. ILPs are known to increase glucose membrane permeability and promote the conversion of glucose into glycogen (Semaniuk et al. 2021). Because AKH counteracts ILP activity by enhancing glycogen phosphorylase to mobilize glycogen storage in the fat body to increase glucose levels in circulation (Gäde and Auerswald 2003; Kim and Rulifson 2004), insulin signaling activity has to be increased, and AKH signaling activity has to be decreased for effective energy storage.

According to careful genetic analyses in flies, *ILP1–5* mutant flies appear to show decreased TAG levels, whereas *ILP6* and *ILP7* mutant flies do not show obvious metabolic defects (Zhang et al. 2009). In *D. melanogaster*, ILP2 appears to decrease glycogen phosphorylase activity, which in turn reduces glycogen breakdown and enhances glycogen deposition (Post et al. 2018). In contrast, ILP3 reduces circulating trehalose levels and glycogen accumulation, particularly under low-protein food conditions, whereas ILP5 increases glycogen and TAG levels (Semaniuk et al. 2018). ILP7 appears to decrease glycogen but increase TAG levels under low-protein food conditions, suggesting that the roles of ILPs in metabolic homeostasis are complex and vary under different nutritional conditions (Semaniuk et al. 2018).

9.2.2 Nutrient Mobilization and Adipokinetic Hormone (AKH)

In organisms, maintaining organismal and cellular functions under negative energy balance is crucial. Under such conditions, nutrient-storing and nutrient-consuming organs communicate with each other, and a broad range of hormones is released from several nutrient-sensing cells to mobilize stored energy reserves and produce circulating sugars and lipids to sustain cellular activity in peripheral organs. The primary hormone for nutrient mobilization in insects is AKH, which mobilizes stored nutrients and increases the levels of circulating sugars and lipids. AKH is produced in and secreted from the CC (Lee and Park 2004). Once released, AKH activates its receptor AKH-R mainly in the fat body. This is consistent with the notion that the primary role of AKH/AKH-R signaling is the mobilization of stored energy (Park et al. 2002; Staubli et al. 2002), particularly lipolysis in *D. melanogaster* (Galsgaard et al. 2019; Toprak 2020). In developing larvae, although overexpression of *AKH* reduces fat storage, disruption of *AKH* does not influence fat storage under normal diet conditions (Lee and Park 2004). In adults, disruption of *AKH* reduces the rate of lipid mobilization under starved conditions, whereas in many studies, AKH has been reported to have no or little effect on glycogen breakdown (Grönke et al. 2007; Gáliková et al. 2015; Yamada et al. 2018). This lipid-mobilizing effect of AKH in adults leads to a rapid loss of stored energy under starved conditions and reduced starvation resistance (Lee and Park 2004; Grönke et al. 2007; Bharucha et al. 2008; Gáliková et al. 2015).

In *D. melanogaster*, AKH is encoded by a single gene, which is expressed in and released from the CC (Schaffer et al. 1990; Noyes et al. 1995). In contrast, there are two *AKH* genes in some other species (Stone et al. 1976; Gäde 1985; Siegert et al. 1985; Schooneveld et al. 1987). AKH is produced as prepro-AKH and becomes biologically active AKH through the cleavage of a signal peptide, pyroglutamylation of the N-terminus, and amidation of the C-terminus (Rhea et al. 2010; Sidyelyeva et al. 2010). Primarily, AKH secretion from the CC into circulation is induced in a cell-autonomous manner in response to low hemolymph sugar levels (Kim and Rulifson 2004; Kim and Neufeld 2015) as well as by signals from different cells/organs, as discussed below. Once hemolymph trehalose levels drop too low, cytoplasmic glucose levels in the CC are affected, which in turn reduces ATP production and increases AMP levels. This change in the AMP-to-ATP ratio is detected by the AMP-activated protein kinase (AMPK) complex, which increases intracellular Ca^{2+} levels, triggering AKH secretion (Braco et al. 2012). These intracellular mechanisms are analogous to glucagon secretion from pancreatic islet α-cells in mammals (Gromada et al. 2007). Interestingly, hypertrehalosemia also induces AKH secretion in *D. melanogaster* larvae (Kim and Rulifson 2004; Kim and Neufeld 2015; Song et al. 2017), suggesting that AKH secretion is sensitive to both low and high concentrations of circulating trehalose in a biphasic manner. Similar biphasic responses to high and low glucose levels are observed in glucagon secretion in mammals (Salehi et al. 2006), indicating that the biphasic response of CC to trehalose levels is an evolutionarily conserved mechanism for energy homeostasis.

9.2.3 Factors That Regulate IPC Activity

In organisms, various cells and organs detect fluctuations in a wide range of internal and external conditions. They maintain metabolic homeostasis by communicating with the rest of the body through the exchange of inter-organ signals. The gut, fat body, and nervous system, in particular, are known to release numerous signaling factors in response to environmental fluctuations. Specifically, in *D. melanogaster*, the fat body senses internal nutritional conditions through mechanistic target of rapamycin (mTOR) signaling and releases several humoral factors to remotely control the expression and secretion of several ILPs from the IPCs (Colombani et al. 2003; Géminard et al. 2009) (see also Sect. 7.3.3).

Some of such factors (adipokines) released from the fat body are sensitive to intracellular amino acid levels, which are sensed through mTOR signaling. The IPCs directly receive the signals, such as Eiger (Egr) and Stunted (Sun), and potentially growth-blocking peptide (GBP) 1 and 2 (Fig. 9.2; also see Sect. 7.3.3). Egr, the *D. melanogaster* orthologue of tumor necrosis factor α (TNF-α), is released from larval fat body in response to low intracellular amino acid concentrations (Agrawal

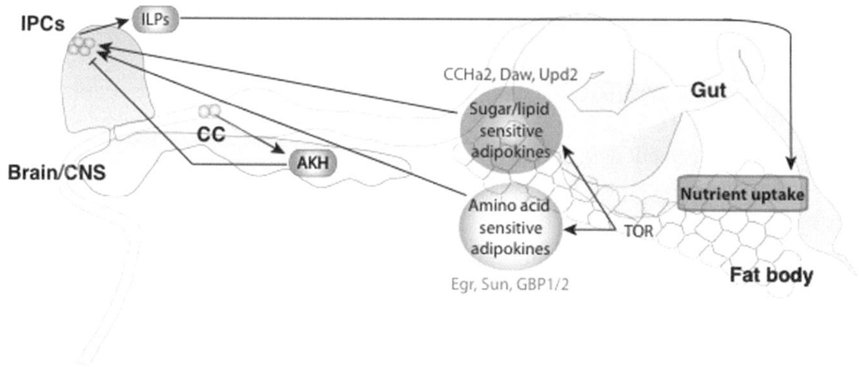

Fig. 9.2 Hormonal circuits that regulate IPC activities. *ILPs* Insulin-like peptides, *AKH* adipokinetic hormone, *IPCs* insulin-producing cells, *CC* corpora cardiaca, *CNS* central nervous system, *Daw* dawdle, *Upd2* unpaired 2, *Egr* eiger, *Sun* stunted, *GBP* growth blocking peptide (see Sect. 9.2.3 for details)

et al. 2016). This signal acts in an inhibitory manner via its receptor Grindelwald (Grnd) in the IPCs, activating the Jun Kinase cascade and suppressing *ILP* expression under low amino acid conditions. Sun is released from the larval fat body under high intracellular amino acid levels through mTOR signaling, activating Methuselah (Mth), a member of the secretin-incretin receptor subfamily, in the IPCs (Delanoue et al. 2016). GBP1/2 are similarly sensitive to high intercellular amino acid levels through mTOR (Koyama and Mirth 2016) and directly stimulate the IPCs through Methuselah-like receptor-10 (Mthl10) (Sung et al. 2017). It has recently been shown that GBP2 stimulates Mthl10 to regulate *GBP1* expression, forming a complex cytokine network that finetunes ILP secretion from the IPCs (Ono et al. 2024). In addition, GBP1/2 are proposed to stimulate a subpopulation of inhibitory IPC-connecting neurons (ICNs) through epidermal growth factor receptor (EGFR). Upon GBP binding, ICN-mediated inhibition of IPC function is relieved, promoting ILP secretion (Meschi et al. 2019).

Several fat body-derived signals, such as CCHa2 and Dawdle (Daw), are sensitive to sugar consumption. CCHa2 (see Sect. 2.3.5) responds to nutritional sugars and lipids via mTOR signaling and directly stimulates the IPCs for ILP secretion through its CCHa2 receptor (Sano et al. 2015). The Activin-like factor Daw is also secreted from the larval fat body in response to dietary glucose and promotes ILP release from the IPCs, although this may be an indirect effect (Ghosh and O'Connor 2014). In addition, Daw regulates the expression of key metabolic enzymes in the tricarboxylic acid (TCA) cycle (Ghosh and O'Connor 2014) and acts on the midgut to reduce the digestive enzyme amylase in response to dietary sugars, a process known as glucose repression, which prevents feeding excess glucose (Chng et al. 2014). Daw is also involved in sugar tolerance, pH balance regulation, and mitochondrial metabolism (Ghosh and O'Connor 2014). Therefore, Daw is a regulator of sugar homeostasis as well as a factor involved in regulating ILP secretion. It is important to note that most of the IPC regulation factors described above may function differently in adult flies.

9.2.4 Factors That Regulate CC Activity

As seen in regulation of ILP secretion, AKH secretion is also tightly regulated in insects. In adult *D. melanogaster*, AKH appears to regulate its own expression through negative feedback: the loss of functional AKH peptide leads to increased *AKH* mRNA expression (Fig. 9.3) (Gáliková et al. 2015). In addition to AKH's self-feedback regulation, several humoral factors are known to regulate AKH secretion. Interestingly, some of these factors regulate both AKH and ILP secretion by acting on the CC and IPCs, respectively. These factors are mentioned in the following section.

In adult *D. melanogaster*, bursicon-α (burs)-producing gut enteroendocrine cells (EECs) act through a neuronal relay to reduce AKH signaling under starved conditions (Scopelliti et al. 2019). Similarly, gut-derived allatostatin C (AST-C), which is homologous to mammalian somatostatin (see Sect. 2.3.3), has also been identified as a regulator of AKH secretion. The secretion of gut-derived AST-C is sensitive to dietary nutrition via mTOR signaling. Loss of *AST-C* from the gut suppresses AKH secretion via one of its receptors, AST-C-receptor 2 (AST-C-R2), which is highly expressed in the CC. This suppression results in strong starvation resistance, a hallmark of reduced AKH signaling activity (Kubrak et al. 2022). In addition, the loss of *AST-C* in the EECs or AST-C-R2 in the CC impairs lipid and carbohydrate mobilization under starved conditions (Kubrak et al. 2022). In other insects, different peptide hormones are characterized to regulate AKH secretion. In the migratory locust *Locusta migratoria*, AKH is well studied due to its role in facilitating long-distance flight (Bogerd et al. 1995). In this species, there are three prepro-AKH precursors presumably derived from three different genes, and expression levels of all three transcripts increase after flight (Bogerd et al. 1995) to promote energy mobilization (van Marrewijk et al. 1986). One of the peptides modulating AKH release is the insect tachykinin-related peptide, locustatachykinin I (Nässel et al.

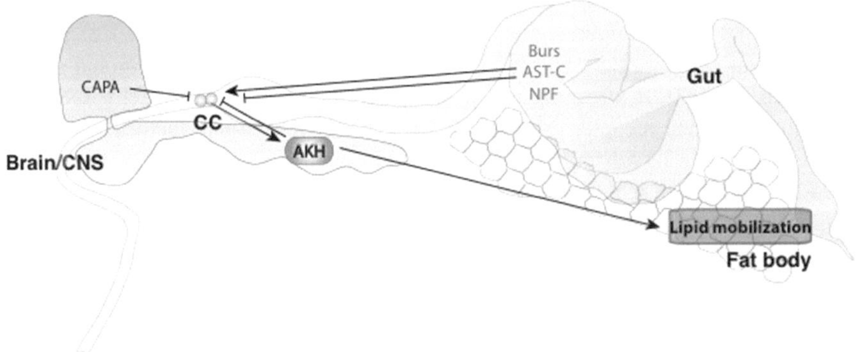

Fig. 9.3 Hormonal circuits that regulate CC activities. *AKH* Adipokinetic hormone, *CC* corpora cardiaca, *CNS* central nervous system, *Burs* bursicon, *AST-C* allatostatin C, *NPF* neuropeptide F (see Sect. 9.2.4 for details)

1995), although it is unclear what environmental conditions trigger the secretion of this peptide. Besides locustatachykinin I, FMRFamide (Vullings et al. 1998), proctolin (Clark et al. 2006), several ASTs (Clark et al. 2008), crustacean cardioactive peptide (Harthoorn et al. 2002), octopamine (Pannabecker and Orchard 1986), and melatonin (Huybrechts et al. 2005) are also known to affect AKH secretion in this species, suggesting that although what environmental stimuli trigger the secretion of these peptides, the CC appear to receive a broad range of signals that are under control of different internal and environmental stimuli.

9.2.5 Feedback Regulation of Insulin and AKH Pathways

Although AKH and ILPs play complementary roles in metabolic homeostasis in *D. melanogaster*, there is no substantial evidence to suggest that ILPs directly regulate *AKH* expression or AKH secretion. Rather, insulin signaling indirectly influences *AKH* transcription. Although ILP2 affects *AKH* expression, this regulation is dependent on ILP1 (Post et al. 2018). In contrast, AKH appears to affect the expression of *ILP*. Silencing *AKH-R* in the IPCs dramatically increases ILP3 accumulation in the IPCs in response to a sugar-rich diet, suggesting that reduced AKH signaling activity in the IPCs leads to the suppression of ILP3 secretion in developing larvae of *D. melanogaster*, at least in particular nutritional conditions (Fig. 9.2), although a similar observation has not been reported in adults (Kim and Neufeld 2015).

For effective energy homeostasis, it is crucial to control energy uptake and mobilization, and therefore several neuropeptides regulate both ILP and AKH secretion, thereby controlling the regulatory circuits that govern energy homeostasis. In *D. melanogaster*, both IPCs and CC express AST-A-R2. Reduction of AST-A results in accumulation of lipids, suggesting that AST-A signaling is key for balanced ILP/AKH secretion (Fig. 9.4) (Hentze et al. 2015; Deveci et al. 2019). Indeed, expression of *AST-A* and *AST-A-R2* is regulated differentially by dietary carbohydrates and protein, and neuronal activities of AST-A-producing cells modulate consumption of sugars and protein (Hentze et al. 2015). Taken together, AST-A appears to regulate adaptive metabolic responses and feeding decisions by regulating ILP/AKH in response to the dietary sugar-to-protein ratio.

Short neuropeptide F (sNPF) is also known to regulate ILP/AKH. In *D. melanogaster* larvae and adults, sNPF is produced in the brain, and its secretion is stimulated by pre-starvation treatments to enhance feeding and increase *ILP* expression (Lee et al. 2004, 2008; Kapan et al. 2012; Carlsson et al. 2013; Oh et al. 2019; Sudhakar et al. 2020). This enhanced feeding is due to a feedback regulation circuit between the sNPF-producing cells and ILPs, whereby insulin signaling stimulates production of sNPF for the increase in feeding in response to short-term starvation (Sudhakar et al. 2020). In adult *D. melanogaster*, sNPF-producing cells sense circulating sugar levels to promote sNPF release in response to high sugar levels. The released sNPF induces secretion of ILPs from the IPCs

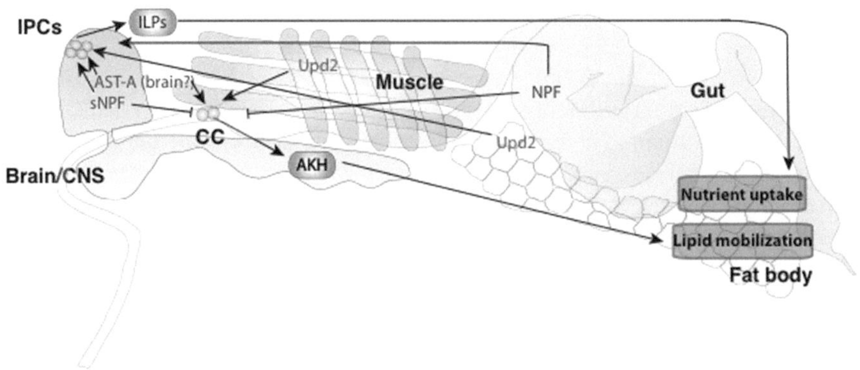

Fig. 9.4 Hormonal circuits that regulate both IPC and CC activities. *ILPs* Insulin-like peptides, *AKH* adipokinetic hormone, *IPCs* insulin-producing cells, *CC* corpora cardiaca, *CNS* central nervous system *AST-A* allatostatin A, *sNPF* short neuropeptide F, *Upd2* unpaired 2 (see Sect. 9.2.5 for details)

because the sNPF receptor (sNPF-R) is associated with stimulatory G-proteins. In contrast, because the same sNPF-R in the CC is presumably associated with inhibitory G-proteins, sNPF reduces AKH secretion from the CC (Oh et al. 2019). As a result, in response to dietary sugar, sNPF increases insulin signaling activity and decreases AKH signaling activity, thereby enhancing uptake of circulating sugar and reducing energy mobilization from stored nutrients without disturbing further feeding.

Besides the adipokines mentioned above (see Sect. 9.2.3), one of the adipokines that stimulate ILP secretion in response to sugars and lipids is Unpaired 2 (Upd2). Upd2 is structurally and functionally similar to mammalian leptin. Leptin is secreted from adipose cells in response to their lipid content and acts on the brain to affect appetite and metabolism (Konner et al. 2009). Similar to leptin in mammals, Upd2 is released from the fat body in response to dietary sugars and lipids, stimulating ILP secretion via its receptor Domeless expressed on presynaptic GABAergic neurons (Rajan and Perrimon 2012). In addition, Upd2 is produced and secreted from the adult muscle to stimulate AKH secretion from the CC (Zhao and Karpac 2017). This muscle-CC axis enables the coordination of diurnal energy demands with fat turnover, which may be crucial for controlling flight-associated high energy demands. Taken together, Upd2 also appears to fine-tune ILP/AKH secretion to coordinate energy storage and mobilization.

Besides central nervous system–derived neuropeptides and adipokines, gut EECs also contribute to balanced ILP/AKH secretion. One of these factors is gut-derived NPF. NPF is released from EECs in response to dietary sugar and suppresses AKH secretion, leading to sugar satiety and an increase in consumption of protein-rich foods (Malita et al. 2022). NPF also appears to stimulate ILP secretion from the IPCs (Yoshinari et al. 2021). Taken together, NPF is a sugar-sensitive key regulator that controls energy uptake by regulating dual endocrine axes.

9.3 Hormonal Regulation of Ion and Water Balance

9.3.1 Diuresis and Diuretic Hormones (DHs)

Insects utilize a wide variety of ecological niches and colonize the whole range of possible habitats. One of the most critical reasons for this success is, at least in part, their ability to fine-tune the homeostatic control of water balance. Indeed, primary urine production is tightly controlled by several humoral factors that are sensitive to various environmental stimuli (Dow et al. 2018; Koyama et al. 2023). Some of these factors are secreted from the Malpighian tubules (MTs) and act in a paracrine manner, while other factors are derived from the central nervous system and gut EECs and act in an endocrine manner.

The MTs are the primary organs responsible for secreting excess body fluid and metabolic wastes, as well as producing primary urine, and therefore are key organs for osmoregulation (Wigglesworth 1931a, b; Halberg and Denholm 2024). *R. prolixus* is known to have one of the most effective MTs that can secrete its own volume of fluid every 15 s (Maddrell 1991). Presumably, this fast-secreting epithelium has adapted to a large volume of a bloodmeal. While the MTs of basal exopterygote insects, such as *R. prolixus*, consist of a single secretory cell type, the tubules of higher endopterygote insects, including *D. melanogaster*, contain two distinct secretory cell types that differ both physiologically and morphologically. These two cell types are the principal cells (PCs) and the secondary, or stellate, cells (SCs). Each cell type responds to different signals and is responsible for separately transporting cations and anions (Dow 2012). Briefly, the PCs are energized by the H^+ electrochemical gradient in the apical membrane established by the large H^+ motive V-ATPase complex (Fig. 9.5). In turn, this gradient results in the secondary active K^+ secretion through a colocalized H^+/K^+ exchanger. The basolateral entry of K^+ into the PCs is controlled via several routes, including the Na^+/K^+ ATPase (Dow 2012; Halberg and Denholm 2024). The net movement of K^+ across the epithelium in the PCs leads to the movement of positive charges into the lumen. In the SCs, the flux of Cl^- is mediated by two chloride channels, chloride channel-a (ClC-a) in the basal membrane and the secretory chloride channel (SecCl) in the apical membrane (O'Donnell et al. 1996; Cabrero et al. 2014; Feingold et al. 2019). This Cl^- secretion in the SCs drives the movement of osmotically obliged water from the hemolymph to the lumen through the apical and basolateral aquaporins, Prip and Drip, respectively (Dow 2012; Halberg and Denholm 2024). Finally, although both PCs and SCs have apical microvilli, those in PCs are much longer, are stabilized by the cell adhesion molecule Fasciclin2, and contain mitochondria to support the high activity of V-ATPase (Halberg et al. 2016).

Although both PCs and SCs function as a unit to transport water together, they are under independent neuroendocrine control. In *D. melanogaster*, the PCs are regulated by calcitonin-like diuretic hormone (CT-DH, also known as DH31 in *D. melanogaster*), CRF-DH (DH44), and CAP2b (also known as CAPA in *D. melanogaster*) (Coast et al. 2001; Cabrero et al. 2002; Terhzaz et al. 2012), whereas SCs are controlled by kinin (also known as leucokinin, LK, in

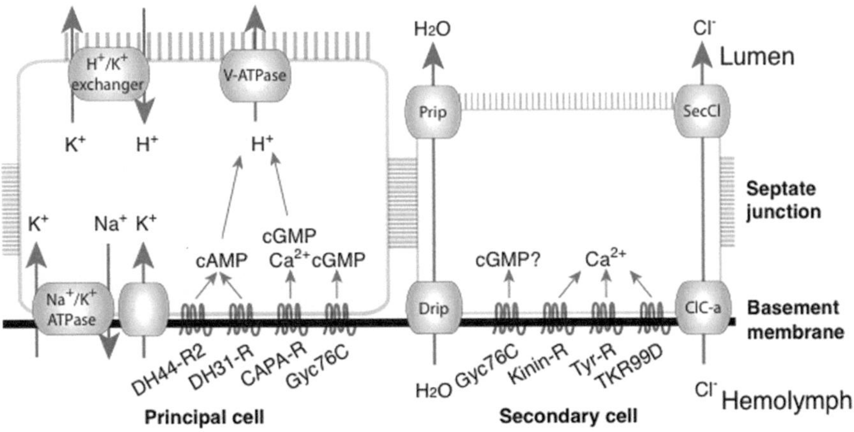

Fig. 9.5 Schematic diagram of classic two-cell-type model of *D. melanogaster* Malpighian tubule functions. DH44-R2 is one of the receptors for CRF-DH (DH44), DH31-R is a receptor for CT-DH (DH31), CAPA-R is a receptor for CAPA, Gyc76C is a receptor for ITP, Kinin-R is a receptor for kinin, Tyr-R is a receptor for tyramine, and TKR99D is one of the receptors for TK in *D. melanogaster* (see Sect. 9.3.1 for details)

D. melanogaster) and tyramine (Terhzaz et al. 1999; Cabrero et al. 2013). Although homologous neuropeptides are well conserved and these receptors are expressed in specific cell types in a broad range of species, a secondary loss of neuropeptides and altered expression pattern of these receptors are observed in some species, which may be correlated with evolutionary adaptation for specific niches (Halberg et al. 2015; Koyama et al. 2021a, 2023; Beaven et al. 2023) as explained below.

9.3.2 CRF-DH (DH44)

In *D. melanogaster*, CRF-DH is known as DH44, which consists of 44 amino acid residues. CRF-DH is encoded by the *DH44* gene and produced in and released from the brain in this species (Fig. 9.6) (Cabrero et al. 2002). CRF-DH is produced in three pairs of DH44-producing neurons in the *pars intercerebralis* (see Sect. 4.2.1) and is secreted into the circulation via the CC and intestinal neurons (Cabrero et al. 2002; Dus et al. 2015). Once secreted, it stimulates one of its receptors, DH44 receptor 2 (DH44-R2), which is highly expressed in the PCs of MTs, and induces renal secretion (Hector et al. 2009; Cannell et al. 2016). CRF-DH has been isolated and purified in several different species and has been shown to stimulate fluid secretion by increasing intracellular cyclic AMP (cAMP) concentrations (Kataoka et al. 1989; Kay et al. 1991, 1992; Lehmberg et al. 1991; Furuya et al. 1995). In *D. melanogaster*, silencing *DH44* or *DH44-R2* or ablation of DH44-producing cells shows a strong desiccation resistance ability, due to reduced renal excretion (Cannell et al. 2016).

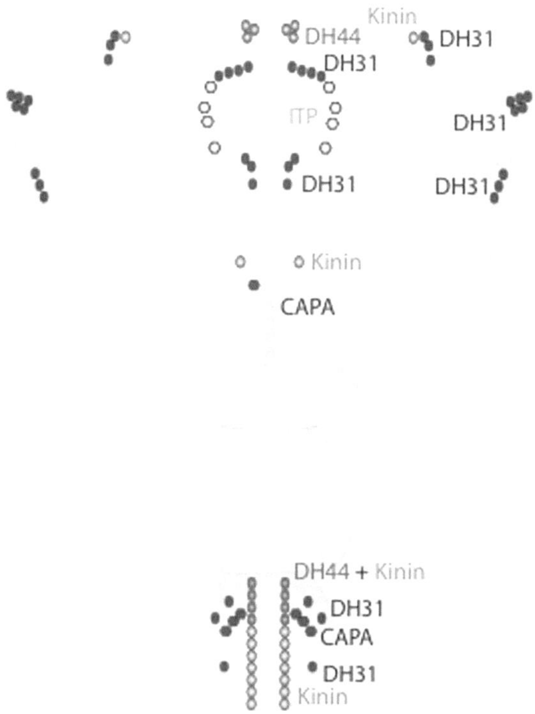

Fig. 9.6 Schematic diagram of DH- and ADH-producing cells in the adult brain-ventral nerve code complex of *D. melanogaster*. *DH44* Diuretic hormone 44, *DH31* diuretic hormone 31, *ITP* ion transport peptide (see Sects. 9.3.2, 9.3.3, 9.3.4, 9.3.5, and 9.3.6 for details)

In Coleoptera, a single gene, *Urinate* (*Urn8*), gives rise to two different CRF-DHs, DH37 and DH47, through alternative splicing (Furuya et al. 1995, 1998; Koyama et al. 2021a). Both of these peptides exert their effects through the receptor Urinate receptor (Urn8-R) (Koyama et al. 2021a). Interestingly, in the red flour beetle *Tribolium castaneum*, Urn8-R is specifically expressed in the SCs (Koyama et al. 2021a) instead of the PCs as seen in many other species, suggesting that this change in receptor localization acts as a key to the evolutionary success of this beetle in colonizing arid habitats.

Besides osmoregulation, CRF-DH/DH44 also controls behaviors that can indirectly affect water homeostasis. The DH44-producing cells are activated in response to nutritive sugars in *D. melanogaster* (Dus et al. 2015; Yang et al. 2018; Oh et al. 2021), and the secreted DH44 promotes sugar consumption and gut peristalsis to stimulate waste excretion (Dus et al. 2015). Besides sugars, DH44-producing cells also directly sense three different amino acids, L-glutamate, L-alanine, and L-aspartate, and activated cells secrete DH44 to promote food consumption (Yang et al. 2018). These facts suggest that the DH44 pathway coordinates feeding behavior in response to specific macronutrients, which can partially contribute to whole-body osmoregulation in flies. In addition to macronutrients, the neuronal networks involved in the regulation of circadian rhythms control DH44-producing cells. Adult females of *D. melanogaster* show clear diurnal rhythms. Wild-type flies deposit a higher number of more water-rich (less concentrated) feces during photophases, whereas they deposit less number of more solid (highly concentrated) feces during scotophases (Koyama et al. 2023). This is presumably related to the fact that flies sleep at night and feed freely during the day. Perhaps, DH44 signaling contributes, at least partially, to the control of diurnal rhythms in insects. Taken together, although DH44 directly contributes to water homeostasis by acting on the MTs via DH44-R2 to stimulate renal secretion, this signaling pathway also seems to contribute indirectly to water homeostasis by controlling multiple behaviors.

9.3.3 CT-DH (DH31)

Calcitonin-like DH (CT-DH; see Sect. 2.3.8), also acts on the PCs and signals through cAMP (Coast et al. 2001). In *D. melanogaster*, CT-DH is also known as DH31, which consists of 31 amino acids, and is produced in several clusters of neurosecretory cells in the brain (Fig. 9.6) (Kunst et al. 2014). DH31 is secreted into the hemolymph through the CC and acts through its receptor DH31 receptor (DH31-R) in the PCs (Johnson et al. 2005). DH31 exhibits a moderate fluid secretion effect when synthetic DH31 is applied on the MTs ex vivo. In the PCs, DH31-R leads to activation of apically localized V-ATPase (Coast et al. 2001). Interestingly, a subset of corazonin (CRZ)-producing cells co-expresses DH44-R1 and DH31-R in the *D. melanogaster* adult brains, indicating that these two DH pathways converge on CRZ-producing cells and act as a large osmoregulatory network (Johnson et al. 2005; Zandawala et al. 2021).

DH31 signaling also contributes to the regulation of behavior in this species. The DH31 circuit is a part of a sleep and daytime arousal-controlling neuroendocrine hormone network (Kunst et al. 2014), suggesting that similar to CRF-DH/DH44 signaling, the CT-DH/DH31 circuit may also contribute to the regulation of daily rhythms of diurnal excretion from the MTs, as seen in the mammalian kidney (Papper and Rosenbaum 1952). In addition to brain-derived DH31, DH31 is also produced in and released from gut EECs (Guo et al. 2019). Although the functional divisions between the brain- and gut-derived DH31 in the regulation of diuresis are not clear, it might be important to explore the function of gut-derived DH31 because of spatial proximity of the MTs and the gut (Koyama et al. 2023).

9.3.4 CAP2b (CAPA)

The evolutional and functional homolog of mammalian neuromedin U, CA2b (also known as CAPA especially in flies; see Sect. 2.3.13), also regulates diuresis from the MTs in a broad range of species (Fig. 9.6) (Tublitz et al. 1992; Davies et al. 1997; Wiehart et al. 2002; Halberg et al. 2015; Koyama et al. 2021b). In the case of

D. melanogaster, the *Capability* (*CAPA*) gene encodes two CAPA peptides (CAPA-1 and CAPA-2) as well as structurally unrelated CAPA-3 (also known as pyrokinin-1, PK-1) (Kean et al. 2002). These two CAPA peptides are expressed in two distinct cell clusters in the central nervous system: one pair of subesophageal ganglion neurons and three pairs of ventroabdominal (Va) neurons (Predel et al. 2004; Wegener et al. 2006; Koyama et al. 2021b). Once secreted from the Va neurons, both CAPA peptides circulate and activate the CAPA receptor (CAPA-R) (Santos et al. 2006; Koyama et al. 2021b), which is highly expressed in the PCs of the MTs (Iversen et al. 2002; Terhzaz et al. 2012). Activated CAPA-R in the PCs in turn stimulates primary urine production from the MTs by increasing Ca^{2+} concentrations, synthesizing nitric oxide, and increasing cGMP concentration in *D. melanogaster* as well as a broad range of species (Tublitz et al. 1992; Davies et al. 1995, 1997; Rosay et al. 1997; Terhzaz et al. 2012; Koyama et al. 2021b). In contrast, CAPA is also suggested to act as an ADH in several species. This antidiuretic function of CAPA/CAP2b is discussed below (see Sect. 9.3.2). Due to the functions of CAPA/CAPA-R in osmoregulation, CAPA signaling is also linked to chill coma recovery and desiccation resistance presumably by regulating internal water and salt balance (Terhzaz et al. 2012, 2015; MacMillan et al. 2018). Indeed, CAPA secretion is primarily sensitive to internal water abundance in *D. melanogaster*. While desiccation leads to inhibition of CAPA secretion, rehydration after desiccation is sufficient to induce CAPA secretion from the CAPA-producing Va neurons (Terhzaz et al. 2012, 2015; Koyama et al. 2021b). However, this effect of rehydration on CAPA secretion is transient (Koyama et al. 2021b). In addition, CAPA secretion is also sensitive to temperature shifts. Although the CAPA-producing Va neurons retain CAPA peptides when adult males are under cold stress, CAPA peptide secretion is enhanced following cold shock recovery (Terhzaz et al. 2012, 2015; MacMillan et al. 2018).

Besides osmoregulation, CAPA regulates multiple physiological processes. The CAPA-producing Va neurons have been shown to stimulate gut peristalsis in response to internal sugar concentrations, promoting nutrient absorption, and this effect is long-lasting (Koyama et al. 2021b). This observation indicates that CAPA signaling is involved in the homeostatic control of metabolism. Furthermore, CAPA-R is highly expressed in the CC, and this CAPA-CAPA-R axis suppresses AKH secretion from the CC to reduce energy mobilization from the fat body (Koyama et al. 2021b). Taken together, CAPA orchestrates post-ingestive physiology by regulating the functions of several different organs through complex interactions between water and metabolic regulations.

9.3.5 Kinin (Leucokinin) and Biogenic Amines

Two different endocrine factors are shown to stimulate diuresis acting through the SCs—kinin and tyramine (Fig. 9.6). Although both kinin and tyramine act through distinct receptors, kinin receptor (kinin-R) and tyramine receptor (tyramine-R), respectively, their downstream responses are indistinguishable, and both of these receptors are similarly expressed in the SCs (Radford et al. 2002; Blumenthal 2003; Herman and Blumenthal 2006; Cabrero et al. 2013). In adult *D. melanogaster*, kinin is produced in several distinct clusters of neurons in the brain and ventral nerve cord (VNC) (Terhzaz et al. 1999; Zandawala et al. 2018a). These different clusters respond to different environmental stimuli in different manners. Kinin-producing cells in the brain are activated by hyperosmotic conditions to promote the formation of water-rewarded memories (Senapati et al. 2019). In contrast, kinin-producing cells in the VNC are activated by signals associated with water ingestion, stimulating water secretion from the MTs (Zandawala et al. 2018a). Therefore, although both brain and VNC kinin-producing cells are likely sensitive to osmotic pressure, the nature of osmotic stimuli controlling kinin-producing cells in these two clusters is contradictory, and kinins derived from two distinct clusters coordinate systemic water homeostasis differently.

In the *D. melanogaster* MTs, kinin-R is only found in the SCs (Radford et al. 2002). Kinin-R-mediated stimulation of SCs in the MTs is consistent in most of holometabolous insects, except for some coleopterans, which have lost kinin signaling secondarily (Halberg et al. 2015). Activation of kinin-R increases intracellular Ca^{2+} concentrations in the SCs rapidly, which leads to activation of two types of chloride channels, apical ClC-a and basolateral SecCl, to increase transepithelial Cl^- flux and collapse of the lumen positive potential, resulting in stimulating fluid secretion (O'Donnell et al. 1996; Cabrero et al. 2014a; Feingold et al. 2019a). As mentioned above, tyramine-R also activates the SCs through the same pathway. However, kinin is a long-distance and systemic humoral factor. In contrast, tyramine is produced in the PCs by tyrosine decarboxylase, which converts tyrosine to tyramine (Blumenthal 2009), and acts on the neighboring SCs in a paracrine manner to activate the chloride shunt conductance (Herman and Blumenthal 2006). These observations suggest that the two pathways work closely together to regulate post-ingestive physiology—tyramine triggers an initial diuretic response in a tissue-autonomous manner through paracrine signaling, and kinin acts slowly but persistently via a long-distance inter-organ communication program.

Another biogenic amine, serotonin (also known as 5-hydroxytryptamine), is also reported as a factor that controls water balance. In *R. prolixus*, serotonin acts as a diuretic factor and stimulates excretion after a bloodmeal (Maddrell et al. 1991; Orchard 2006; Gioino et al. 2014). Similarly, serotonin stimulates diuresis in a wide range of insects (Pietrantonio et al. 2001; Coast 2011). In the house cricket, *Acheta domesticus*, serotonin increases intracellular Ca^{2+} concentrations without affecting cAMP concentrations, which is similar to the downstream pathway of kinin signaling (Coast 2011). However, in *R. prolixus* and the mosquito *Aedes aegypti*, serotonin stimulates increases in cAMP levels as well as intracellular Ca^{2+} concentrations (Pietrantonio et al. 2001; Gioino et al. 2014). The different modes of action of serotonin in different species may be due to different serotonin receptors functioning in the MTs or different downstream components involved.

9.3.6 Water Reabsorption and Antidiuretic Hormones

Although the distal (upper) MTs produce primary urine, the proximal (lower) MTs are involved in water and ion reabsorption (Wigglesworth 1931a; Irvine 1969; O'Donnell and Maddrell 1995; Koyama et al. 2021a). However, it is widely recognized that ion and water reabsorption are also carried out from the hindgut and rectum. Therefore, antidiuretic factors must act either to inhibit fluid secretion from the distal part of the MTs or to enhance water reabsorption at the proximal regions of the MTs, rectum, or hindgut.

The presence of antidiuretic hormones has long been widely believed, and an active compound was first partially purified from the house cricket, *A. domesticus* (Spring et al. 1988). This partially purified antidiuretic factor is secreted from the CC and inhibits fluid secretion by the MTs without affecting water reabsorption from the hindgut. Similarly, several antidiuretic factors have also been partially purified in beetles (Eigenheer et al. 2002, 2003; Wiehart et al. 2002). However, most of these partially purified antidiuretic factors remain incompletely characterized. Additionally, receptors that could potentially mediate their antidiuretic functions have yet to be identified. Therefore, although antidiuretic activity is widely seen in a broad range of insects, antidiuretic factors are less well described in comparison to DHs.

One of the chemically characterized antidiuretic factors is neuroparsin, the first ADH identified in insects. It was isolated from the CC of *L. migratoria* and stimulates water reabsorption from the hindgut (Fournier and Girardie 1988; Girardie et al. 1989). However, due to the secondary loss of homologous genes in the genetic model insect *D. melanogaster* and lack of characterization of its receptor, the molecular, biochemical, and cellular actions of neuroparsin are not well studied. Another ADH, ion transport peptide (ITP), was originally identified in the desert locust *Schistocerca gregaria* and is widely distributed in other species (Dai et al. 2007; Drexler et al. 2007), including *D. melanogaster* (Dircksen et al. 2008). ITP also promotes water reabsorption in the hindgut (Fig. 9.6) (Audsley et al. 1992a, b).

ITP is encoded by a single *ITP* gene, and usually this gene produces one ITP and multiple ITP-like (ITPL) peptides via alternative splicing. ITP belongs to the crustacean hyperglycemic, molt-inhibiting, and vitellogenesis-inhibiting family of peptides (Keller 1992; Lacombe et al. 1999; Chan et al. 2003; Chen et al. 2005; Dai et al. 2007; Drexler et al. 2007; Dircksen et al. 2008) (see Sect. 2.5.1). The ITP and ITPL sequences do not share homology with neuroparsins. Accumulating evidence has revealed that although ITP is originally described as ADH, it is also involved in water and food intake, foraging behavior, fecundity, and metabolic regulation in both AKH-dependent and -independent manners. These findings suggest that ITP is a key hormone that orchestrates adult physiology, including metabolic and water homeostasis. Therefore, the functions of ITP are described separately below (see Sect. 9.3.7).

Besides these antidiuretic factors, CAPA peptides are suggested to act as an antidiuretic signal in *D. melanogaster* (MacMillan et al. 2018) when CAPA concentration is extremely low (femtomolar order). Similar antidiuretic functions of

CAPA have also been reported in the mealworm *Tenebrio molitor* (Wiehart et al. 2002), *R. prolixus* (Quinlan et al. 1997; Quinlan and O'Donnell 1998), and *A. aegypti* (Ionescu and Donini 2012). However, the molecular and cellular mechanisms underlying this putative regulation remain unknown. In *D. melanogaster* adult males, NPF and sNPF also inhibit fluid secretion by inhibiting adenylate cyclase (Chintapalli et al. 2012). This sex specificity is due to male-specific enrichment of their receptors in the MTs. Interestingly, NPF regulates water-seeking and feeding behaviors by regulating thirst and hunger (Landayan et al. 2021). Because NPF and ITP are co-expressed in certain neurons in the brain (Johard et al. 2009), NPF and ITP may be a part of anti-diuretic osmoregulatory circuit in male flies.

9.3.7 A Key Regulator for Metabolism and Water Balance: Ion Transport Peptides (ITP)

As mentioned above, the *ITP* gene encodes ITP and several ITPLs. Although there is no genetic evidence, in vitro studies suggest that ITPLs antagonize the function of ITP (Phillips et al. 1998; Wang et al. 2000; Dircksen et al. 2008; Tsutsui 2021). In *D. melanogaster*, ITPL2 expression is found in a broad range of tissues, and the antidiuretic and metabolic functions of ITP may be counteracted and fine-tuned by ITPLs. Therefore, although several studies using *D. melanogaster* have shown that knockdown and overexpression of *ITP* result in similar rather than opposite phenotypes, it is worth pointing out that these results may not reflect accurate ITP functions, as all RNAi lines available in public target both *ITP* and all *ITPLs*.

Besides the antidiuretic effect, ITP induces thirst-promoting signals in *D. melanogaster* (Gáliková et al. 2018), suggesting that ITP not only directly coordinates adult water physiology by regulating diuresis but also indirectly through regulating behaviors that can positively contribute to water retention. Interestingly, in this species, ITP secretion is under the control of circadian pacemaker neurons, which coordinate its secretion in a rhythmical manner (Johard et al. 2009). Consistent with this notion, in *clock* mutant flies, this daily ITP secretion rhythm is reduced (Hermann-Luibl et al. 2014). Therefore, like diuretic peptides, ITP may also contribute to daily diuretic rhythms.

Beyond controlling water homeostasis, ITP also increases energy expenditure and inhibits feeding behavior (Gáliková and Klepsatel 2022). Some of these processes are mediated by AKH, which induces lipolysis and increases hemolymph glucose and trehalose concentrations, whereas other processes, such as regulating feeding and foraging behaviors, are not mediated by AKH (Gáliková and Klepsatel 2022). Thus, ITP functions as a key regulator for metabolism and behavior, which may indirectly influence water homeostasis.

ITP receptors are identified in the silkworm *Bombyx mori* (Nagai et al. 2014). Among the orphan neuropeptide G protein-coupled receptors of *B. mori* (BNGRs), BNGR-A2 and BNGR-A34 were found to be ITP receptors and BNGR-A24 an

ITPL receptor. Upon binding to the ligands, activated receptors increase cGMP levels presumably via guanylyl cyclase at 76C (Gyc76C) (Nagai et al. 2014). In *B. mori*, *BNGR-A2* mutation leads to approximately 50% higher excretion in the larval stage compared to wild-type larvae. Interestingly, loss of *BNGR-A2* reduces expression of multiple insulin signaling-related genes as well as the concentrations of the molting steroid hormone ecdysone (Sun et al. 2020), suggesting that ITP signaling strongly influences development and growth as well as physiology in *B. mori*. Furthermore, the ITP expression pattern differs between adult males and females, and the male-specific ITP-positive neurons innervate the accessory glands, suggesting a potential function of ITP in male-specific reproductive organ maturation in this species (Klöcklerová et al. 2023). In contrast, in *T. castaneum*, knocking down *ITPLs* seems to affect ovarian maturation (Begum et al. 2009). These results suggest that although it is not consistent whether ITP/ITPLs affect males or females, they appear to affect reproductive organ maturation in several different species.

In contrast to *B. mori*, a recent study has suggested that Gyc76C may potentially function as a receptor for ITP in *D. melanogaster* (Gera et al. 2024). *Gyc76C* knockdown in the MTs abolishes the inhibitory effect of ITP on fluid secretion. In addition, knocking down *Gyc76C* affects feeding behavior and energy homeostasis, similar to the ITP knockdown flies. Taken together, ITP signaling systemically orchestrates metabolic and osmotic homeostasis in insects.

9.4 Homeostatic Regulation Circuits for Metabolism and Osmolarity

As described above, ITP appears to be a key neuropeptide that integrates metabolic and osmotic homeostasis, fine-tuning overall adult physiology and various other processes. Presumably, this is because different osmotic perturbations caused by different environmental conditions may require distinct combinations of osmoregulatory responses (Koyama et al. 2023). Similarly, different metabolic perturbations resulting from different environmental conditions may require distinct combinations of metabolic responses. Therefore, understanding how different hormonal circuits integrate both osmotic and metabolic homeostasis is crucial for understanding the functions of larger systemic homeostatic regulation circuits.

Although numerous neuropeptides coordinate osmotic homeostasis, these neuropeptides also influence various other biological processes. In *D. melanogaster*, CAPA and ITP are known to regulate AKH secretion from the CC, whereas other DHs or ADHs do not (Fig. 9.7), suggesting that only these two peptides, in addition to their roles in osmoregulation, mobilize stored nutrients in the fat body via AKH signaling (Koyama et al. 2021b; Gáliková and Klepsatel 2022). In addition, courtship behavior and reproductive dormancy are regulated by DH31, but not by other DHs or ADHs (Lin et al. 2022; Kurogi et al. 2023). Since reproduction is a highly energy-demanding biological process, different neuropeptides may regulate distinct

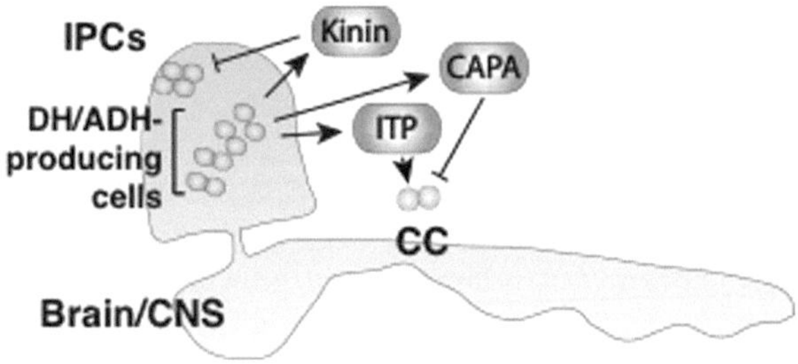

Fig. 9.7 Diuretic and antidiuretic hormones that regulate IPC and CC activities. *IPCs* Insulin-producing cells, *CC* corpora cardiaca, *CNS* central nervous system, *DH* diuretic hormone, *ADH* antidiuretic hormone, *ITP* Ion transport peptide (see Sect. 9.4 for details)

biological processes to allocate limited internal resources effectively. Indeed, in *S. gregaria*, injection of locust CRF-DH induces satiety, suppresses oocyte growth, and lowers ecdysone concentrations in the circulation and ovaries (Van Wielendaele et al. 2012).

In *D. melanogaster*, internal sugar perturbations or dietary sugar affects secretion of DH44, CAPA, and kinin, but not DH31 or ITP, suggesting that different macronutrients modulate osmoregulation through different osmoregulatory circuits (Dus et al. 2015; Senapati et al. 2019; Koyama et al. 2021b) and showing that feeding different diets activates different osmoregulatory circuits. These observations reveal that the homeostatic regulation of metabolism and water balance is tightly interconnected, with different environmental challenges activating different cocktails of DHs and ADHs to implement the appropriate homeostatic program. Consistent with this notion, the IPCs in adult *D. melanogaster* also receive neuronal inputs via kinin, which reduce *ILP* expression and secretion (Fig. 9.7) (Zandawala et al. 2018b). Kinin also controls adult behaviors associated with metabolism, including food preference and locomotion activity (Lopez-Arias et al. 2011). Besides DHs and ADHs, both brain- and gut-derived hormones also play key roles in the homeostatic regulations of metabolism and water balance and ILP secretion. Tachykinin (TK)-producing neurons in the brain directly innervate the IPCs and activate the TK receptor TKR99D in the IPCs presumably via a paracrine mechanism (Agard et al. 2024). This signal is known to regulate the transcription of *ILP2* and *ILP3* (Birse et al. 2011; Song et al. 2014). As a result, the loss of *TKR99D* in the IPCs causes rapid sugar depletion under starvation, leading in increased starvation sensitivity (Birse et al. 2011). Interestingly, kinin is also produced in gut EECs and regulates lipid metabolism both in the gut and whole body in response to dietary yeast (Song et al. 2014). Furthermore, gut-derived TK has been reported to control selective appetite regulation and AKH secretion in *D. melanogaster* (Ahrentløv et al. 2025). Gut-derived TK is released in response to protein-rich food consumption and acts on CC to promote AKH secretion. AKH suppresses protein appetite and promotes sugar hunger. Therefore, TK appears to be crucial for proper regulation of food choice. Interestingly, a recent study has shown that TK also acts as a DH (Agard et al. 2024). In the SCs, TKR99D is expressed, and TK stimulates fluid secretion from the MTs, possibly through conserved Ca^{2+} signaling as seen in the tyramine-R and kinin-R pathways. Taken together, several peptide hormones orchestrate systemic homeostatic responses, adjusting both osmotic and metabolic balance to optimize whole-body physiology.

The presence of multiple signaling pathways regulating water balance may provide an advantage for adaptation to different environmental conditions. In different species, different osmotic and metabolic regulatory circuits must be activated under similar circumstances. For instance, in blood-sucking insects, extremely large amounts of water and salts must be excreted upon feeding, whereas dry grain–feeding insects may not take sufficient water for their survival upon feeding. Therefore, even the "same" feeding behaviors must be associated with different osmoregulatory responses in different species, depending on their food qualities and ecological niches. If different osmoregulatory circuits are associated with different metabolic outcomes, species that are adapted differently may utilize different DHs or ADHs, which could be key to the evolutionary success of insects.

Conversely, different metabolic responses also likely affect the outcomes of osmoregulation. A number of classic studies have shown that, in general, animals exposed to arid environments exhibit much higher metabolic rates compared to those in normal humidity. In insects and other animals, this elevated metabolic rate is thought to generate metabolic (oxidative) water to counterbalance increased body-water evaporation (Buxton 1930; Mellanby 1958; Wharton et al. 1965), suggesting that some insects may enhance metabolic water production as a part of an osmoregulation mechanism to counteract the effect of desiccation. In *D. melanogaster*, recent studies have revealed that reducing the water sensor *pickpocket 28* switches metabolic processes by promoting AKH secretion, resulting in the stimulation of water production to compensate for the perceived lack of water availability (Waterson et al. 2014). Consistent with this notion, CAPA-producing cells become activated by rehydration and silencing their neuronal activity stimulates fluid secretion and inhibits AKH secretion (Koyama et al. 2021b), suggesting that excessive internal water inhibits energy mobilization, reducing the production of metabolic water, similar to how antidiuretic vasopressin mediates glucagon expression in mammals (Kim et al. 2021).

In insects, there are two types of hygroreceptors: one responds to increased humidity (moist receptor) and the other to decreased humidity (dry receptor) (Yokohari and Tateda 1976; Tichy 1979; Itoh et al. 1984). In *D. melanogaster*, *water witch* is required to detect moist air, and *nanchung* is involved in detecting dry air (Liu et al. 2007). Interestingly, in the brains of *D. melanogaster*, four neurons in the interoceptive subesophageal zone neurons (ISNs) have been identified as a switch that modulates hunger and thirst signals, ultimately determining sugar and water consumption, respectively (Jourjine et al. 2016). These neurons in the ISNs co-express *nanchung* and *AKH-R*, sensing osmolarity via the dry receptor Nanchung and nutritional conditions directly via AKH-R. In addition, this subset of neurons appears to sense insulin indirectly. Importantly, both low osmolarity and AKH stimulate the ISNs to promote sugar consumption and insulin secretion. In contrast, high osmolarity and the absence of AKH reduce their neuronal activity, encouraging water consumption (Jourjine et al. 2016). This suggests that osmotic and metabolic homeostasis are tightly coupled, orchestrating eating and drinking behaviors, at least partially, through a common neuronal network.

9.5 Conclusions

Although the homeostatic regulation of both metabolism and water balance has traditionally been studied separately, accumulating evidence over the last several decades indicates that these processes are tightly coupled and many neuroendocrine factors appear to regulate multiple physiological processes, including metabolism and behavior, enabling effective systemic control to optimize organismal outcomes. Accordingly, analyzing the multidimensional functions of each neuropeptide is crucial for uncovering their integrated roles at the organismal level. Especially, different insects appear to utilize distinct neuroendocrine circuits in a very complex manner to effectively control systemic homeostasis of metabolism and water balance, depending on their food sources and ecological niches. Recent cutting-edge techniques such as next-generation sequencing, RNA interference, and CRISPR-mediated genome editing will be invaluable in addressing these questions across a wide range of insect species.

Acknowledgments I thank Dr. Kenneth Veland Halberg for fruitful discussions and comments on an earlier version of this manuscript and Ms. Anna Garger for her assistance in correcting the manuscript. In addition, I thank two reviewers for critical comments and useful suggestions.

References

Agard MA, Zandawala M, Paluzzi JV (2024) Another fly diuretic hormone: tachykinins increase fluid and ion transport by adult *Drosophila melanogaster* Malpighian 'renal' tubules. J Exp Biol 227(19). https://doi.org/10.1242/jeb.247668

Agrawal N, Delanoue R, Mauri A, Basco D, Pasco M, Thorens B et al (2016) The *Drosophila* TNF Eiger is an adipokine that acts on insulin-producing cells to mediate nutrient response. Cell Metab 23(4):675–684. https://doi.org/10.1016/j.cmet.2016.03.003

Ahrentløv N, Kubrak O, Lassen M, Malita A, Koyama T, Frederiksen AS et al (2025) Protein-responsive gut hormone tachykinin directs food choice and impacts lifespan. Nat Metab 7(6): 1223–1245. https://doi.org/10.1038/s42255-025-01267-0

Antonova Y, Arik AJ, Moore W, Riehle MA, Brown MR (2012) 2—insulin-like peptides: structure, signaling, and function.InLI Gilbert (Ed.), Insect endocrinology (pp. 63–92). San Diego: Academic Press

Arquier N, Geminard C, Bourouis M, Jarretou G, Honegger B, Paix A et al (2008) *Drosophila* ALS regulates growth and metabolism through functional interaction with insulin-like peptides. Cell Metab 7(4):333–338. https://doi.org/10.1016/j.cmet.2008.02.003

Audsley N, McIntosh C, Phillips JE (1992a) Action of ion-transport peptide from locust corpus cardiacum on several hindgut transport processes. J Exp Biol 173(1):275–288. https://doi.org/10.1242/jeb.173.1.275

Audsley N, McIntosh C, Phillips JE (1992b) Isolation of a neuropeptide from locust corpus cardiacum which influences ileal transport. J Exp Biol 173:261–274. https://doi.org/10.1242/jeb.173.1.261

Bader R, Sarraf-Zadeh L, Peters M, Moderau N, Stocker H, Kohler K et al (2013) The IGFBP7 homolog imp-L2 promotes insulin signaling in distinct neurons of the *drosophila* brain. J Cell Sci 126(Pt 12):2571–2576. https://doi.org/10.1242/jcs.120261

Bai H, Kang P, Tatar M (2012) Drosophila insulin-like peptide-6 (*dilp6*) expression from fat body extends lifespan and represses secretion of drosophila insulin-like peptide-2 from the brain. Aging Cell 11(6):978–985. https://doi.org/10.1111/acel.12000

Beaven R, Halberg KV, Denholm B (2023) The insect cryptonephridial complex. Curr Biol 33(14): R748–R749. https://doi.org/10.1016/j.cub.2023.05.060

Begum K, Li B, Beeman RW, Park Y (2009) Functions of ion transport peptide and ion transport peptide-like in the red flour beetle *Tribolium castaneum*. Insect Biochem Mol Biol 39(10):717–725. https://doi.org/10.1016/j.ibmb.2009.08.005

Benoit JB, McCluney KE, DeGennaro MJ, Dow JAT (2023) Dehydration dynamics in terrestrial arthropods: from water sensing to trophic interactions. Annu Rev Entomol 68:129–149. https://doi.org/10.1146/annurev-ento-120120-091609

Bharucha KN, Tarr P, Zipursky SL (2008) A glucagon-like endocrine pathway in *drosophila* modulates both lipid and carbohydrate homeostasis. J Exp Biol 211(Pt 19):3103–3110. https://doi.org/10.1242/jeb.016451

Birse RT, Söderberg JAE, Luo J, Winther AME, Nässel DR (2011) Regulation of insulin-producing cells in the adult *drosophila* brain *via* the tachykinin peptide receptor DTKR. J Exp Biol 214 (Pt 24):4201–4208. https://doi.org/10.1242/jeb.062091

Blumenthal EM (2003) Regulation of chloride permeability by endogenously produced tyramine in the *drosophila* Malpighian tubule. Am J Physiol Cell Physiol 284(3):C718–C728. https://doi.org/10.1152/ajpcell.00359.2002

Blumenthal EM (2009) Isoform- and cell-specific function of tyrosine decarboxylase in the *drosophila* Malpighian tubule. J Exp Biol 212(Pt 23):3802–3809. https://doi.org/10.1242/jeb.035782

Bogerd J, Kooiman FP, Pijnenburg MA, Hekking LH, Oudejans RC, Van der Horst DJ (1995) Molecular cloning of three distinct cDNAs, each encoding a different adipokinetic hormone precursor, of the migratory locust, *Locusta migratoria*. Differential expression of the distinct adipokinetic hormone precursor genes during flight activity. J Biol Chem 270(39): 23038–23043. https://doi.org/10.1074/jbc.270.39.23038

Braco JT, Gillespie EL, Alberto GE, Brenman JE, Johnson EC (2012) Energy-dependent modulation of glucagon-like signaling in *drosophila* via the AMP-activated protein kinase. Genetics 192(2):457–466. https://doi.org/10.1534/genetics.112.143610

Brogiolo W, Stocker H, Ikeya T, Rintelen F, Fernandez R, Hafen E (2001) An evolutionarily conserved function of the *drosophila* insulin receptor and insulin-like peptides in growth control. Curr Biol 11(4):213–221. https://doi.org/10.1016/s0960-9822(01)00068-9

Buxton PA (1930) Evaporation from the meal worm (*Tenebrio*: Coleoptera) and atmospheric humidity. Proc R Soc B Biol Sci 106(747):560–577. https://doi.org/10.1098/rspb.1930.0045

Cabrero P, Radford JC, Broderick KE, Costes L, Veenstra JA, Spana EP et al (2002) The *Dh* gene of *Drosophila melanogaster* encodes a diuretic peptide that acts through cyclic AMP. J Exp Biol 205(Pt 24):3799–3807. https://doi.org/10.1242/jeb.205.24.3799

Cabrero P, Richmond L, Nitabach M, Davies SA, Dow JA (2013) A biogenic amine and a neuropeptide act identically: tyramine signals through calcium in *drosophila* tubule stellate cells. Proc Biol Sci 280(1757):20122943. https://doi.org/10.1098/rspb.2012.2943

Cabrero P, Terhzaz S, Romero MF, Davies SA, Blumenthal EM, Dow JA (2014) Chloride channels in stellate cells are essential for uniquely high secretion rates in neuropeptide-stimulated *drosophila* diuresis. Proc Natl Acad Sci USA 111(39):14301–14306. https://doi.org/10.1073/pnas.1412706111

Cannell E, Dornan AJ, Halberg KA, Terhzaz S, Dow JAT, Davies SA (2016) The corticotropin-releasing factor-like diuretic hormone 44 (DH44) and kinin neuropeptides modulate desiccation and starvation tolerance in *Drosophila melanogaster*. Peptides 80:96–107. https://doi.org/10.1016/j.peptides.2016.02.004

Carlson SH, Beitz A, Osborn JW (1997) Intragastric hypertonic saline increases vasopressin and central Fos immunoreactivity in conscious rats. Am J Phys 272(3 Pt 2):R750–R758. https://doi.org/10.1152/ajpregu.1997.272.3.R750

Carlsson MA, Enell LE, Nassel DR (2013) Distribution of short neuropeptide F and its receptor in neuronal circuits related to feeding in larval *drosophila*. Cell Tissue Res 353(3):511–523. https://doi.org/10.1007/s00441-013-1660-4

Chan SM, Gu PL, Chu KH, Tobe SS (2003) Crustacean neuropeptide genes of the CHH/MIH/GIH family: implications from molecular studies. Gen Comp Endocrinol 134(3):214–219. https://doi.org/10.1016/s0016-6480(03)00263-6

Chen SH, Lin CY, Kuo CM (2005) *In silico* analysis of crustacean hyperglycemic hormone family. Mar Biotechnol (NY) 7(3):193–206. https://doi.org/10.1007/s10126-004-0020-5

Chiang RG, Davey KG (1988) A novel receptor capable of monitoring applied pressure in the abdomen of an insect. Science 241(4873):1665–1667. https://doi.org/10.1126/science.241.4873.1665

Chiang RG, Chiang JA, Davey KG (1990) Structure of the abdominal receptor responsive to internally applied pressure in the blood-feeding insect, *Rhodnius prolixus*. Cell Tissue Res 261(3):583–587. https://doi.org/10.1007/BF00313538

Chintapalli VR, Terhzaz S, Wang J, Al Bratty M, Watson DG, Herzyk P et al (2012) Functional correlates of positional and gender-specific renal asymmetry in *drosophila*. PLoS One 7(4): e32577. https://doi.org/10.1371/journal.pone.0032577

Chng WB, Sleiman MS, Schupfer F, Lemaitre B (2014) Transforming growth factor beta/Activin signaling functions as a sugar-sensing feedback loop to regulate digestive enzyme expression. Cell Rep 9(1):336–348. https://doi.org/10.1016/j.celrep.2014.08.064

Clark L, Zhang JR, Tobe S, Lange AB (2006) Proctolin: a possible releasing factor in the corpus cardiacum/corpus allatum of the locust. Peptides 27(3):559–566. https://doi.org/10.1016/j.peptides.2005.06.028

Clark L, Lange AB, Zhang JR, Tobe SS (2008) The roles of Dippu-allatostatin in the modulation of hormone release in Locusta migratoria. J Insect Physiol 54(6):949–958. https://doi.org/10.1016/j.jinsphys.2008.03.007

Clements J, Hens K, Francis C, Schellens A, Callaerts P (2008) Conserved role for the *drosophila* Pax6 homolog eyeless in differentiation and function of insulin-producing neurons. Proc Natl Acad Sci USA 105(42):16183–16188. https://doi.org/10.1073/pnas.0708330105

Coast G (2011) Serotonin has kinin-like activity in stimulating secretion by Malpighian tubules of the house cricket *Acheta domesticus*. Peptides 32(3):500–508. https://doi.org/10.1016/j.peptides.2010.07.021

Coast GM, Webster SG, Schegg KM, Tobe SS, Schooley DA (2001) The *Drosophila melanogaster* homologue of an insect calcitonin-like diuretic peptide stimulates V-ATPase activity in fruit fly Malpighian tubules. J Exp Biol 204(Pt 10):1795–1804. https://doi.org/10.1242/jeb.204.10.1795

Colombani J, Raisin S, Pantalacci S, Radimerski T, Montagne J, Leopold P (2003) A nutrient sensor mechanism controls *drosophila* growth. Cell 114(6):739–749. https://doi.org/10.1016/s0092-8674(03)00713-x

Dai L, Zitnan D, Adams ME (2007) Strategic expression of ion transport peptide gene products in central and peripheral neurons of insects. J Comp Neurol 500(2):353–367. https://doi.org/10.1002/cne.21192

Davies SA, Huesmann GR, Maddrell SH, O'Donnell MJ, Skaer NJ, Dow JA et al (1995) CAP2b, a cardioacceleratory peptide, is present in *drosophila* and stimulates tubule fluid secretion via cGMP. Am J Phys 269(6 Pt 2):R1321–R1326. https://doi.org/10.1152/ajpregu.1995.269.6.R1321

Davies SA, Stewart EJ, Huesmann GR, Skaer NJ, Maddrell SH, Tublitz NJ et al (1997) Neuropeptide stimulation of the nitric oxide signaling pathway in *Drosophila melanogaster* Malpighian tubules. Am J Phys 273(2 Pt 2):R823–R827. https://doi.org/10.1152/ajpregu.1997.273.2.R823

Delanoue R, Meschi E, Agrawal N, Mauri A, Tsatskis Y, McNeill H et al (2016) *Drosophila* insulin release is triggered by adipose stunted ligand to brain methuselah receptor. Science 353(6307): 1553–1556. https://doi.org/10.1126/science.aaf8430

Deveci D, Martin FA, Leopold P, Romero NM (2019) AstA signaling functions as an evolutionary conserved mechanism timing juvenile to adult transition. Curr Biol 29(5):813–822 e814. https://doi.org/10.1016/j.cub.2019.01.053

Dircksen H, Tesfai LK, Albus C, Nassel DR (2008) Ion transport peptide splice forms in central and peripheral neurons throughout postembryogenesis of *drosophila melanogaster*. J Comp Neurol 509(1):23–41. https://doi.org/10.1002/cne.21715

Dow JAT (2012) The versatile stellate cell—more than just a space-filler. J Insect Physiol 58(4): 467–472. https://doi.org/10.1016/j.jinsphys.2011.12.003

Dow JAT, Halberg KA, Terhzaz S, Davies SA (2018) *Drosophila* as a model for neuroendocrine control of renal homeostasis. In: Model Animals in Neuroendocrinology: From Worm to Mouse to Man, pp 81–100. https://doi.org/10.1002/9781119391128.ch4

Drexler AL, Harris CC, dela Pena MG, Asuncion-Uchi M, Chung S, Webster S et al (2007) Molecular characterization and cell-specific expression of an ion transport peptide in the tobacco hornworm, *Manduca sexta*. Cell Tissue Res 329(2):391–408. https://doi.org/10.1007/s00441-007-0391-9

Dus M, Lai JS, Gunapala KM, Min S, Tayler TD, Hergarden AC et al (2015) Nutrient sensor in the brain directs the action of the brain-gut axis in *drosophila*. Neuron 87(1):139–151. https://doi.org/10.1016/j.neuron.2015.05.032

Eigenheer RA, Nicolson SW, Schegg KM, Hull JJ, Schooley DA (2002) Identification of a potent antidiuretic factor acting on beetle Malpighian tubules. Proc Natl Acad Sci USA 99(1):84–89. https://doi.org/10.1073/pnas.012436199

Eigenheer RA, Wiehart UM, Nicolson SW, Schoofs L, Schegg KM, Hull JJ et al (2003) Isolation, identification and localization of a second beetle antidiuretic peptide. Peptides 24(1):27–34. https://doi.org/10.1016/s0196-9781(02)00273-5

Feingold D, Knogler L, Starc T, Drapeau P, O'Donnell MJ, Nilson LA et al (2019) *secCl* is a cys-loop ion channel necessary for the chloride conductance that mediates hormone-induced fluid secretion in *drosophila*. Sci Rep 9(1):7464. https://doi.org/10.1038/s41598-019-42849-9

Folk DG, Bradley TJ (2004) The evolution of recovery from desiccation stress in laboratory-selected populations of *Drosophila melanogaster*. J Exp Biol 207(Pt 15):2671–2678. https://doi.org/10.1242/jeb.01048

Fournier B, Girardie J (1988) A new function for the locust neuroparsins: stimulation of water reabsorption. J Insect Physiol 34(4):309–313. https://doi.org/10.1016/0022-1910(88)90141-2

Furuya K, Schegg KM, Wang H, King DS, Schooley DA (1995) Isolation and identification of a diuretic hormone from the mealworm *Tenebrio molitor*. Proc Natl Acad Sci USA 92(26): 12323–12327. https://doi.org/10.1073/pnas.92.26.12323

Furuya K, Schegg KM, Schooley DA (1998) Isolation and identification of a second diuretic hormone from *Tenebrio molitor*. Peptides 19(4):619–626. https://doi.org/10.1016/s0196-9781(97)00475-0

Gäde G (1985) Mode of action of the hypertrehalosaemic peptides from the American cockroach. Z Naturforsch 40c:670–676. https://doi.org/10.1515/znc-1985-9-1015

Gäde G, Auerswald L (2003) Mode of action of neuropeptides from the adipokinetic hormone family. Gen Comp Endocrinol 132(1):10–20. https://doi.org/10.1016/s0016-6480(03)00159-x

Gáliková M, Klepsatel P (2022) Ion transport peptide regulates energy intake, expenditure, and metabolic homeostasis in *drosophila*. Genetics. https://doi.org/10.1093/genetics/iyac150

Gáliková M, Diesner M, Klepsatel P, Hehlert P, Xu Y, Bickmeyer I et al (2015) Energy homeostasis control in *drosophila* adipokinetic hormone mutants. Genetics 201(2):665–683. https://doi.org/10.1534/genetics.115.178897

Gáliková M, Dircksen H, Nassel DR (2018) The thirsty fly: ion transport peptide (ITP) is a novel endocrine regulator of water homeostasis in *drosophila*. PLoS Genet 14(8):e1007618. https://doi.org/10.1371/journal.pgen.1007618
Galsgaard KD, Pedersen J, Knop FK, Holst JJ, Albrechtsen NJW (2019) Glucagon receptor signaling and lipid metabolism. Front Physiol 10:413. https://doi.org/10.3389/fphys.2019.00413
Géminard C, Rulifson EJ, Léopold P (2009) Remote control of insulin secretion by fat cells in *drosophila*. Cell Metab 10(3):199–207. https://doi.org/10.1016/j.cmet.2009.08.002
Gera J, Agard M, Nave H, Sajadi F, Thorat L, Kondo S et al (2024) Anti-diuretic hormone ITP signals via a guanylate cyclase receptor to modulate systemic homeostasis in *drosophila*. elife 13:RP97043. https://doi.org/10.7554/elife.97043.1
Ghosh AC, O'Connor MB (2014) Systemic Activin signaling independently regulates sugar homeostasis, cellular metabolism, and pH balance in *Drosophila melanogaster*. Proc Natl Acad Sci USA 111(15):5729–5734. https://doi.org/10.1073/pnas.1319116111
Gioino P, Murray BG, Ianowski JP (2014) Serotonin triggers cAMP and PKA-mediated intracellular calcium waves in Malpighian tubules of *Rhodnius prolixus*. Am J Physiol Regul Integr Comp Physiol 307(7):R828–R836. https://doi.org/10.1152/ajpregu.00561.2013
Girardie J, Girardie A, Huet JC, Pernollet JC (1989) Amino acid sequence of locust neuroparsins. FEBS Lett 245(1–2):4–8. https://doi.org/10.1016/0014-5793(89)80179-6
González Segarra AJ, Pontes G, Jourjine N, Del Toro A, Scott K (2023) Hunger- and thirst-sensing neurons modulate a neuroendocrine network to coordinate sugar and water ingestion. elife 12. https://doi.org/10.7554/eLife.88143
Gromada J, Franklin I, Wollheim CB (2007) α-Cells of the endocrine pancreas: 35 years of research but the enigma remains. Endocr Rev 28(1):84–116. https://doi.org/10.1210/er.2006-0007
Grönke S, Muller G, Hirsch J, Fellert S, Andreou A, Haase T et al (2007) Dual lipolytic control of body fat storage and mobilization in *drosophila*. PLoS Biol 5(6):e137. https://doi.org/10.1371/journal.pbio.0050137
Grönke S, Clarke DF, Broughton S, Andrews TD, Partridge L (2010) Molecular evolution and functional characterization of drosophila insulin-like peptides. PLoS Genet 6(2):e1000857. https://doi.org/10.1371/journal.pgen.1000857
Guo X, Yin C, Yang F, Zhang Y, Huang H, Wang J et al (2019) The cellular diversity and transcription factor code of *drosophila* enteroendocrine cells. Cell Rep 29(12):4172–4185. https://doi.org/10.1016/j.celrep.2019.11.048
Halberg KV, Denholm B (2024) Mechanisms of systemic osmoregulation in insects. Annu Rev Entomol 69:415–438. https://doi.org/10.1146/annurev-ento-040323-021222
Halberg KA, Terhzaz S, Cabrero P, Davies SA, Dow JA (2015) Tracing the evolutionary origins of insect renal function. Nat Commun 6:6800. https://doi.org/10.1038/ncomms7800
Halberg KA, Rainey SM, Veland IR, Neuert H, Dornan AJ, Klambt C et al (2016) The cell adhesion molecule Fasciclin2 regulates brush border length and organization in *drosophila* renal tubules. Nat Commun 7:11266. https://doi.org/10.1038/ncomms11266
Harthoorn LF, Oudejans RC, Diederen JH, Van der Horst DJ (2002) Coherence between biosynthesis and secretion of insect adipokinetic hormones. Peptides 23(4):629–634. https://doi.org/10.1016/s0196-9781(01)00672-6
Hector CE, Bretz CA, Zhao Y, Johnson EC (2009) Functional differences between two CRF-related diuretic hormone receptors in *drosophila*. J Exp Biol 212(19):3142–3147. https://doi.org/10.1242/jeb.033175
Hentze JL, Carlsson MA, Kondo S, Nassel DR, Rewitz KF (2015) The neuropeptide allatostatin a regulates metabolism and feeding decisions in *drosophila*. Sci Rep 5:11680. https://doi.org/10.1038/srep11680
Herman AM, Blumenthal EM (2006) Identification of the tyramine receptor in the *drosophila* Malpighian tubule. FASEB J 20(4):A345–A346. https://doi.org/10.1096/fasebj.20.4.A345-c

Hermann-Luibl C, Yoshii T, Senthilan PR, Dircksen H, Helfrich-Forster C (2014) The ion transport peptide is a new functional clock neuropeptide in the fruit fly *Drosophila melanogaster*. J Neurosci 34(29):9522–9536. https://doi.org/10.1523/JNEUROSCI.0111-14.2014

Huybrechts J, De Loof A, Schoofs L (2005) Melatonin-induced neuropeptide release from isolated locust corpora cardiaca. Peptides 26(1):73–80. https://doi.org/10.1016/j.peptides.2004.07.012

Ikeya T, Galic M, Belawat P, Nairz K, Hafen E (2002) Nutrient-dependent expression of insulin-like peptides from neuroendocrine cells in the CNS contributes to growth regulation in *drosophila*. Curr Biol 12(15):1293–1300. https://doi.org/10.1016/s0960-9822(02)01043-6

Ionescu A, Donini A (2012) *Aedes*CAPA-PVK-1 displays diuretic and dose dependent antidiuretic potential in the larval mosquito *Aedes aegypti* (Liverpool). J Insect Physiol 58(10):1299–1306. https://doi.org/10.1016/j.jinsphys.2012.07.002

Irvine HB (1969) Sodium and potassium secretion by isolated insect Malpighian tubules. Am J Phys 217(5):1520–1527. https://doi.org/10.1152/ajplegacy.1969.217.5.1520

Itoh T, Yokohari F, Tominaga Y (1984) Two types of antennal hygro- and thermoreceptive sensilla of the cricket, *Gryllus bimaculatus* (de Geer). Zool Sci 1(4):533–543. https://doi.org/10.34425/zs000058

Iversen A, Cazzamali G, Williamson M, Hauser F, Grimmelikhuijzen CJ (2002) Molecular cloning and functional expression of a *drosophila* receptor for the neuropeptides capa-1 and -2. Biochem Biophys Res Commun 299(4):628–633. https://doi.org/10.1016/s0006-291x(02)02709-2

Johard HA, Yoishii T, Dircksen H, Cusumano P, Rouyer F, Helfrich-Forster C et al (2009) Peptidergic clock neurons in *drosophila*: ion transport peptide and short neuropeptide F in subsets of dorsal and ventral lateral neurons. J Comp Neurol 516(1):59–73. https://doi.org/10.1002/cne.22099

Johnson EC, Shafer OT, Trigg JS, Park J, Schooley DA, Dow JA et al (2005) A novel diuretic hormone receptor in *drosophila*: evidence for conservation of CGRP signaling. J Exp Biol 208 (Pt 7):1239–1246. https://doi.org/10.1242/jeb.01529

Jourjine N, Mullaney BC, Mann K, Scott K (2016) Coupled sensing of hunger and thirst signals balances sugar and water consumption. Cell 166(4):855–866. https://doi.org/10.1016/j.cell.2016.06.046

Kapan N, Lushchak OV, Luo J, Nassel DR (2012) Identified peptidergic neurons in the *drosophila* brain regulate insulin-producing cells, stress responses and metabolism by coexpressed short neuropeptide F and corazonin. Cell Mol Life Sci 69(23):4051–4066. https://doi.org/10.1007/s00018-012-1097-z

Kataoka H, Troetschler RG, Li JP, Kramer SJ, Carney RL, Schooley DA (1989) Isolation and identification of a diuretic hormone from the tobacco hornworm, *Manduca sexta*. Proc Natl Acad Sci USA 86(8):2976–2980. https://doi.org/10.1073/pnas.86.8.2976

Kay I, Coast GM, Cusinato O, Wheeler CH, Totty NF, Goldsworthy GJ (1991) Isolation and characterization of a diuretic peptide from *Acheta domesticus*. Evidence for a family of insect diuretic peptides. Biol Chem Hoppe Seyler 372(7):505–512. https://doi.org/10.1515/bchm3.1991.372.2.505

Kay I, Patel M, Coast GM, Totty NF, Mallet AI, Goldsworthy GJ (1992) Isolation, characterization and biological activity of a CRF-related diuretic peptide from *Periplaneta americana* L. Regul Pept 42(3):111–122. https://doi.org/10.1016/0167-0115(92)90091-8

Kean L, Cazenave W, Costes L, Broderick KE, Graham S, Pollock VP et al (2002) Two nitridergic peptides are encoded by the gene *capability* in *Drosophila melanogaster*. Am J Physiol Regul Integr Comp Physiol 282(5):R1297–R1307. https://doi.org/10.1152/ajpregu.00584.2001

Keller R (1992) Crustacean neuropeptides: structures, functions and comparative aspects. Experientia 48(5):439–448. https://doi.org/10.1007/BF01928162

Kim J, Neufeld TP (2015) Dietary sugar promotes systemic TOR activation in *drosophila* through AKH-dependent selective secretion of Dilp3. Nat Commun 6:6846. https://doi.org/10.1038/ncomms7846

Kim SK, Rulifson EJ (2004) Conserved mechanisms of glucose sensing and regulation by *drosophila* corpora cardiaca cells. Nature 431(7006):316–320. https://doi.org/10.1038/nature02897

Kim A, Knudsen JG, Madara JC, Benrick A, Hill TG, Abdul Kadir L et al (2021) Arginine-vasopressin mediates counter-regulatory glucagon release and is diminished in type 1 diabetes. elife 10. https://doi.org/10.7554/eLife.72919

Klöcklerová V, Gáliková Z, Roller L, Žitňan D (2023) Differential expression of ITP and ITPL indicate multiple functions in the silkworm *Bombyx mori*. Cell Tissue Res 392(3):715–731. https://doi.org/10.1007/s00441-023-03752-y

Konner AC, Klockener T, Bruning JC (2009) Control of energy homeostasis by insulin and leptin: targeting the arcuate nucleus and beyond. Physiol Behav 97(5):632–638. https://doi.org/10.1016/j.physbeh.2009.03.027

Koyama T, Mirth CK (2016) Growth-blocking peptides as nutrition-sensitive signals for insulin secretion and body size regulation. PLoS Biol 14(2):e1002392. https://doi.org/10.1371/journal.pbio.1002392

Koyama T, Mirth CK (2018) Unravelling the diversity of mechanisms through which nutrition regulates body size in insects. Curr Opin Insect Sci 25(Supplement C):1–8. https://doi.org/10.1016/j.cois.2017.11.002

Koyama T, Texada MJ, Halberg KA, Rewitz K (2020) Metabolism and growth adaptation to environmental conditions in *drosophila*. Cell Mol Life Sci 77(22):4523–4551. https://doi.org/10.1007/s00018-020-03547-2

Koyama T, Naseem MT, Kolosov D, Vo CT, Mahon D, Jakobsen ASS et al (2021a) A unique Malpighian tubule architecture in *Tribolium castaneum* informs the evolutionary origins of systemic osmoregulation in beetles. Proc Natl Acad Sci USA 118(14). https://doi.org/10.1073/pnas.2023314118

Koyama T, Terhzaz S, Naseem MT, Nagy S, Rewitz K, Dow JAT et al (2021b) A nutrient-responsive hormonal circuit mediates an inter-tissue program regulating metabolic homeostasis in adult *drosophila*. Nat Commun 12(1):5178. https://doi.org/10.1038/s41467-021-25445-2

Koyama T, Rana DW, Halberg KV (2023) Managing fuels and fluids: network integration of osmoregulatory and metabolic hormonal circuits in the polymodal control of homeostasis in insects. Bioessays 45(9):e2300011. https://doi.org/10.1002/bies.202300011

Kubrak O, Koyama T, Ahrentlov N, Jensen L, Malita A, Naseem MT et al (2022) The gut hormone Allatostatin C/somatostatin regulates food intake and metabolic homeostasis under nutrient stress. Nat Commun 13(1):692. https://doi.org/10.1038/s41467-022-28268-x

Kunst M, Hughes ME, Raccuglia D, Felix M, Li M, Barnett G et al (2014) Calcitonin gene-related peptide neurons mediate sleep-specific circadian output in *drosophila*. Curr Biol 24(22):2652–2664. https://doi.org/10.1016/j.cub.2014.09.077

Kuramochi G, Kobayashi I (2000) Regulation of the urine concentration mechanism by the oropharyngeal afferent pathway in man. Am J Nephrol 20(1):42–47. https://doi.org/10.1159/000013554

Kurogi Y, Imura E, Mizuno Y, Hoshino R, Nouzova M, Matsuyama S et al (2023) Female reproductive dormancy in *drosophila* is regulated by DH31-producing neurons projecting into the corpus allatum. Development 150(10):dev201186. https://doi.org/10.1242/dev.201186

Lacombe C, Greve P, Martin G (1999) Overview on the sub-grouping of the crustacean hyperglycemic hormone family. Neuropeptides 33(1):71–80. https://doi.org/10.1054/npep.1999.0016

Landayan D, Wang BP, Zhou J, Wolf FW (2021) Thirst interneurons that promote water seeking and limit feeding behavior in *drosophila*. elife 10. https://doi.org/10.7554/eLife.66286

Lechner SG, Markworth S, Poole K, Smith ES, Lapatsina L, Frahm S et al (2011) The molecular and cellular identity of peripheral osmoreceptors. Neuron 69(2):332–344. https://doi.org/10.1016/j.neuron.2010.12.028

Lee G, Park JH (2004) Hemolymph sugar homeostasis and starvation-induced hyperactivity affected by genetic manipulations of the adipokinetic hormone-encoding gene in *Drosophila melanogaster*. Genetics 167(1):311–323. https://doi.org/10.1534/genetics.167.1.311

Lee KS, You KH, Choo JK, Han YM, Yu K (2004) *Drosophila* short neuropeptide F regulates food intake and body size. J Biol Chem 279(49):50781–50789. https://doi.org/10.1074/jbc.M407842200

Lee KS, Kwon OY, Lee JH, Kwon K, Min KJ, Jung SA et al (2008) *Drosophila* short neuropeptide F signalling regulates growth by ERK-mediated insulin signalling. Nat Cell Biol 10(4): 468–475. https://doi.org/10.1038/ncb1710

Lehmberg E, Ota RB, Furuya K, King DS, Applebaum SW, Ferenz HJ et al (1991) Identification of a diuretic hormone of *Locusta migratoria*. Biochem Biophys Res Commun 179(2):1036–1041. https://doi.org/10.1016/0006-291x(91)91923-z

Lin S, Owald D, Chandra V, Talbot C, Huetteroth W, Waddell S (2014) Neural correlates of water reward in thirsty *drosophila*. Nat Neurosci 17(11):1536–1542. https://doi.org/10.1038/nn.3827

Lin HH, Kuang MC, Hossain I, Xuan Y, Beebe L, Shepherd AK et al (2022) A nutrient-specific gut hormone arbitrates between courtship and feeding. Nature 602(7898):632–638. https://doi.org/10.1038/s41586-022-04408-7

Liu L, Li Y, Wang R, Yin C, Dong Q, Hing H et al (2007) *Drosophila* hygrosensation requires the TRP channels water witch and nanchung. Nature 450(7167):294–298. https://doi.org/10.1038/nature06223

Lopez-Arias B, Dorado B, Herrero P (2011) Blockade of the release of the neuropeptide leucokinin to determine its possible functions in fly behavior: chemoreception assays. Peptides 32(3): 545–552. https://doi.org/10.1016/j.peptides.2010.07.002

MacMillan HA, Nazal B, Wali S, Yerushalmi GY, Misyura L, Donini A et al (2018) Anti-diuretic activity of a CAPA neuropeptide can compromise *drosophila* chill tolerance. J Exp Biol 221 (Pt 19). https://doi.org/10.1242/jeb.185884

Maddrell SHP (1991) The fastest fluid-secreting cell known: the upper malpighian tubule of *Rhodnius*. BioEssays 13(7):357–362. https://doi.org/10.1002/bies.950130710

Maddrell SH, Herman WS, Mooney RL, Overton JA (1991) 5-Hydroxytryptamine: a second diuretic hormone in *Rhodnius prolixus*. J Exp Biol 156:557–566. https://doi.org/10.1242/jeb.156.1.557

Malita A, Kubrak O, Koyama T, Ahrentløv N, Texada MJ, Nagy S et al (2022) A gut-derived hormone suppresses sugar appetite and regulates food choice in *drosophila*. Nat Metab 4(11): 1532–1550. https://doi.org/10.1038/s42255-022-00672-z

Mellanby K (1958) Water content and insect metabolism. Nature 181(4620):1403–1403. https://doi.org/10.1038/1811403a0

Meschi E, Leopold P, Delanoue R (2019) An EGF-responsive neural circuit couples insulin secretion with nutrition in *drosophila*. Dev Cell 48(1):76–86 e75. https://doi.org/10.1016/j.devcel.2018.11.029

Nagai C, Mabashi-Asazuma H, Nagasawa H, Nagata S (2014) Identification and characterization of receptors for ion transport peptide (ITP) and ITP-like (ITPL) in the silkworm *Bombyx mori*. J Biol Chem 289(46):32166–32177. https://doi.org/10.1074/jbc.M114.590646

Nässel DR, Zandawala M (2019) Recent advances in neuropeptide signaling in *drosophila*, from genes to physiology and behavior. Prog Neurobiol 179:101607. https://doi.org/10.1016/j.pneurobio.2019.02.003

Nässel DR, Passier PC, Elekes K, Dircksen H, Vullings HG, Cantera R (1995) Evidence that locustatachykinin I is involved in release of adipokinetic hormone from locust corpora cardiaca. Regul Pept 57(3):297–310. https://doi.org/10.1016/0167-0115(95)00043-B

Noyes BE, Katz FN, Schaffer MH (1995) Identification and expression of the *drosophila* adipokinetic hormone gene. Mol Cell Endocrinol 109(2):133–141. https://doi.org/10.1016/0303-7207(95)03492-p

O'Donnell MJ, Maddrell SH (1995) Fluid reabsorption and ion transport by the lower Malpighian tubules of adult female *drosophila*. J Exp Biol 198(Pt 8):1647–1653. https://doi.org/10.1242/jeb.198.8.1647

O'Donnell MJ, Dow JA, Huesmann GR, Tublitz NJ, Maddrell SH (1996) Separate control of anion and cation transport in Malpighian tubules of *Drosophila melanogaster*. J Exp Biol 199(Pt 5): 1163–1175. https://doi.org/10.1242/jeb.199.5.1163

Oh Y, Lai JS, Mills HJ, Erdjument-Bromage H, Giammarinaro B, Saadipour K et al (2019) A glucose-sensing neuron pair regulates insulin and glucagon in *drosophila*. Nature 574(7779): 559–564. https://doi.org/10.1038/s41586-019-1675-4

Oh Y, Lai JS, Min S, Huang HW, Liberles SD, Ryoo HD et al (2021) Periphery signals generated by Piezo-mediated stomach stretch and Neuromedin-mediated glucose load regulate the *drosophila* brain nutrient sensor. Neuron 109(12):1979–1995 e1976. https://doi.org/10.1016/j.neuron.2021.04.028

Ohhara Y, Kobayashi S, Yamakawa-Kobayashi K, Yamanaka N (2018) Adult-specific insulin-producing neurons in *Drosophila melanogaster*. J Comp Neurol 526(8):1351–1367. https://doi.org/10.1002/cne.24410

Okamoto N, Nishimura T (2015) Signaling from glia and cholinergic neurons controls nutrient-dependent production of an insulin-like peptide for *drosophila* body growth. Dev Cell 35(3): 295–310. https://doi.org/10.1016/j.devcel.2015.10.003

Okamoto N, Yamanaka N, Yagi Y, Nishida Y, Kataoka H, O'Connor MB et al (2009) A fat body-derived IGF-like peptide regulates postfeeding growth in *drosophila*. Dev Cell 17(6):885–891. https://doi.org/10.1016/j.devcel.2009.10.008

Okamoto N, Nakamori R, Murai T, Yamauchi Y, Masuda A, Nishimura T (2013) A secreted decoy of InR antagonizes insulin/IGF signaling to restrict body growth in *drosophila*. Genes Dev 27(1):87–97. https://doi.org/10.1101/gad.204479.112

Ono M, Matsumura T, Sung EJ, Koyama T, Ochiai M, Shears SB et al (2024) *Drosophila* cytokine GBP2 exerts immune responses and regulates *GBP1* expression through GPCR receptor Mthl10. Insect Biochem Mol Biol 167:104086. https://doi.org/10.1016/j.ibmb.2024.104086

Orchard I (2006) Serotonin: a coordinator of feeding-related physiological events in the blood-gorging bug, *Rhodnius prolixus*. Comp Biochem Physiol A Mol Integr Physiol 144(3):316–324. https://doi.org/10.1016/j.cbpa.2005.11.010

Pannabecker T, Orchard I (1986) Octopamine and cyclic AMP mediate release of adipokinetic hormone I and II from isolated locust neuroendocrine tissue. Mol Cell Endocrinol 48(2–3): 153–159. https://doi.org/10.1016/0303-7207(86)90037-7

Papper S, Rosenbaum JD (1952) Diurnal variation in the diuretic response to ingested water. J Clin Invest 31(4):401–405. https://doi.org/10.1172/JCI102622

Park Y, Kim YJ, Adams ME (2002) Identification of G protein-coupled receptors for *drosophila* PRXamide peptides, CCAP, corazonin, and AKH supports a theory of ligand-receptor coevolution. Proc Natl Acad Sci USA 99(17):11423–11428. https://doi.org/10.1073/pnas.162276199

Phillips JE, Meredith J, Audsley N, Richardson N, Macins A, Ring M (1998) Locust ion transport peptide (ITP): a putative hormone controlling water and ionic balance in terrestrial insects. Am Zool 38(3):461–470. https://doi.org/10.1093/icb/38.3.461

Pietrantonio PV, Jagge C, McDowell C (2001) Cloning and expression analysis of a 5HT7-like serotonin receptor cDNA from mosquito *Aedes aegypti* female excretory and respiratory systems. Insect Mol Biol 10(4):357–369. https://doi.org/10.1046/j.0962-1075.2001.00274.x

Post S, Tatar M (2016) Nutritional geometric profiles of insulin/IGF expression in *Drosophila melanogaster*. PLoS One 11(5):e0155628. https://doi.org/10.1371/journal.pone.0155628

Post S, Karashchuk G, Wade JD, Sajid W, De Meyts P, Tatar M (2018) *Drosophila* insulin-like peptides DILP2 and DILP5 differentially stimulate cell signaling and glycogen phosphorylase to regulate longevity. Front Endocrinol (Lausanne) 9:245. https://doi.org/10.3389/fendo.2018.00245

Predel R, Wegener C, Russell WK, Tichy SE, Russell DH, Nachman RJ (2004) Peptidomics of CNS-associated neurohemal systems of adult *Drosophila melanogaster*: a mass spectrometric survey of peptides from individual flies. J Comp Neurol 474(3):379–392. https://doi.org/10.1002/cne.20145

Quinlan MC, O'Donnell MJ (1998) Anti-diuresis in the blood-feeding insect *Rhodnius prolixus* Stål: antagonistic actions of cAMP and cGMP and the role of organic acid transport. J Insect Physiol 44(7–8):561–568. https://doi.org/10.1016/s0022-1910(98)00047-x

Quinlan MC, Tublitz NJ, O'Donnell MJ (1997) Anti-diuresis in the blood-feeding insect *Rhodnius prolixus* Stal: the peptide CAP2b and cyclic GMP inhibit Malpighian tubule fluid secretion. J Exp Biol 200(Pt 17):2363–2367. https://doi.org/10.1242/jeb.200.17.2363

Radford JC, Davies SA, Dow JA (2002) Systematic G-protein-coupled receptor analysis in *Drosophila melanogaster* identifies a leucokinin receptor with novel roles. J Biol Chem 277(41):38810–38817. https://doi.org/10.1074/jbc.M203694200

Rajan A, Perrimon N (2012) *Drosophila* cytokine unpaired 2 regulates physiological homeostasis by remotely controlling insulin secretion. Cell 151(1):123–137. https://doi.org/10.1016/j.cell.2012.08.019

Rhea JM, Wegener C, Bender M (2010) The proprotein convertase encoded by *amontillado* (*amon*) is required in *drosophila* corpora cardiaca endocrine cells producing the glucose regulatory hormone AKH. PLoS Genet 6(5):e1000967. https://doi.org/10.1371/journal.pgen.1000967

Rosay P, Davies SA, Yu Y, Sozen MA, Kaiser K, Dow JA (1997) Cell-type specific calcium signalling in a *drosophila* epithelium. J Cell Sci 110(Pt 15):1683–1692. https://doi.org/10.1242/jcs.110.15.1683

Rosenbaum T, Islas LD (2023) Molecular physiology of TRPV channels: controversies and future challenges. Annu Rev Physiol 85:293–316. https://doi.org/10.1146/annurev-physiol-030222-012349

Ruaud AF, Lam G, Thummel CS (2011) The *drosophila* NR4A nuclear receptor DHR38 regulates carbohydrate metabolism and glycogen storage. Mol Endocrinol 25(1):83–91. https://doi.org/10.1210/me.2010-0337

Salehi A, Vieira E, Gylfe E (2006) Paradoxical stimulation of glucagon secretion by high glucose concentrations. Diabetes 55(8):2318–2323. https://doi.org/10.2337/db06-0080

Sano H, Nakamura A, Texada MJ, Truman JW, Ishimoto H, Kamikouchi A et al (2015) The nutrient-responsive hormone CCHamide-2 controls growth by regulating insulin-like peptides in the brain of Drosophila melanogaster. PLoS Genet 11(5):e1005209. https://doi.org/10.1371/journal.pgen.1005209

Santos JG, Pollak E, Rexer KH, Molnar L, Wegener C (2006) Morphology and metamorphosis of the peptidergic Va neurons and the median nerve system of the fruit fly, *Drosophila melanogaster*. Cell Tissue Res 326(1):187–199. https://doi.org/10.1007/s00441-006-0211-7

Sarraf-Zadeh L, Christen S, Sauer U, Cognigni P, Miguel-Aliaga I, Stocker H et al (2013) Local requirement of the drosophila insulin binding protein imp-L2 in coordinating developmental progression with nutritional conditions. Dev Biol. https://doi.org/10.1016/j.ydbio.2013.06.008

Schaffer MH, Noyes BE, Slaughter CA, Thorne GC, Gaskell SJ (1990) The fruitfly *Drosophila melanogaster* contains a novel charged adipokinetic-hormone-family peptide. Biochem J 269(2):315–320. https://doi.org/10.1042/bj2690315

Schooneveld H, van Herp F, van Minnen J (1987) Demonstration of substances immunologically related to the identified arthropod neuropeptides AKH/RPCH in the CNS of several invertebrate species. Brain Res 406(1–2):224–232. https://doi.org/10.1016/0006-8993(87)90786-4

Scopelliti A, Bauer C, Yu Y, Zhang T, Kruspig B, Murphy DJ et al (2019) A neuronal relay mediates a nutrient responsive gut/fat body axis regulating energy homeostasis in adult *drosophila*. Cell Metab 29(2):269–284e210. https://doi.org/10.1016/j.cmet.2018.09.021

Semaniuk UV, Gospodaryov DV, Feden'ko KM, Yurkevych IS, Vaiserman AM, Storey KB et al (2018) Insulin-like peptides regulate feeding preference and metabolism in *drosophila*. Front Physiol 9:1083. https://doi.org/10.3389/fphys.2018.01083

Semaniuk U, Piskovatska V, Strilbytska O, Strutynska T, Burdyliuk N, Vaiserman A et al (2021) *Drosophila* insulin-like peptides: from expression to functions—a review. Entomol Exp Appl 169(2):195–208. https://doi.org/10.1111/eea.12981

Senapati B, Tsao CH, Juan YA, Chiu TH, Wu CL, Waddell S et al (2019) A neural mechanism for deprivation state-specific expression of relevant memories in *drosophila*. Nat Neurosci 22(12):2029–2039. https://doi.org/10.1038/s41593-019-0515-z

Sidyelyeva G, Wegener C, Schoenfeld BP, Bell AJ, Baker NE, McBride SM et al (2010) Individual carboxypeptidase D domains have both redundant and unique functions in *drosophila*

development and behavior. Cell Mol Life Sci 67(17):2991–3004. https://doi.org/10.1007/s00018-010-0369-8
Siegert K, Morgan P, Mordue W (1985) Primary structures of locust adipokinetic hormones II. Biol Chem Hoppe Seyler 366(8):723–727. https://doi.org/10.1515/bchm3.1985.366.2.723
Slaidina M, Delanoue R, Gronke S, Partridge L, Leopold P (2009) A *drosophila* insulin-like peptide promotes growth during nonfeeding states. Dev Cell 17(6):874–884. https://doi.org/10.1016/j.devcel.2009.10.009
Song W, Veenstra JA, Perrimon N (2014) Control of lipid metabolism by tachykinin in *drosophila*. Cell Rep 9(1):40–47. https://doi.org/10.1016/j.celrep.2014.08.060
Song W, Cheng D, Hong S, Sappe B, Hu Y, Wei N et al (2017) Midgut-derived activin regulates glucagon-like action in the fat body and glycemic control. Cell Metab 25(2):386–399. https://doi.org/10.1016/j.cmet.2017.01.002
Spring JH, Morgan AM, Hazelton SR (1988) A novel target for antidiuretic hormone in insects. Science 241(4869):1096–1098. https://doi.org/10.1126/science.241.4869.1096
Staubli F, Jørgensen TJ, Cazzamali G, Williamson M, Lenz C, Søndergaard L et al (2002) Molecular identification of the insect adipokinetic hormone receptors. Proc Natl Acad Sci USA 99(6):3446–3451. https://doi.org/10.1073/pnas.052556499
Stone JV, Mordue W, Batley KE, Morris HR (1976) Structure of locust adipokinetic hormone, a neurohormone that regulates lipid utilisation during flight. Nature 263(5574):207–211. https://doi.org/10.1038/263207a0
Sudhakar SR, Pathak H, Rehman N, Fernandes J, Vishnu S, Varghese J (2020) Insulin signalling elicits hunger-induced feeding in drosophila. Dev Biol 459(2):87–99. https://doi.org/10.1016/j.ydbio.2019.11.013
Sun L, Zhang Z, Zhang R, Yu Y, Yang F, Tan A (2020) Molecular disruption of ion transport peptide receptor results in impaired water homeostasis and developmental defects in *Bombyx mori*. Front Physiol 11:424. https://doi.org/10.3389/fphys.2020.00424
Sung EJ, Ryuda M, Matsumoto H, Uryu O, Ochiai M, Cook ME et al (2017) Cytokine signaling through *drosophila* Mthl10 ties lifespan to environmental stress. Proc Natl Acad Sci USA 114(52):13786–13791. https://doi.org/10.1073/pnas.1712453115
Terhzaz S, O'Connell FC, Pollock VP, Kean L, Davies SA, Veenstra JA et al (1999) Isolation and characterization of a leucokinin-like peptide of *Drosophila melanogaster*. J Exp Biol 202 (Pt 24):3667–3676. https://doi.org/10.1242/jeb.202.24.3667
Terhzaz S, Cabrero P, Robben JH, Radford JC, Hudson BD, Milligan G et al (2012) Mechanism and function of *drosophila* capa GPCR: a desiccation stress-responsive receptor with functional homology to human neuromedinU receptor. PLoS One 7(1):e29897. https://doi.org/10.1371/journal.pone.0029897
Terhzaz S, Teets NM, Cabrero P, Henderson L, Ritchie MG, Nachman RJ et al (2015) Insect capa neuropeptides impact desiccation and cold tolerance. Proc Natl Acad Sci USA 112(9): 2882–2887. https://doi.org/10.1073/pnas.1501518112
Texada MJ, Koyama T, Rewitz K (2020) Regulation of body size and growth control. Genetics 216(2):269–313. https://doi.org/10.1534/genetics.120.303095
Tichy H (1979) Hygro- and thermoreceptive triad in antennal sensillum of the stick insect, *Carausius morosus*. J Comp Physiol 132(2):149–152. https://doi.org/10.1007/BF00610718
Toprak U (2020) The role of peptide hormones in insect lipid metabolism. Front Physiol 11:434. https://doi.org/10.3389/fphys.2020.00434
Tsutsui N (2021) Chapter 77—ion transport peptide. In: Ando H, Ukena K, Nagata S (eds) Handbook of hormones, 2nd edn. Academic Press, San Diego, pp 735–737
Tublitz NJ, Allen AT, Cheung CC, Edwards KK, Kimble DP, Loi PK et al (1992) Insect cardioactive peptides: regulation of hindgut activity by cardioacceleratory peptide 2 (CAP2) during wandering behaviour in *Manduca sexta* larvae. J Exp Biol 165:241–264. https://doi.org/10.1242/jeb.165.1.241

Van der Horst DJ (2003) Insect adipokinetic hormones: release and integration of flight energy metabolism. Comp Biochem Physiol B Biochem Mol Biol 136(2):217–226. https://doi.org/10.1016/s1096-4959(03)00151-9

Van der Horst DJ, Van Marrewijk WJ, Diederen JH (2001) Adipokinetic hormones of insect: release, signal transduction, and responses. Int Rev Cytol 211:179–240. https://doi.org/10.1016/s0074-7696(01)11019-3

van Marrewijk WJ, van den Broek AT, Beenakkers AM (1986) Hormonal control of fat-body glycogen mobilization for locust flight. Gen Comp Endocrinol 64(1):136–142. https://doi.org/10.1016/0016-6480(86)90039-0

Van Wielendaele P, Dillen S, Marchal E, Badisco L, Vanden Broeck J (2012) CRF-like diuretic hormone negatively affects both feeding and reproduction in the desert locust, *Schistocerca gregaria*. PLoS One 7(2):e31425. https://doi.org/10.1371/journal.pone.0031425

Vullings HG, Ten Voorde SE, Passier PC, Diederen JH, Van Der Horst DJ, Nassel DR (1998) A possible role of SchistoFLRFamide in inhibition of adipokinetic hormone release from locust corpora cardiaca. J Neurocytol 27(12):901–913. https://doi.org/10.1023/a:1006901123566

Wang YJ, Zhao Y, Meredith J, Phillips JE, Theilmann DA, Brock HW (2000) Mutational analysis of the C-terminus in ion transport peptide (ITP) expressed in *drosophila* Kc1 cells. Arch Insect Biochem Physiol 45(3):129–138. https://doi.org/10.1002/1520-6327(200011)45:3<129::AID-ARCH4>3.0.CO;2-L

Wang S, Tulina N, Carlin DL, Rulifson EJ (2007) The origin of islet-like cells in *drosophila* identifies parallels to the vertebrate endocrine axis. Proc Natl Acad Sci USA 104(50): 19873–19878. https://doi.org/10.1073/pnas.0707465104

Waterson MJ, Chung BY, Harvanek ZM, Ostojic I, Alcedo J, Pletcher SD (2014) Water sensor *ppk28* modulates *drosophila* lifespan and physiology through AKH signaling. Proc Natl Acad Sci USA 111(22):8137–8142. https://doi.org/10.1073/pnas.1315461111

Wegener C, Reinl T, Jansch L, Predel R (2006) Direct mass spectrometric peptide profiling and fragmentation of larval peptide hormone release sites in *Drosophila melanogaster* reveals tagma-specific peptide expression and differential processing. J Neurochem 96(5):1362–1374. https://doi.org/10.1111/j.1471-4159.2005.03634.x

Wharton DR, Wharton ML, Lola J (1965) Blood volume and water content of the male American cockroach, *Periplaneta Americana* L.-methods and the influence of age and starvation. J Insect Physiol 11:391–404. https://doi.org/10.1016/0022-1910(65)90046-6

Wiehart UI, Nicolson SW, Eigenheer RA, Schooley DA (2002) Antagonistic control of fluid secretion by the Malpighian tubules of *Tenebrio molitor*: effects of diuretic and antidiuretic peptides and their second messengers. J Exp Biol 205(Pt 4):493–501. https://doi.org/10.1242/jeb.205.4.493

Wigglesworth VB (1931a) The physiology of excretion in a blood-sucking insect, *Rhodnius prolixus* (Hemiptera, Reduviidae): II. Anatomy and histology of the excretory system. J Exp Biol 8(4):428–441. https://doi.org/10.1242/jeb.8.4.428

Wigglesworth VB (1931b) The physiology of excretion in a blood-sucking insect, *Rhodnius prolixus* (Hemiptera, Reduviidae): III. The mechanism of uric acid excretion. J Exp Biol 8(4): 443–451. https://doi.org/10.1242/jeb.8.4.443

Wigglesworth VB (1949) Insect biochemistry. Annu Rev Biochem 18:595–614. https://doi.org/10.1146/annurev.bi.18.070149.003115

Yamada T, Habara O, Kubo H, Nishimura T (2018) Fat body glycogen serves as a metabolic safeguard for the maintenance of sugar levels in *drosophila*. Development 145(6). https://doi.org/10.1242/dev.158865

Yang Z, Huang R, Fu X, Wang G, Qi W, Mao D et al (2018) A post-ingestive amino acid sensor promotes food consumption in *drosophila*. Cell Res. https://doi.org/10.1038/s41422-018-0084-9

Yokohari F, Tateda H (1976) Moist and dry hygroreceptors for relative humidity of the cockroach, *Periplaneta americana* L. J Comp Physiol 106(2):137–152. https://doi.org/10.1007/BF00620495

Yoshinari Y, Kosakamoto H, Kamiyama T, Hoshino R, Matsuoka R, Kondo S et al (2021) The sugar-responsive enteroendocrine neuropeptide F regulates lipid metabolism through glucagon-like and insulin-like hormones in *Drosophila melanogaster*. Nat Commun 12(1):4818. https://doi.org/10.1038/s41467-021-25146-w

Zandawala M, Marley R, Davies SA, Nassel DR (2018a) Characterization of a set of abdominal neuroendocrine cells that regulate stress physiology using colocalized diuretic peptides in *drosophila*. Cell Mol Life Sci 75(6):1099–1115. https://doi.org/10.1007/s00018-017-2682-y

Zandawala M, Yurgel ME, Texada MJ, Liao S, Rewitz KF, Keene AC et al (2018b) Modulation of *drosophila* post-feeding physiology and behavior by the neuropeptide leucokinin. PLoS Genet 14(11):e1007767. https://doi.org/10.1371/journal.pgen.1007767

Zandawala M, Nguyen T, Balanya Segura M, Johard HAD, Amcoff M, Wegener C et al (2021) A neuroendocrine pathway modulating osmotic stress in *drosophila*. PLoS Genet 17(3):e1009425. https://doi.org/10.1371/journal.pgen.1009425

Zhang H, Liu J, Li CR, Momen B, Kohanski RA, Pick L (2009) Deletion of *drosophila* insulin-like peptides causes growth defects and metabolic abnormalities. Proc Natl Acad Sci USA 106(46): 19617–19622. https://doi.org/10.1073/pnas.0905083106

Zhao X, Karpac J (2017) Muscle directs diurnal energy homeostasis through a Myokine-dependent hormone module in *drosophila*. Curr Biol 27(13):1941–1955 e1946. https://doi.org/10.1016/j.cub.2017.06.004

Chapter 10
Regulation of Insect Reproduction

Yosuke Mizuno and Ryusuke Niwa

Abstract Reproductive success is crucial to the persistence of insect and other animal populations. The basic processes of reproduction common to almost all insects can be divided into three categories: (1) gametogenesis to prepare mature sperm and eggs, (2) attraction of males through female pheromones, and (3) sperm transfer through mating and induction of female post-mating responses. Previous studies have shown that multiple crucial peptide hormones regulate the reproductive processes. This chapter focuses on the function of these identified peptide hormones, particularly those required for the proper control of egg maturation, pheromone biosynthesis, and female post-mating responses. The peptide hormones described in this chapter are neuroparsin, ovary-maturating parsin, ovary ecdysteroidogenic hormone, neuropeptide F, short neuropeptide F, insulin-like peptides, allatostatin-C, calcitonin-like diuretic hormone, corazonin, ecdysis-triggering hormone, trypsin-modulating oostatic factor, pheromone biosynthesis-activating neuropeptide, and sex peptide. These hormones are crucial, at least in part, for regulating reproduction in response to external environmental cues, such as temperature, photoperiod, nutrition, and mating stimuli.

Keywords Reproduction · Oogenesis · Egg maturation · Pheromone · Post-mating response · Ovary ecdysteroidogenic hormone (OEH) · Neuroparsin · Insulin-like peptides (ILPs) · Neuropeptide F (NPF) · Vitellogenesis

Y. Mizuno
Degree Programs in Life and Earth Sciences, Graduate School of Science and Technology, University of Tsukuba, Tsukuba, Japan

R. Niwa (✉)
Life Science Center for Survival Dynamics, Tsukuba Advanced Research Alliance (TARA), University of Tsukuba, Tsukuba, Japan
e-mail: ryusuke-niwa@tara.tsukuba.ac.jp

A. Mizoguchi, N. Okamoto (eds.), *Insect Peptide Hormones and Neuropeptides*, Entomology Monographs, https://doi.org/10.1007/978-981-95-7510-7_10

10.1 Introduction

Reproduction in animals is essential for producing offspring. This is no exception in insects. Among insect species, there are considerable variations in processes of gamete formation, the strategies that males and females use to encounter each other, the sequence of mating behaviors, and the effects of mating. Nevertheless, the basic processes common to all insect species can be summarized as follows: (i) gametogenesis to prepare mature sperm and eggs, (ii) attraction of males through female pheromones, and (iii) sperm transfer through mating and the induction of female post-mating responses (PMRs). Historically, many studies indicated that two of the most important insect hormones, i.e., juvenile hormones (JHs) and ecdysteroids, are deeply involved in controlling these processes, although there are species-specific differences. The roles of JHs and ecdysteroids in insect reproduction are not the focus of this chapter; therefore, we leave the overview of their roles to the published review articles (Roy et al. 2018; Song and Zhou 2020; Khalid et al. 2021; Leyria 2024; Leyria et al. 2025).

In addition to JHs and ecdysteroids, some of these reproductive processes are regulated, at least in part, by multiple crucial peptide hormones and neuropeptides (PHs/NPs). This chapter focuses on the function of these identified PHs/NPs, particularly those required for the proper control of egg maturation, pheromone biosynthesis, and female PMRs. All the hormones mentioned in this chapter are listed in Table 10.1.

10.2 Peptide Hormones/Neuropeptides Regulating Mature Egg Formation

Insect oogenesis is a complex process involving several stages (Büning 1994). This process begins with the germline stem cells (GSCs) in the ovaries, each of which divides asymmetrically to produce a daughter GSC and a cystoblast. Cystoblasts undergo several rounds of division to form a cystocyte complex, consisting of one oocyte and multiple nurse cells. This complex is enveloped by a layer of follicular epithelial cells. Oogenesis progresses through various stages, from pre-vitellogenesis to vitellogenesis, during which yolk accumulation occurs. Oocytes undergo significant growth during oogenesis, primarily due to the transfer of cytoplasm from the nurse cells. Growth is accompanied by the synthesis and accumulation of proteins, lipids, and other macromolecules necessary for embryogenesis after fertilization. This process is tightly regulated and involves the coordination of various cellular and molecular mechanisms to ensure the proper development of mature eggs ready for fertilization. Several PHs/NPs regulate the biosynthesis of JHs and ecdysteroids, both of which are essential for mature egg formation in most insects. Additional PHs/NPs directly affect the gonads to control gametogenesis in a wide variety of insects, as described below. A schematic diagram

Table 10.1 Peptides mentioned in Chap. 10

Peptide	Amino acid sequence (N-terminus > C = terminus)	Species	Production site
AST-C	pEVRYRQCYFNPISCF	*Drosophila melanogaster*	Brain
CRZ	pyroglutamicQTFQYSRGWTN-amide	*Drosophila melanogaster*	Brain
DILP2 A chain	QGIVERCCKKSCDMKALREYCSVVRN	*Drosophila melanogaster*	Brain
DILP2 B chain	TLCSEKLNEVLSMVCEEYNPVIPH	*Drosophila melanogaster*	Brain
DH31	TVDFGLARGYSGTQEAK HRMGLAAANFAGGP-amide	*Drosophila melanogaster*	Brain
ETH	DDSSPGFFLKITKNVPRL-amide	*Drosophila melanogaster*	Inka cell
MIP1	AWQSLQSSW-amide	*Drosophila melanogaster*	Subesophageal ganglion
DILP8 A chain	DHSSRSYNNIPYCCLNQCEEEFFC	*Drosophila melanogaster*	Ovary, imagi-nal discs
DILP8 B chain	SFCSLERMKKFAMEA CEHLFQADEGARR	*Drosophila melanogaster*	Ovary, imagi-nal discs
Neuroparsin A	NPISRSCEGANCVVDLTRCEY GDVTDFFGRKVCAKGPGDKCGGPYE LHGKCGVGMDCRCGLCSGCSLHN LQCFFFEGGLPSSC	*Locust migratoria*	Brain
NPF	SNSRPPRKNDVNTMADAYKFLQDLD TYYGDRARVRF-amide	*Drosophila melanogaster*	Brain, Midgut
OEH	QPTNVLEIRCKLYSGPAVQNTGECVH GAELNPCGKLSCLKGVGDKCGESTAG IIMSGKCASGLMCCGGQCVGCK NGICDHRLCPPRLTMNHHPFGLGLMAGSPQQ QQPVGVFPSLYKMFDYYSSESA	*Aedes aegypti*	Brain
OMP	YYEAPPDGRHLLLQPAPAAPAVAPA APASWPHQQRRQALDEFAAAAAAAADAQFQDEE EDGGRRV	*Locust migratoria*	Brain
PBAN	LSDDMPATPADQEMYRQ DPEQIDSRTKYFSPRL-amide	*Helicoverpa zea*	Subesophageal ganglion
sNPF	ARGPQLRLRF-amide	*Leptinotarsa decemlineata*	Brain
SP	WEWPWNRKPTKFPIPSPN PRDKWCRLNLGPAWGGRC	*Drosophila melanogaster*	Accessory gland
TMOF	YDPAPPPPPP	*Aedes aegypti*	Ovary
TMOF	NPTNLH	*Neobellieria bullata*	Ovary

The peptide sequence information and tissues of presence are listed for only one representative species

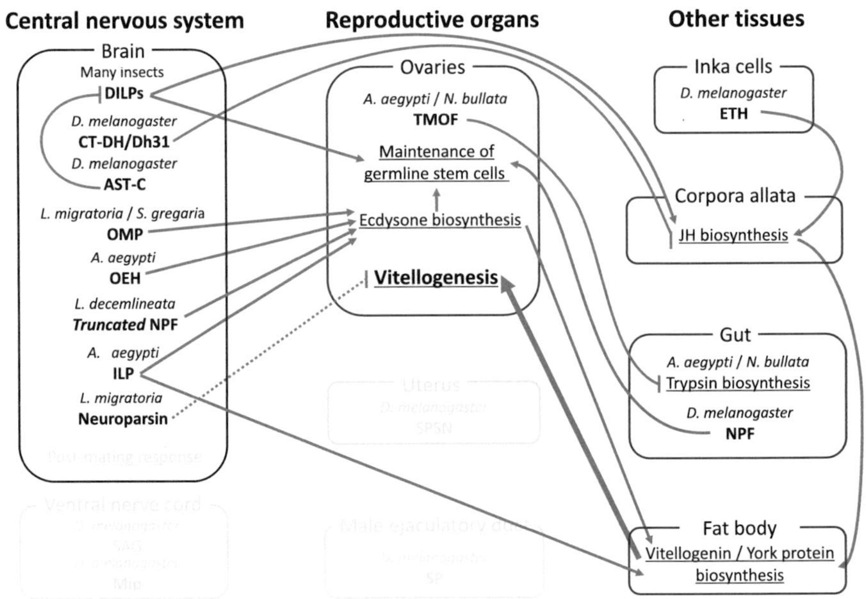

Fig. 10.1 Schematic diagram of hormone-mediated interorgan communications to regulate oogenesis in the ovaries. Abbreviations: DILPs *Drosophila* insulin-like peptides, CT-DH calcitonin-like diuretic hormone, DH31 diuretic hormone 31, PISCF/AST PISCF-related allatostatin, AST-C allatostatin C, OMP ovary maturating parsin, OEH ovary ecdysteroidogenic hormone, TMOF trypsin-modulating oostatic factor, ETH ecdysis-triggering hormone, JH juvenile hormone, NPF neuropeptide F

of the PH/NP-mediated interorgan communications that regulates oogenesis in the ovaries is shown in Fig. 10.1.

10.2.1 Neuroparsin

In the early days of identifying the PHs involved in vitellogenesis and oogenesis, biochemical and physiological studies using locusts made essential contributions. To our knowledge, neuroparsin (see Sect. 2.5.2), which was originally isolated from the corpora cardiaca (CC) of the migratory locust *Locusta migratoria* (Girardie et al. 1987b, 1990), is the first PH clearly shown to be involved in ovarian development. A subsequent study confirmed that neuroparsin is present in the pars intercerebralis (PI) region of the brain of *L. migratoria*.

Both bioassays and gene-silencing approaches have indicated that neuroparsin plays a role in suppressing ovarian development in multiple species. After determining the primary structure of neuroparsin, it was found that its injection exerts anti-JH effects, likely by acting on the central region of the PI, inhibiting vitellogenesis in *L. migratoria*. In contrast, the injection of anti-neuroparsin antiserum promotes vitellogenesis in locusts (Girardie et al. 1987a). RNA interference (RNAi) of neuroparsin in the desert locust *Schistocerca gregaria* enhances the transcription of the *vitellogenin* gene—which encodes an essential egg yolk component—and oocyte growth (Badisco et al. 2011). In addition, RNAi targeting *ecdysone receptor* (*EcR*) and its heterodimeric partner, *retinoid X receptor* (*RXR*), leads to a reduction in the transcript levels of two of the four *neuroparsin* genes (Lenaerts et al. 2019), consistent with the fact that ecdysone signaling promotes ovarian development. The crucial role of neuroparsin in reproduction has recently been recognized in social ants and bees, including *Harpegnathos venator*, *Atta cephalotes*, and *Monomorium pharaonis* (Nagel et al. 2020; Zhang et al. 2024; Gilbert et al. 2025). Proper caste differentiation to produce individuals with reproductive and non-reproductive roles, such as queens and workers, respectively, is crucial for maintaining their social order in colonies in these social insects. The *neuroparsin A* gene plays an essential role in differentiation, as it is highly expressed in the brains of non- or reduced-reproductive individuals, such as workers. Mechanistically, at least in *H. venator*, neuroparsin A presumably acts through the putative neuroparsin A receptor, Venus kinase receptor (VKR; see Sect. 10.2.2), to regulate the expression of *JH binding protein* positively and the ecdysone biosynthesis enzyme gene *shadow* negatively, suppressing oogenesis. Conversely, downregulation of neuroparsin A induces elevated oogenesis (Gilbert et al. 2025).

Neuroparsin genes are conserved among many insect species and even in some crustaceans, whereas they are absent in the genomes of the fruit fly *Drosophila melanogaster* and closely related species (Veenstra 2010). In the mud crab *Scylla paramamosain*, administration of recombinant neuroparsin peptides leads to a reduction in *vitellogenin* mRNA levels in the hepatopancreas, similar to effects observed in insects (Liu et al. 2020). In some species, the *neuroparsin* gene is duplicated within their genomes.

Notably, neuroparsin has a high degree of N-terminal amino acid sequence similarity with human insulin-like growth factor (IGF)-binding protein 7 (IGFBP7) (Badisco et al. 2007). The N-terminal regions of IGFBPs are critical for IGF binding. Reflecting this sequence similarity, a binding assay using *S. gregaria* insulin-related peptide (IRP) and neuroparsin showed that the two polypeptides could physically interact with each other (Badisco et al. 2008). This result suggests that, similar to IGFBPs, neuroparsin may function by binding to IRP. Further studies to identify the in vivo targets of neuroparsin will shed light on the mechanisms by which it inhibits vitellogenesis.

10.2.2 Ovary Ecdysteroidogenic Hormone (OEH)

OEH (see Sect. 2.5.2) is a PH with 92 amino acids identified in mosquitoes and shows high amino acid sequence homology with neuroparsin (Brown et al. 1998). Although OEH belongs to the neuroparsin superfamily, its physiological functions are significantly different from those of neuroparsin. Therefore, we present this as a different hormone.

The reproductive cycle of female mosquitoes is activated by ingestion of blood from vertebrate hosts. A substance in the mosquito brain promotes egg maturation after the ingestion of a blood meal. Evidence for this comes from a study (Lea 1967) that demonstrated that surgical removal of the neurosecretory cells in the PI region of the brain (PI-NSCs) immediately after eclosion inhibits egg maturation, even following blood feeding, in the yellow fever mosquito *Aedes aegypti* and other aedine mosquitoes. Notably, the arrested egg development in these operated mosquitoes is rescued by the transplantation of the surgically isolated PI-NSCs. Furthermore, transplantation of PI-NSCs failed to restore egg development arrested by removal of the corpora allata (CA), and transplantation of the CA did not restore egg development arrested by PI-NSC ablation, suggesting that JH and the substances produced in the mosquito PI-NSCs regulate different processes of egg maturation (Lea 1967). Later, a study reported the isolation of several ecdysteroidogenic peptides from the heads of female *A. aegypti* using conventional liquid chromatography and a bioassay for gonadotropic activity using autogenous *A. atropalpus* (Matsumoto et al. 1989). In the bioassay, injection of one of the purified peptides induced yolk accumulation in *A. atropalpus* females decapitated during eclosion. Furthermore, the purified peptide stimulated ecdysteroid biosynthesis in the ovaries of *A. aegypti*. Based on these results, the purified peptide was designated "ovarian ecdysteroidogenic hormone (OEH)," and the entire amino acid sequence of mature OEH from *A. aegypti* was determined (Brown et al. 1998).

OEH is produced by PI-NSCs and released into the hemolymph. The release of OEH is triggered by blood feeding, which is consistent with the fact that blood feeding stimulates ovarian ecdysteroid biosynthesis (Brown and Cao 2001). OEH in the hemolymph is received by VKR, a subtype of receptor tyrosine kinases (RTKs), which acts as an OEH receptor and is primarily present in ovarian follicular epithelial cells (Vogel et al. 2015). It is noteworthy that the strategy used to identify the OEH receptor was elegant. As we mentioned above, neuroparsin family members are absent in *D. melanogaster*. Therefore, researchers have hypothesized that a receptor for OEH is present in most insect genomes, but not in the *D. melanogaster* genome. Phylogenetic analysis revealed that a *RTK* gene similar to the *insulin receptor* (*InR*) gene is present in the genomes of three mosquito species and the cactophilic fruit fly *D. mojavensis*, but not in *D. melanogaster*. An RNAi experiment confirmed that the knockdown of the RTK (VKR) disabled OEH-mediated egg formation in *A. aegypti*, suggesting that VKR is indeed an OEH receptor. Moreover, upon ligand binding, activated VKR induces the phosphorylation of the intracellular signal transduction molecule AKT (Vogel et al. 2015), which seems to activate ovarian ecdysteroid biosynthesis and promote egg maturation.

OEH is the only neuroparsin family member present in the genomes of the mosquitoes examined, including *A. aegypti* (Dhara et al. 2013). As described above, the function of mosquito OEH differs from that of neuroparsin, as OEH and neuroparsin affect ovarian development positively and negatively, respectively. Therefore, during insect evolution, mosquito "neuroparsin" has acquired an opposite effect on regulating ovarian development. However, the evolutionary and physiological significance of this process is not fully understood.

10.2.3 Ovary Maturating Parsin (OMP)

Biochemical and physiological studies of the CC of the migratory locust *L. migratoria* have contributed to the identification of OMP, another hormone that regulates ovarian development.

In one study (Couillaud et al. 1989), 20 chromatographic fractions of *L. migratoria* CC extracts were collected and injected into females at various developmental stages, including the adult stages, to screen for their effects on oocyte growth. While some fractions contained neuroparsin, which exhibits an antagonistic effect on oocyte growth, another fraction stimulated ovarian maturation. This fraction did not influence metamorphosis, JH biosynthesis, or JH-dependent body coloration, suggesting that the bioactive compound in this fraction enhances ovarian development independent of JH (Couillaud et al. 1989). Later, a peptide with 65 amino acids, namely, OMP, was isolated from this fraction (Girardie et al. 1991, 1992). The primary structure of OMP is entirely different from that of neuroparsin.

OMP is also released from PI-NSCs (Tamarelle et al. 2000). Like OEH, OMP is thought to induce the expression of *vitellogenin* by promoting ovarian ecdysteroid biosynthesis, thereby inducing ovarian maturation (Girardie and Girardie 1996).

OMP has also been identified in the desert locust *S. gregaria* and exhibits gonadotropic activity (Girardie et al. 1998). However, OMP is conserved in only a few orthopteran species (Richard et al. 1994).

10.2.4 Neuropeptide F (NPF)

NPF is an invertebrate homolog of mammalian NPY and is widely distributed in animals (Nässel and Wegener 2011) (see Sect. 2.3.11). NPF was first discovered in *D. melanogaster* (Brown et al. 1999). Following this discovery, several intensive genetic studies using *D. melanogaster* have demonstrated that NPF is essential for diverse physiological processes, such as learning, stress responses, feeding, courtship behavior, and reproduction (Nässel and Wegener 2011; Hoshino and Niwa 2021). Among these functions, NPF promotes oogenesis in the fruit fly *D. melanogaster*; however, NPF-mediated stimulation of ovarian ecdysteroid biosynthesis has not been detected in fruit flies (Ameku et al. 2018; Hoshino et al. 2023).

NPF is produced in some brain neurons and in a subpopulation of enteroendocrine cells (EECs) in insects (Brown et al. 1999; Nässel and Wegener 2011). Tissue-specific RNAi experiments in *D. melanogaster* demonstrated that gut-derived NPF is essential for oogenesis and influences mating-induced germline stem cell (GSC) proliferation (Ameku et al. 2018). In this case, NPF release from EECs in females was affected by mating stimuli and sugar feeding. Upon mating, dietary glucose is actively converted to fructose by an evolutionarily conserved metabolic pathway known as the polyol pathway (Hoshino et al. 2023). De novo synthesized fructose circulates in the hemolymph and is then received by the fructose taste receptor Gr43a present in NPF-positive EECs, leading to the secretion of gut NPF. Circulating NPF is received by the NPF receptor expressed on ovarian somatic cells, activating niche signals and stimulating the proliferation of female GSCs (Hoshino et al. 2023). In addition to *D. melanogaster*, the role of NPF in the reproduction of other insects has not been thoroughly studied.

10.2.5 Short Neuropeptide F (sNPF)

Prior to the discovery of *D. melanogaster* NPF (Brown et al. 1999), NPF-related peptides were isolated and fully characterized from the cerebral ganglia of the Colorado potato beetle *Leptinotarsa decemlineata* (Spittaels et al. 1996). Since *L. decemlineata* NPF-related peptide is shorter than NPF, this peptide is referred to as truncated NPF and currently formally known as sNPF, which stands for "short" NPF (see Sect. 2.3.15).

The first reported in vivo function of sNPF was egg maturation in locusts. A previous study reported that heterologous injection of sNPF from the Colorado potato beetle accelerated egg development in *L. migratoria* but not in the gray fleshfly *Neobellieria bullata* (Cerstiaens et al. 1999). A subsequent study also revealed that the injection of sNPF into adult female *S. gregaria* results in higher ovarian ecdysteroid titers and accelerates the appearance of ecdysteroid peaks in the hemolymph (Van Wielendaele et al. 2013a). In addition, RNAi targeting *S. gregaria sNPF* gene produced significantly smaller oocytes and reduced both ovarian and hemolymph ecdysteroid titers (Van Wielendaele et al. 2013a). These results suggested that sNPF stimulates ovarian ecdysteroid synthesis, leading to egg maturation in locusts.

Additionally, sNPF is also involved in regulating male reproductive processes, such as testis accretion, seminal vesicle accretion, and post-copulation fertility in *S. gregaria*; however, the detailed mechanisms have not yet been investigated (Van Wielendaele et al. 2013b).

10.2.6 Insulin-Like Peptides (ILPs)

The insulin superfamily of peptides is widely distributed in animals. In invertebrates, members of this superfamily are collectively referred to as insulin-like peptides (ILPs; see Sect. 2.2.1). The first identified invertebrate ILPs were isolated from the brain of *B. mori* (Nagasawa et al. 1984) (see Sect. 1.1.4). Similar to vertebrate insulin and IGFs, insect ILPs play a central role in regulating nutrient-dependent growth and metabolism. The effects of insect ILPs are described in Chaps. 7 and 9.

Insect ILPs are essential for regulating reproduction (Badisco et al. 2013). The first evidence of the reproductive function of ILPs was obtained from a study using the mosquito *A. aegypti* (Graf et al. 1997). Later molecular genetic studies in *D. melanogaster* provided critical evidence that ILPs are essential for reproduction, as oogenesis and fecundity were severely impaired in mutants of *chico* and *InR*, the *D. melanogaster* orthologs of the genes for the insulin receptor substrate and insulin receptor, respectively (Böhni et al. 1999; Drummond-Barbosa and Spradling 2001; Tatar et al. 2001). Similar ILP functions have been described in many other species, including the red flour beetle *Tribolium castaneum*, cockroach *Blattella germanica*, bloodsucking bug *Rhodnius prolixus*, and eusocial ants *Ooceraea biroi* and *Harpegnathos saltator* (Chandra et al. 2018; Yan et al. 2022; Leyria 2024).

Genetic studies using *D. melanogaster* have revealed that ILP signaling influences oogenesis through activation in three key sites: (1) in ovarian somatic cells to activate oogenesis (Hsu and Drummond-Barbosa 2009; Shimada et al. 2011) and stimulate ovarian ecdysteroid biosynthesis; (2) in fat body cells to activate the production of egg yolk components; and (3) in the CA to activate JH biosynthesis, leading to the active production of egg yolk components (Tatar et al. 2001; Tu et al. 2005). The role of ILP signaling in oogenesis is, at least in part, essentially conserved in many other insects (LaFever and Drummond-Barbosa 2005; Hsu and Drummond-Barbosa 2009), and it is worth noting that these three actions are not necessarily used equally in all insects.

In *D. melanogaste*r, ILPs are also known as DILPs. Nutrient-rich conditions stimulate the production and secretion of DILPs in adult females, thereby activating the insulin signaling pathway in ovarian follicle cells (Shimada et al. 2011; Burn et al. 2014). Conversely, insulin signaling in follicle cells, which regulates both microtubule and P-body organization in the underlying germline cells, is suppressed under nutrient-poor conditions, thereby easing the progression of oogenesis. This starvation response in germline cells is thought to be adaptive because it can suppress the energy-demanding process of oogenesis when sufficient nutrients are unavailable to females. In *D. melanogaster*, insulin signaling in the ovary plays a crucial role in maintaining the cap cells, which are the niche cells of female GSCs. Activation of insulin signaling leads to an increase in the number of GSCs, whereas suppression of insulin signaling has the opposite effect (LaFever and Drummond-Barbosa 2005; Hsu and Drummond-Barbosa 2009). Additionally, DILPs are crucial for maintaining GSCs in males (Ueishi et al. 2009; McLeod et al. 2010).

Insulin signaling influences oogenesis in *D. melanogaster*, by acting on the CA to regulate JH biosynthesis. Specific *InR* mutations severely impair vitellogenesis and JH biosynthesis. In these animals, vitellogenesis is substantially recovered when *InR* mutants are treated with methoprene, a JH analog (Tatar et al. 2001). In addition, insulin signaling–mediated regulation of JH biosynthesis influences the induction of reproductive diapause, also known as reproductive dormancy, at least in *D. melanogaster* (Ojima et al. 2018; Kurogi et al. 2021). Reproductive dormancy is a process in which oogenesis is strongly suppressed to conserve energy under adverse environmental conditions such as cold temperatures and short days (Kurogi et al. 2021). DILP production and secretion are suppressed during reproductive dormancy induction. Therefore, insufficient DILPs cannot stimulate the CA, leading to the suppression of JH biosynthesis and, subsequently, the suppression of oogenesis (Ojima et al. 2018). However, it should be noted that the interplay between insulin signaling and JH signaling in the regulation of vitellogenesis varies dramatically between insects (Zhu et al. 2020).

The function of ILPs in adult mosquitoes is similar to that of OEH. Both OEH and ILPs are produced in PI-NSCs and act on ovarian cells to stimulate the production and secretion of ovarian ecdysteroids, which are essential for vitellogenesis. However, unlike OEH, ILPs act on more than just ovarian cells. For example, ILPs stimulate fat body cells after a blood meal to activate *vitellogenin* gene expression. One study demonstrated that mammalian insulin can stimulate *A. aegypti* fat bodies and promote insulin signaling (Roy et al. 2007). In this study, insulin in combination with 20-hydroxyecdysone induced the expression of the *vitellogenin* gene, which was inhibited in the presence of chemical inhibitors of insulin signaling molecules. Moreover, RNAi targeting mosquito *InR* and its downstream signaling component genes inhibited insulin-induced *vitellogenin* gene expression (Roy et al. 2007).

ILP signaling plays a central role in coordinating reproduction and longevity in ants. The expression of the ILP gene, *ilp2*, is upregulated in the reproductives in the clonal raider ant *Ooceraea biroi*, and larval signals suppress reproduction by inhibiting *ilp2*, thereby generating a colony reproductive cycle reminiscent of ancestral subsociality (Chandra et al. 2018). However, elevated ILP2 levels can overcome larval suppression and stabilize the reproductive division of labor, suggesting a mechanism by which nutritional and larval cues shape the evolution of eusociality. In *Harpegnathos* ants, caste switching from workers to gamergates, which are reproductive queen-like individuals, links ILP signaling to both fecundity and lifespan (Yan et al. 2022). Increased ILP signaling via the MAP kinase pathway promotes vitellogenesis and egg production. In contrast, concurrent ovarian production of the insulin antagonist Imp-L2 downregulates AKT/FOXO signaling in the fat body, a mechanism associated with the remarkable longevity of gamergates. Together, these studies illustrate how differential modulation of insulin signaling underlies both the division of reproductive roles and the paradoxical combination of high fecundity and extended life span in eusocial ants.

Among the ILP family members, DILP8 from *D. melanogaster* and its ortholog, gonadulin, from *L. migratoria* and *R. prolixus*, are essential for reproduction in these insects. DILP8/gonadulin is biochemically distinct from other ILPs because the DILP8 receptor is a leucine-rich repeat-containing G protein-coupled receptor 3 (LGR3), rather than an RTK (Gontijo and Garelli 2018). In adult *D. melanogaster* females, ovarian follicle cell–derived DILP8 is important for ovulation (Liao and Nässel 2020). DILP8 acts locally on *Lgr3*-expressing follicle cells and systemically on *Lgr3*-positive efferent neurons in the abdominal ganglia, projecting to the oviduct musculature. Accordingly, DILP8 is likely to coordinate follicle rupture and oviduct contractions during ovulation (Liao and Nässel 2020). Thus, it is expected that DILP8 controls ovulation in adult females by regulating LGR3-positive neurons in the ventral nerve cord. In other fruit flies, *Bactrocera dorsalis* and *Drosophila mercatorum*, *Ilp8* is specifically expressed in the ovaries and essential for female fecundity. Behavioral analysis has demonstrated that the

Ilp8 mutation impairs female sexual attractiveness in fruit flies, resulting in decreased mating success and is likely the cause of fecundity reduction (Li et al. 2023). In the locust *L. migratoria*, gonadulin influences vitellogenesis and oocyte growth (Veenstra et al. 2021). In the bloodsucking bug *R. prolixus*, gonadulin is involved in ovulation and oviposition (Leyria et al. 2022).

ILPs have essential reproductive regulatory functions in dipterans and lepidopterans. Knockout of the *Bombyx mori* IGF-like peptide (BIGFLP) using the CRISPR-Cas9 system significantly reduced pupal ovary size and the number of eggs laid by adult females (Fujinaga et al. 2019). Furthermore, transplantation of ovaries from *BIGFLP* knockout animals into wild-type animals restores the development of *BIGFLP* mutant ovaries, indicating that BIGFLP in the hemolymph plays an important role in ovarian development (Fujinaga et al. 2019). Given that ovarian development in *B. mori* progresses during the pupal stage and that the expression level of *BIGFLP* is significantly elevated during the pupal stage, it is likely that BIGFLP is important for ovarian development in the pupal stage (Mizoguchi and Okamoto 2013).

10.2.7 Allatostatin-C (AST-C)

AST-C, also known as PISCF-related allatostatin (PISCF/AST), is well-known for its role in suppressing JH biosynthesis in the CA (see Sect. 2.3.3). This PH was initially isolated from the heads of the tobacco hornworm *Manduca sexta* (Kramer et al. 1991).

Studies using *D. melanogaster* have shown that AST-C plays multiple roles in regulating female reproduction. One role of AST-C in reproduction is that a subpopulation of AST-C-producing neurons is essential for gating the initiation of vitellogenesis after eclosion and re-initiation after mating (Zhang et al. 2022). In virgin females, AST-C-producing neurons are activated, resulting in the suppression of JH biosynthesis and, subsequently, the suppression of vitellogenesis. In contrast, in mated females, AST-C-producing neurons are deactivated; thus, JH-mediated vitellogenesis actively progresses (Zhang et al. 2022).

AST-C also controls diurnal rhythmicity during oogenesis, which is dependent on circadian clock pacemaker neurons (Zhang et al. 2021). Indeed, AST-C is present in a subset of pacemaker neurons, designated as DN1p (see Chap. 13), which send inhibitory inputs to brain ILP-producing cells expressing AST-C receptors. As a result, rhythmic changes in AST-C-producing DN1p activity generate an oogenic rhythm by regulating JH biosynthesis and vitellogenesis via insulin signaling (Zhang et al. 2021).

Besides DN1p, AST-C is present in other pacemaker neurons called DN3s, which are crucial for recovery from cold-induced reproductive dormancy (Meiselman et al. 2022). Notably, AST-C levels are suppressed by cold in *D. melanogaster*, other *Drosophila* species, and some mosquitoes. Moreover, AST-C in DN3s stimulates egg production, facilitating recovery from reproductive dormancy (Meiselman et al. 2022).

10.2.8 Calcitonin-Like Diuretic Hormone (CT-DH)

CT-DH, also known as diuretic hormone 31 (DH31; see Sect. 2.3.8), was originally isolated from the cockroach *D. punctata* as a diuretic hormone crucial for controlling water balance (Furuya et al. 2000) (see Sect. 9.3.3).

A recent study using *D. melanogaster* demonstrated that CT-DH is also essential for inducing reproductive dormancy by suppressing JH biosynthesis under cold conditions (Kurogi et al. 2023). Since CT-DH-producing neurons in the brain project into the CA, thereby suppressing JH biosynthesis in adult flies, Knockdown of *CT-DH* in these CA-projecting neurons or *CT-DH receptor* in the CA induces a decrease in JH titers, which should occur under dormancy-inducing conditions, leading to abnormal yolk accumulation in the ovaries (Kurogi et al. 2023). The cell bodies of CA-projecting CT-DH-producing neurons are present in the lateral protocerebral region of the *D. melanogaster* brain. Notably, the lateral protocerebral neurons projecting to the CA are also required to induce reproductive dormancy (diapause) in other insect species (Kurogi et al. 2021). It would be intriguing to examine whether CA-projecting neurons in other insects are CT-DH-positive.

10.2.9 Corazonin (CRZ)

CRZ, homologous to vertebrate gonadotropin-releasing hormone, is widely distributed across insect species (see Sect. 2.3.1). It is produced by specific groups of neurosecretory neurons (especially in the brain and ventral nerve cord) and affects diverse physiological functions such as metabolism, stress response, and reproductive regulation (Predel et al. 2007).

One of the crucial roles of CRZ in reproduction is the regulation of reproductive diapause. For example, during unfavorable seasons (such as winter), the bean bug *Riptortus pedestris* suspends reproduction in a process called diapause to conserve energy and improve survival chances (Xi et al. 2025). CRZ is thought to act as a molecular signal in the brain in response to photoperiodic cues. CRZ levels rise under short-day conditions, leading to suppressed ovarian development and enhanced fat accumulation (Xi et al. 2025).

In the eusocial ant *H. saltator* and other eusocial ant species, CRZ is involved in natural reproductive plasticity, where adult workers can transition to reproductive gamergates (Nagel et al. 2020). *CRZ* is downregulated as workers become gamergates. *CRZ* is preferentially expressed in workers and foragers. Notably, injection of CRZ into transitioning individuals suppresses *vitellogenin* expression and stimulates worker-like hunting behaviors while inhibiting gamergate behaviors, including egg deposition (Nagel et al. 2020).

In male fruit flies such as *D. melanogaster* and *B. dorsalis*, CRZ signaling is essential for normal mating behavior, sperm transfer, and seminal fluid provision (Tayler et al. 2012; Hou et al. 2018).

10.2.10 Ecdysis-Triggering Hormone (ETH)

ETH was initially identified as a hormone that triggers specific behavioral sequences of ecdysis through its direct action on the nervous system of *M. sexta* (Zitnan et al. 1996). ETH is now recognized as a well-conserved insect PH produced in the epitracheal gland cells, called Inka cells, in all insects examined to date (Roller et al. 2010) (see Sect. 2.4.1).

ETH has versatile physiological roles in insects, including the promotion of JH biosynthesis. It has been revealed that the ETH receptor is highly expressed in larval CA of *B. mori*, and ETH is crucial for JH biosynthesis in pharate adults of *A. aegypti* (Yamanaka et al. 2008; Areiza et al. 2014). In *D. melanogaster*, ETH is essential for JH biosynthesis in the adult stage and stimulates vitellogenesis through JH biosynthesis (Meiselman et al. 2017).

10.2.11 Trypsin-Modulating Oostatic Factor (TMOF)

All the PHs/NPs discussed thus far were produced outside the ovary. In contrast, TMOF is a unique peptide produced in and released from the ovary and influences oogenesis.

TMOF was initially isolated from the ovaries of the mosquito *A. aegypti* as a peptide that suppresses vitellogenesis (Borovsky 1985). The injection of a partially purified fraction containing TMOF suppressed egg development (Borovsky 1988). Interestingly, TMOF did not appear to act directly on the ovary; instead, it inhibited the biosynthesis of trypsin, a digestive enzyme present in the midgut. The inhibition of trypsin biosynthesis suppresses blood meal digestion, thereby suppressing vitellogenesis. The primary structure of TMOF, which is NH_2-YDPAPPPPP-COOH, was subsequently determined (Borovsky et al. 1990). Further studies revealed that *A. aegypti* TMOF is released after blood feeding and acts on the intestine to suppress the expression of the *trypsin* gene (Borovsky 2003).

A subsequent study reported that bioactive peptides that inhibit midgut trypsin and suppress oogenesis are also present in the grey fleshfly *Neobellieria bullata*. Even though the function of the peptides of *N. bullata* is very similar to that of *A. aegypti* TMOF, peptide sequencing revealed that the *N. bullata* "TMOF" structure is NH_2-NPTNLH-COOH (Bylemans et al. 1994). Surprisingly, these two peptides with entirely different sequences have almost the same "TMOF" activity in each species. Currently, the receptor for TMOF has not been identified, and future research is needed to determine the receptors used in these different species.

10.3 Peptide Hormones Regulating Pheromone Biosynthesis

Pheromones are chemical substances produced by organisms that are released into the environment to stimulate other individuals of the same species. The numerous pheromones produced by females include those that attract males and enable encounters that ultimately result in mating. Thus, pheromones are crucial for reproductive success. Efforts have been made to identify the pheromone biosynthetic pathway and PHs that stimulate it. To date, only a brain-derived PH that controls pheromone biosynthesis has been identified in lepidopterans.

10.3.1 Pheromone Biosynthesis Activating Neuropeptide (PBAN)

PBAN (see Sect. 2.3.13) controls pheromone biosynthesis in lepidopterans. A classical experiment showed that head ligation immediately after eclosion reduced pheromone synthesis in the pheromone glands in female corn earworm *Helicoverpa zea* (Raina and Klun 1984). Pheromone levels in their bodies were restored when the brain was transplanted into head-ligated newly eclosed adults. These results strongly suggest the presence of a brain factor that promotes pheromone biosynthesis in pheromone glands (Raina and Klun 1984). Subsequently, bioactive peptides, later known as PBAN, were isolated from *H. zea* and *B. mori*, and their amino acid sequences were determined (Raina et al. 1989; Kitamura et al. 1989). PBAN is generated from a precursor protein common to multiple peptides, including diapause hormone (DH; see Sect. 8.2.1.) (Sato et al. 1993). PBAN is thought to be released from neurons in the subesophageal ganglion (Ma et al. 1994, 1998). The PBAN receptor has also been identified and is highly expressed in the pheromone glands (Choi et al. 2003; Hull et al. 2004).

PBAN belongs to the PRXamide NP family (see Sect. 2.3.13), which is evolutionarily conserved in bilaterian animals. Comparative informatics has revealed that *PBAN* genes are widely present in many insect species. However, their role in pheromone biosynthesis has only been demonstrated in a few lepidopteran species. *PBAN* genes are lost in higher dipteran species, including *D. melanogaster* (Farris 2023). Pheromones in *Drosophila* species are mainly biosynthesized in oenocytes. Although the PH that promotes pheromone biosynthesis in oenocytes has not been identified, ecdysteroids and JHs have been shown to stimulate pheromone biosynthesis in oenocytes (Bilen et al. 2013; Baron et al. 2018).

10.4 Bioactive Peptides Regulating Female Post-mating Responses

Mating between males and females is necessary for all animals, including insects, except for species capable of parthenogenesis. In insects, a series of physiological changes, called post-mating responses (PMRs), are induced in females after mating. Female PMRs include increased feeding, induction of egg-laying behavior, and refusal to mate. Increased feeding and the induction of egg-laying behavior are essential for females to leave offspring in the next generation. At the same time, refusal to mate again is adaptive for males because it increases the likelihood of siring their own offspring.

The molecular mechanisms regulating PMRs have been extensively studied in *D. melanogaster* (Hopkins and Perry 2022). In particular, identifying sex peptide (SP), a male-derived peptide required to induce female PMRs, is a monumental finding in reproductive biology. A schematic diagram of the interorgan communications in SP-mediated regulation of oogenesis is shown in Fig. 10.2.

10.4.1 Sex Peptide (SP)

SP is a peptide with 36 amino acids that was discovered in *D. melanogaster* and induces a wide variety of female PMRs (Hopkins and Perry 2022). SP is synthesized in the accessory glands of males and is passed from male to female during mating (Chen and Bühler 1970). It is received by neurons expressing the SP receptor (SPR) in the female uterus and induces a series of PMRs (Yapici et al. 2008; Häsemeyer et al. 2009). SP is vital for PMRs such as female egg-laying behavior and mating refusal (Chapman et al. 2003; Liu and Kubli 2003). SP-receptor sensory neurons (SPSN) project to the abdominal ganglion and connect with other neurons that synthesize myoinhibitory peptide (MIP; see Sect. 2.4.4) (Jang et al. 2017). MIP-positive neurons project to nearby neurons (SAG neurons). SAG neurons project to the PI in the brain. This series of neural circuits plays an essential role in the SP-induced female PMRs in *D. melanogaster*. SP also induces a mating-dependent increase in GSCs, which is mediated by ecdysone synthesis in the ovary and the release of NPF from the gut in *D. melanogaster* (Ameku and Niwa 2016; Ameku et al. 2018).

SP genes are well conserved within the genus *Drosophila* species, while no clear orthologs have been found in other insect groups. Surprisingly, SPs in ancestral *Drosophila* species cannot induce female PMRs (Tsuda et al. 2015). Examining whether other components of the seminal fluid are crucial for inducing female PMRs in ancestral *Drosophila* species and other insect species is intriguing. In contrast to the weak evolutionary conservation of *SP* genes, *SPR* genes are highly conserved in insects (Yapici et al. 2008). However, a recent study showed that SPR is not required to suppress remating behavior or induce egg laying in *A. aegypti* (Amaro et al. 2024).

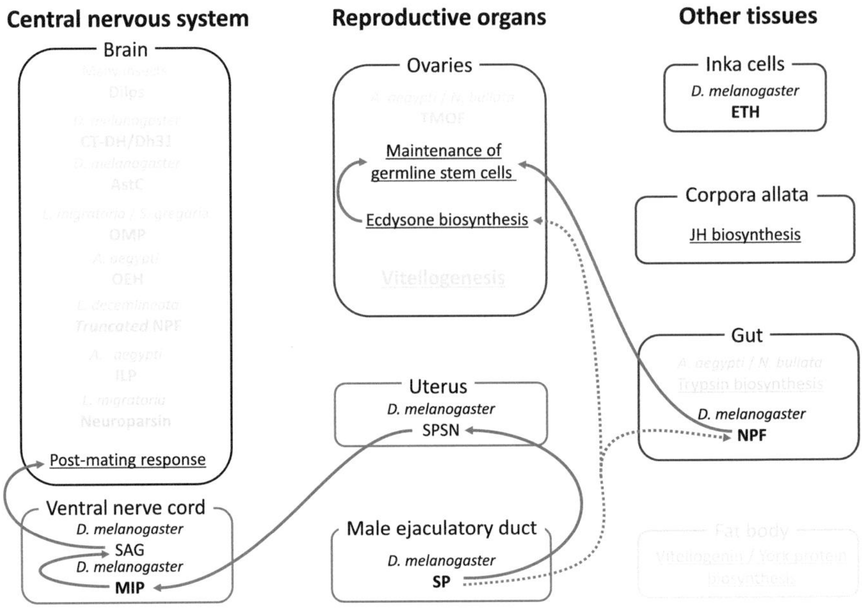

Fig. 10.2 Schematic diagram of interorgan communications of the male sex peptide-mediated regulation of oogenesis. Abbreviations: SAG Sex peptide abdominal ganglion neuron, MIP myoinhibitory peptide, SPSN Sex peptide sensory neuron, SP Sex peptide

10.5 Future Perspectives

In this chapter, we outlined the PHs/NPs that control insect reproduction. Many of these PHs/NPs have been identified in various insect species, and functional analysis of these peptides has also been actively conducted in each species. However, some peptides are poorly conserved among species. In addition, there are instances where the peptide itself is widely conserved but the function initially discovered in one species is not observed in other species. It is hypothesized that these peptides may have independently gained or lost their function during insect evolution, and future studies are needed to test this hypothesis in more detail.

It should be noted that specific receptors for some of the peptides discussed in this chapter have yet to be identified. The identification and functional analysis of receptors are expected to progress by utilizing informatics and genetic techniques, such as next-generation sequencing, genome editing, and RNAi methods, which have been drastically developed in recent years.

References

Amaro IA, Wohl MP, Pitcher S, Alfonso-Parra C, Avila FW, Paige AS et al (2024) Sex peptide receptor is not required for refractoriness to remating or induction of egg laying in *Aedes aegypti*. Genetics 227. https://doi.org/10.1093/genetics/iyae034

Ameku T, Niwa R (2016) Mating-induced increase in germline stem cells via the neuroendocrine system in female *Drosophila*. PLoS Genet 12:e1006123. https://doi.org/10.1371/journal.pgen.1006123

Ameku T, Yoshinari Y, Texada MJ, Kondo S, Amezawa K, Yoshizaki G et al (2018) Midgut-derived neuropeptide F controls germline stem cell proliferation in a mating-dependent manner. PLoS Biol 16:e2005004. https://doi.org/10.1371/journal.pbio.2005004

Areiza M, Nouzova M, Rivera-Perez C, Noriega FG (2014) Ecdysis triggering hormone ensures proper timing of juvenile hormone biosynthesis in pharate adult mosquitoes. Insect Biochem Mol Biol 54:98–105. https://doi.org/10.1016/j.ibmb.2014.09.006

Badisco L, Claeys I, Van Loy T, Van Hiel M, Franssens V, Simonet G et al (2007) Neuroparsins, a family of conserved arthropod neuropeptides. Gen Comp Endocrinol 153:64–71. https://doi.org/10.1016/j.ygcen.2007.03.008

Badisco L, Claeys I, Van Hiel M, Clynen E, Huybrechts J, Vandersmissen T et al (2008) Purification and characterization of an insulin-related peptide in the desert locust, *Schistocerca gregaria*: immunolocalization, cDNA cloning, transcript profiling and interaction with neuroparsin. J Mol Endocrinol 40:137–150. https://doi.org/10.1677/JME-07-0161

Badisco L, Marchal E, Van Wielendaele P, Verlinden H, Vleugels R, Vanden Broeck J (2011) RNA interference of insulin-related peptide and neuroparsins affects vitellogenesis in the desert locust *Schistocerca gregaria*. Peptides 32:573–580. https://doi.org/10.1016/j.peptides.2010.11.008

Badisco L, Van Wielendaele P, Vanden Broeck J (2013) Eat to reproduce: a key role for the insulin signaling pathway in adult insects. Front Physiol 4:202. https://doi.org/10.3389/fphys.2013.00202

Baron A, Denis B, Wicker-Thomas C (2018) Control of pheromone production by ovaries in *Drosophila*. J Insect Physiol 109:138–143. https://doi.org/10.1016/j.jinsphys.2018.07.003

Bilen J, Atallah J, Azanchi R, Levine JD, Riddiford LM (2013) Regulation of onset of female mating and sex pheromone production by juvenile hormone in *Drosophila melanogaster*. Proc Natl Acad Sci USA 110:18321–18326. https://doi.org/10.1073/pnas.1318119110

Böhni R, Riesgo-Escovar J, Oldham S, Brogiolo W, Stocker H, Andruss BF et al (1999) Autonomous control of cell and organ size by CHICO, a *Drosophila* homolog of vertebrate IRS1-4. Cell 97:865–875. https://doi.org/10.1016/s0092-8674(00)80799-0

Borovsky D (1985) Isolation and characterization of highly purified mosquito oostatic hormone. Arch Insect Biochem Physiol 2:333–349. https://doi.org/10.1002/arch.940020402

Borovsky D (1988) Oostatic hormone inhibits biosynthesis of midgut proteolytic enzymes and egg development in mosquitoes. Arch Insect Biochem Physiol 7:187–210. https://doi.org/10.1002/arch.940070305

Borovsky D (2003) Trypsin-modulating oostatic factor: a potential new larvicide for mosquito control. J Exp Biol 206:3869–3875. https://doi.org/10.1242/jeb.00602

Borovsky D, Carlson DA, Griffin PR, Shabanowitz J, Hunt DF (1990) Mosquito oostatic factor: a novel decapeptide modulating trypsin-like enzyme biosynthesis in the midgut. FASEB J 4: 3015–3020. https://doi.org/10.1096/fasebj.4.12.2394318

Brown MR, Cao C (2001) Distribution of ovary ecdysteroidogenic hormone I in the nervous system and gut of mosquitoes. J Insect Sci 1:3. https://doi.org/10.1093/jis/1.1.3

Brown MR, Graf R, Swiderek KM, Fendley D, Stracker TH, Champagne DE et al (1998) Identification of a steroidogenic neurohormone in female mosquitoes. J Biol Chem 273:3967–3971. https://doi.org/10.1074/jbc.273.7.3967

Brown MR, Crim JW, Arata RC, Cai HN, Chun C, Shen P (1999) Identification of a Drosophila brain-gut peptide related to the neuropeptide Y family. Peptides 20:1035–1042. https://doi.org/10.1016/s0196-9781(99)00097-2

Büning J (1994) The insect ovary: ultrastructure, previtellogenic growth and evolution. Springer Science & Business Media
Burn KM, Shimada Y, Ayers K, Lu F, Hudson AM, Cooley L (2014) Somatic insulin signaling regulates a germline starvation response in Drosophila egg chambers. Dev Biol. https://doi.org/10.1016/j.ydbio.2014.11.021
Bylemans D, Borovsky D, Hunt DF, Shabanowitz J, Grauwels L, De Loof A (1994) Sequencing and characterization of trypsin modulating oostatic factor (TMOF) from the ovaries of the grey fleshfly, *Neobellieria (Sarcophaga) bullata*. Regul Pept 50:61–72. https://doi.org/10.1016/0167-0115(94)90192-9
Cerstiaens A, Benfekih L, Zouiten H, Verhaert P, De Loof A, Schoofs L (1999) Led-NPF-1 stimulates ovarian development in locusts. Peptides 20:39–44. https://doi.org/10.1016/s0196-9781(98)00152-1
Chandra V, Fetter-Pruneda I, Oxley PR, Ritger AL, McKenzie SK, Libbrecht R et al (2018) Social regulation of insulin signaling and the evolution of eusociality in ants. Science 361:398–402. https://doi.org/10.1126/science.aar5723
Chapman T, Bangham J, Vinti G, Seifried B, Lung O, Wolfner MF et al (2003) The sex peptide of *Drosophila melanogaster*: female post-mating responses analyzed by using RNA interference. Proc Natl Acad Sci USA 100:9923–9928. https://doi.org/10.1073/pnas.1631635100
Chen PS, Bühler R (1970) Paragonial substance (sex peptide) and other free ninhydrin-positive components in male and female adults of *Drosophila melanogaster*. J Insect Physiol 16:615–627. https://doi.org/10.1016/0022-1910(70)90095-8
Choi M-Y, Fuerst E-J, Rafaeli A, Jurenka R (2003) Identification of a G protein-coupled receptor for pheromone biosynthesis activating neuropeptide from pheromone glands of the moth *Helicoverpa zea*. Proc Natl Acad Sci USA 100:9721–9726. https://doi.org/10.1073/pnas.1632485100
Couillaud F, Girardie A, Girardie J (1989) Identification of gonadotropic and antigonadotropic factors from the nervous part of the corpora cardiaca in the African locust. Invertebr Reprod Dev 16:17–22. https://doi.org/10.1080/07924259.1989.9672056
Dhara A, Eum J-H, Robertson A, Gulia-Nuss M, Vogel KJ, Clark KD et al (2013) Ovary ecdysteroidogenic hormone functions independently of the insulin receptor in the yellow fever mosquito, *Aedes aegypti*. Insect Biochem Mol Biol 43:1100–1108. https://doi.org/10.1016/j.ibmb.2013.09.004
Drummond-Barbosa D, Spradling AC (2001) Stem cells and their progeny respond to nutritional changes during *Drosophila* oogenesis. Dev Biol 231:265–278. https://doi.org/10.1006/dbio.2000.0135
Farris SM (2023) Insect PRXamides: evolutionary divergence, novelty, and loss in a conserved neuropeptide system. J Insect Sci 23:3. https://doi.org/10.1093/jisesa/ieac079
Fujinaga D, Shiomi K, Yagi Y, Kataoka H, Mizoguchi A (2019) An insulin-like growth factor-like peptide promotes ovarian development in the silkmoth *Bombyx mori*. Sci Rep 9:1–12. https://doi.org/10.1038/s41598-019-54962-w
Furuya K, Milchak RJ, Schegg KM, Zhang J, Tobe SS, Coast GM et al (2000) Cockroach diuretic hormones: characterization of a calcitonin-like peptide in insects. Proc Natl Acad Sci USA 97:6469–6474. https://doi.org/10.1073/PNAS.97.12.6469
Gilbert MB, Glastad KM, Fioriti M, Sorek M, Scarpa T, Purnell FS et al (2025) Neuropeptides specify and reprogram division of labor in the leafcutter ant *Atta cephalotes*. Cell 188:3974–3991.e21. https://doi.org/10.1016/j.cell.2025.05.023
Girardie J, Girardie A (1996) Lom OMP, a putative ecdysiotropic factor for the ovary in *Locusta migratoria*. J Insect Physiol 42:215–221. https://doi.org/10.1016/0022-1910(95)00097-6
Girardie J, Boureme D, Couillaud F, Tamarelle M, Girardie A (1987a) Anti-juvenile effect of neuroparsin A, a neuroprotein isolated from locust corpora cardiaca. Insect Biochem 17:977–983. https://doi.org/10.1016/0020-1790(87)90106-5

Girardie J, Boureme D, Girardie A (1987b) Production sites of the three neurosecretory proteins characterized in the corpora cardiaca of the migratory locust. Insect Biochem 17:29–36. https://doi.org/10.1016/0020-1790(87)90140-5

Girardie J, Huet J-C, Pernollet J-C (1990) The locust neuroparsin a: sequence and similarities with vertebrate and insect polypeptide hormones. Insect Biochem 20:659–666. https://doi.org/10.1016/0020-1790(90)90079-A

Girardie J, Richard O, Huet JC, Nespoulous C, Van Dorsselaer A, Pernollet JC (1991) Physical characterization and sequence identification of the ovary maturating parsin. A new neurohormone purified from the nervous corpora cardiaca of the African locust (*Locusta migratoria migratorioides*). Eur J Biochem 202:1121–1126. https://doi.org/10.1111/j.1432-1033.1991.tb16479.x

Girardie J, Richard O, Girardie A (1992) Time-dependent variations in the activity of a novel ovary maturating neurohormone from the nervous corpora cardiaca during oögenesis in the locust, *Locusta migratoria migratorioides*. J Insect Physiol 38:215–221. https://doi.org/10.1016/0022-1910(92)90069-P

Girardie J, Geoffre S, Delbecque J-P, Girardie A (1998) Arguments for two distinct gonadotropic activities triggered by different domains of the ovary maturating parsin of *Locusta migratoria*. J Insect Physiol 44:1063–1071. https://doi.org/10.1016/s0022-1910(98)00066-3

Gontijo AM, Garelli A (2018) The biology and evolution of the Dilp8-Lgr3 pathway: a relaxin-like pathway coupling tissue growth and developmental timing control. Mech Dev 154:44–50. https://doi.org/10.1016/j.mod.2018.04.005

Graf R, Neuenschwander S, Brown MR, Ackermann U (1997) Insulin-mediated secretion of ecdysteroids from mosquito ovaries and molecular cloning of the insulin receptor homologue from ovaries of bloodfed *Aedes aegypti*. Insect Mol Biol 6:151–163. https://doi.org/10.1111/j.1365-2583.1997.tb00083.x

Häsemeyer M, Yapici N, Heberlein U, Dickson BJ (2009) Sensory neurons in the *Drosophila* genital tract regulate female reproductive behavior. Neuron 61:511–518. https://doi.org/10.1016/j.neuron.2009.01.009

Hopkins BR, Perry JC (2022) The evolution of sex peptide: sexual conflict, cooperation, and coevolution. Biol Rev Camb Philos Soc 97:1426–1448. https://doi.org/10.1111/brv.12849

Hoshino R, Niwa R (2021) Regulation of mating-induced increase in female germline stem cells in the fruit Fly *Drosophila melanogaster*. Front Physiol 12:785435. https://doi.org/10.3389/fphys.2021.785435

Hoshino R, Sano H, Yoshinari Y, Nishimura T, Niwa R (2023) Circulating fructose regulates a germline stem cell increase via gustatory receptor–mediated gut hormone secretion in mated *Drosophila*. Sci Adv 9:eadd5551. https://doi.org/10.1126/sciadv.add5551

Hou Q-L, Chen E-H, Jiang H-B, Yu S-F, Yang P-J, Liu X-Q et al (2018) Corazonin signaling is required in the male for sperm transfer in the oriental fruit fly *Bactrocera dorsalis*. Front Physiol 9:660. https://doi.org/10.3389/fphys.2018.00660

Hsu H-J, Drummond-Barbosa D (2009) Insulin levels control female germline stem cell maintenance via the niche in *Drosophila*. Proc Natl Acad Sci USA 106:1117–1121. https://doi.org/10.1073/pnas.0809144106

Hull JJ, Ohnishi A, Moto K, Kawasaki Y, Kurata R, Suzuki MG et al (2004) Cloning and characterization of the pheromone biosynthesis activating neuropeptide receptor from the silkmoth, *Bombyx mori*. Significance of the carboxyl terminus in receptor internalization. J Biol Chem 279:51500–51507. https://doi.org/10.1074/jbc.M408142200

Jang Y-H, Chae H-S, Kim Y-J (2017) Female-specific myoinhibitory peptide neurons regulate mating receptivity in *Drosophila melanogaster*. Nat Commun 8:1630. https://doi.org/10.1038/s41467-017-01794-9

Khalid MZ, Ahmad S, Ngegba PM, Zhong G (2021) Role of endocrine system in the regulation of female insect reproduction. Biology 10:614. https://doi.org/10.3390/biology10070614

Kitamura A, Nagasawa H, Kataoka H, Inoue T, Matsumoto S, Ando T et al (1989) Amino acid sequence of pheromone-biosynthesis-activating neuropeptide (PBAN) of the silkworm, *Bombyx mori*. Biochem Biophys Res Commun 163:520–526. https://doi.org/10.1016/0006-291x(89)92168-2

Kramer SJ, Toschi A, Miller CA, Kataoka H, Quistad GB, Li JP et al (1991) Identification of an allatostatin from the tobacco hornworm *Manduca sexta*. Proc Natl Acad Sci USA 88:9458–9462. https://doi.org/10.1073/pnas.88.21.9458

Kurogi Y, Mizuno Y, Imura E, Niwa R (2021) Neuroendocrine regulation of reproductive dormancy in the fruit fly *Drosophila melanogaster*: a review of juvenile hormone-dependent regulation. Front Ecol Evol 9:715029. https://doi.org/10.3389/FEVO.2021.715029

Kurogi Y, Imura E, Mizuno Y, Hoshino R, Nouzova M, Matsuyama S et al (2023) Female reproductive dormancy in *Drosophila* is regulated by DH31-producing neurons projecting into the corpus allatum. Development 150:dev201186. https://doi.org/10.1242/DEV.201186/310536

LaFever L, Drummond-Barbosa D (2005) Direct control of germline stem cell division and cyst growth by neural insulin in *Drosophila*. Science 309:1071–1073. https://doi.org/10.1126/science.1111410

Lea AO (1967) The medial neurosecretory cells and egg maturation in mosquitoes. J Insect Physiol 13:419–429. https://doi.org/10.1016/0022-1910(67)90082-0

Lenaerts C, Marchal E, Peeters P, Vanden Broeck J (2019) The ecdysone receptor complex is essential for the reproductive success in the female desert locust, *Schistocerca gregaria*. Sci Rep 9:15. https://doi.org/10.1038/s41598-018-36763-9

Leyria J (2024) Endocrine factors modulating vitellogenesis and oogenesis in insects: an update. Mol Cell Endocrinol 587:112211. https://doi.org/10.1016/j.mce.2024.112211

Leyria J, Philip R, Orchard I, Lange AB (2022) Gonadulin: a newly discovered insulin-like peptide involved in ovulation and oviposition in *Rhodnius prolixus*, a vector of Chagas disease. Insect Biochem Mol Biol 150:103848. https://doi.org/10.1016/j.ibmb.2022.103848

Leyria J, Fruttero LL, Paglione PA, Canavoso LE (2025) How insects balance reproductive output and immune investment. Insects 16:311. https://doi.org/10.3390/insects16030311

Li H, Luo X, Li N, Liu T, Zhang J (2023) Insulin-like peptide 8 (Ilp8) regulates female fecundity in flies. Front Cell Dev Biol 11:1103923. https://doi.org/10.3389/fcell.2023.1103923

Liao S, Nässel DR (2020) Drosophila insulin-like peptide 8 (DILP8) in ovarian follicle cells regulates ovulation and metabolism. Front Endocrinol 11:461. https://doi.org/10.3389/fendo.2020.00461

Liu H, Kubli E (2003) Sex-peptide is the molecular basis of the sperm effect in *Drosophila melanogaster*. Proc Natl Acad Sci USA 100:9929–9933. https://doi.org/10.1073/pnas.1631700100

Liu J, Liu A, Liu F, Huang H, Ye H (2020) Role of neuroparsin 1 in vitellogenesis in the mud crab, *Scylla paramamosain*. Gen Comp Endocrinol 285:113248. https://doi.org/10.1016/j.ygcen.2019.113248

Ma PW, Knipple DC, Roelofs WL (1994) Structural organization of the *Helicoverpa zea* gene encoding the precursor protein for pheromone biosynthesis-activating neuropeptide and other neuropeptides. Proc Natl Acad Sci USA 91:6506–6510. https://doi.org/10.1073/pnas.91.14.6506

Ma PWK, Knipple DC, Roelofs WL (1998) Expression of a gene that encodes pheromone biosynthesis activating neuropeptide in the central nervous system of corn earworm, *Helicoverpa zea*. Insect Biochem Mol Biol 28:373–385. https://doi.org/10.1016/S0965-1748(98)00009-5

Matsumoto S, Brown MR, Suzuki A, Lea AO (1989) Isolation and characterization of ovarian ecdysteroidogenic hormones from the mosquito, *Aedes aegypti*. Insect Biochem 19:651–656. https://doi.org/10.1016/0020-1790(89)90100-5

McLeod CJ, Wang L, Wong C, Jones DL (2010) Stem cell dynamics in response to nutrient availability. Curr Biol 20:2100–2105. https://doi.org/10.1016/j.cub.2010.10.038

Meiselman M, Lee SS, Tran R-T, Dai H, Ding Y, Rivera-Perez C et al (2017) Endocrine network essential for reproductive success in *Drosophila melanogaster*. Proc Natl Acad Sci USA 114:E3849–E3858. https://doi.org/10.1073/pnas.1620760114

Meiselman MR, Alpert MH, Cui X, Shea J, Gregg I, Gallio M et al (2022) Recovery from cold-induced reproductive dormancy is regulated by temperature-dependent AstC signaling. Curr Biol 32:1362–1375.e8. https://doi.org/10.1016/J.CUB.2022.01.061

Mizoguchi A, Okamoto N (2013) Insulin-like and IGF-like peptides in the silkmoth *Bombyx mori*: discovery, structure, secretion, and function. Front Physiol 4:217. https://doi.org/10.3389/fphys.2013.00217

Nagasawa H, Kataoka H, Isogai A, Tamura S, Suzuki A, Ishizaki H et al (1984) Amino-terminal amino Acid sequence of the silkworm prothoracicotropic hormone: homology with insulin. Science 226:1344–1345. https://doi.org/10.1126/science.226.4680.1344

Nagel M, Qiu B, Brandenborg LE, Larsen RS, Ning D, Boomsma JJ et al (2020) The gene expression network regulating queen brain remodeling after insemination and its parallel use in ants with reproductive workers. Sci Adv 6:eaaz5772. https://doi.org/10.1126/sciadv.aaz5772

Nässel DR, Wegener C (2011) A comparative review of short and long neuropeptide F signaling in invertebrates: any similarities to vertebrate neuropeptide Y signaling? Peptides 32:1335–1355. https://doi.org/10.1016/j.peptides.2011.03.013

Ojima N, Hara Y, Ito H, Yamamoto D (2018) Genetic dissection of stress-induced reproductive arrest in *Drosophila melanogaster* females. PLoS Genet 14:e1007434. https://doi.org/10.1371/journal.pgen.1007434

Predel R, Neupert S, Russell WK, Scheibner O (2007) Corazonin in insects. Peptides 28:3–10. https://doi.org/10.1016/J.PEPTIDES.2006.10.011

Raina AK, Klun JA (1984) Brain factor control of sex pheromone production in the female corn earworm moth. Science 225:531–533. https://doi.org/10.1126/science.225.4661.531

Raina AK, Jaffe H, Kempe TG, Keim P, Blacher RW, Fales HM et al (1989) Identification of a neuropeptide hormone that regulates sex pheromone production in female moths. Science 244: 796–798. https://doi.org/10.1126/science.244.4906.796

Richard O, Tamarelle M, Geoffre S, Girardie J (1994) Restricted occurrence of *Locusta migratoria* ovary maturing parsin in the brain-corpora cardiaca complex of various insect species. Histochemistry 102:233–239. https://doi.org/10.1007/BF00268900

Roller L, Zitnanová I, Dai L, Simo L, Park Y, Satake H et al (2010) Ecdysis triggering hormone signaling in arthropods. Peptides 31:429–441. https://doi.org/10.1016/j.peptides.2009.11.022

Roy SG, Hansen IA, Raikhel AS (2007) Effect of insulin and 20-hydroxyecdysone in the fat body of the yellow fever mosquito, *Aedes aegypti*. Insect Biochem Mol Biol 37:1317–1326. https://doi.org/10.1016/j.ibmb.2007.08.004

Roy S, Saha TT, Zou Z, Raikhel AS (2018) Regulatory pathways controlling female insect reproduction. Annu Rev Entomol 63:489–511. https://doi.org/10.1146/annurev-ento-020117-043258

Sato Y, Oguchi M, Menjo N, Imai K, Saito H, Ikeda M et al (1993) Precursor polyprotein for multiple neuropeptides secreted from the suboesophageal ganglion of the silkworm *Bombyx mori*: characterization of the cDNA encoding the diapause hormone precursor and identification of additional peptides. Proc Natl Acad Sci USA 90:3251–3255. https://doi.org/10.1073/pnas.90.8.3251

Shimada Y, Burn KM, Niwa R, Cooley L (2011) Reversible response of protein localization and microtubule organization to nutrient stress during *drosophila* early oogenesis. Dev Biol 355: 250–262. https://doi.org/10.1016/j.ydbio.2011.04.022

Song J, Zhou S (2020) Post-transcriptional regulation of insect metamorphosis and oogenesis. Cell Mol Life Sci 77:1893–1909. https://doi.org/10.1007/s00018-019-03361-5

Spittaels K, Verhaert P, Shaw C, Johnston RN, Devreese B, Van Beeumen J et al (1996) Insect neuropeptide F (NPF)-related peptides: isolation from Colorado potato beetle (*Leptinotarsa decemlineata*) brain. Insect Biochem Mol Biol 26:375–382. https://doi.org/10.1016/0965-1748(95)00104-2

Tamarelle M, Coast GM, Veenstra JA (2000) Ovary maturing parsin and diuretic hormone are produced by the same neuroendocrine cells in the migratory locust, *Locusta migratoria*. Peptides 21:737–739. https://doi.org/10.1016/s0196-9781(00)00187-x

Tatar M, Kopelman A, Epstein D, Tu MP, Yin CM, Garofalo RS (2001) A mutant *Drosophila* insulin receptor homolog that extends life-span and impairs neuroendocrine function. Science 292:107–110. https://doi.org/10.1126/science.1057987

Tayler TD, Pacheco DA, Hergarden AC, Murthy M, Anderson DJ (2012) A neuropeptide circuit that coordinates sperm transfer and copulation duration in *Drosophila*. Proc Natl Acad Sci USA 109:20697–20702. https://doi.org/10.1073/pnas.1218246109

Tsuda M, Peyre J-B, Asano T, Aigaki T (2015) Visualizing molecular functions and cross-species activity of sex-peptide in *Drosophila*. Genetics 200:1161–1169. https://doi.org/10.1534/genetics.115.177550

Tu M-P, Yin C-M, Tatar M (2005) Mutations in insulin signaling pathway alter juvenile hormone synthesis in *Drosophila melanogaster*. Gen Comp Endocrinol 142:347–356. https://doi.org/10.1016/j.ygcen.2005.02.009

Ueishi S, Shimizu H, H Inoue Y (2009) Male germline stem cell division and spermatocyte growth require insulin signaling in *Drosophila*. Cell Struct Funct 34:61–69. https://doi.org/10.1247/csf.08042

Van Wielendaele P, Wynant N, Dillen S, Badisco L, Marchal E, Vanden Broeck J (2013a) In vivo effect of Neuropeptide F on ecdysteroidogenesis in adult female desert locusts (*Schistocerca gregaria*). J Insect Physiol 59:624–630. https://doi.org/10.1016/j.jinsphys.2013.03.005

Van Wielendaele P, Wynant N, Dillen S, Zels S, Badisco L, Vanden Broeck J (2013b) Neuropeptide F regulates male reproductive processes in the desert locust, *Schistocerca gregaria*. Insect Biochem Mol Biol 43:252–259. https://doi.org/10.1016/j.ibmb.2012.12.004

Veenstra JA (2010) What the loss of the hormone neuroparsin in the *melanogaster* subgroup of *Drosophila* can tell us about its function. Insect Biochem Mol Biol 40:354–361. https://doi.org/10.1016/j.ibmb.2010.03.001

Veenstra JA, Leyria J, Orchard I, Lange AB (2021) Identification of gonadulin and insulin-like growth factor from migratory locusts and their importance in reproduction in *Locusta migratoria*. Front Endocrinol 12:693068. https://doi.org/10.3389/fendo.2021.693068

Vogel KJ, Brown MR, Strand MR (2015) Ovary ecdysteroidogenic hormone requires a receptor tyrosine kinase to activate egg formation in the mosquito *Aedes aegypti*. Proc Natl Acad Sci USA 112:5057–5062. https://doi.org/10.1073/pnas.1501814112

Xi J, Hamanaka Y, Shiga S (2025) Corazonin mediates photoperiodically induced diapause in the bean bug *Riptortus pedestris*. J Exp Biol 228:jeb250528. https://doi.org/10.1242/jeb.250528

Yamanaka N, Yamamoto S, Zitnan D, Watanabe K, Kawada T, Satake H et al (2008) Neuropeptide receptor transcriptome reveals unidentified neuroendocrine pathways. PLoS One 3:e3048. https://doi.org/10.1371/journal.pone.0003048

Yan H, Opachaloemphan C, Carmona-Aldana F, Mancini G, Mlejnek J, Descostes N et al (2022) Insulin signaling in the long-lived reproductive caste of ants. Science 377:1092–1099. https://doi.org/10.1126/science.abm8767

Yapici N, Kim Y-J, Ribeiro C, Dickson BJ (2008) A receptor that mediates the post-mating switch in *Drosophila* reproductive behaviour. Nature 451:33–37. https://doi.org/10.1038/nature06483

Zhang C, Daubnerova I, Jang Y-H, Kondo S, Žitňan D, Kim Y-J (2021) The neuropeptide allatostatin C from clock-associated DN1p neurons generates the circadian rhythm for oogenesis. Proc Natl Acad Sci USA 118:e2016878118. https://doi.org/10.1073/pnas.2016878118

Zhang C, Kim AJ, Rivera-Perez C, Noriega FG, Kim Y-J (2022) The insect somatostatin pathway gates vitellogenesis progression during reproductive maturation and the post-mating response. Nat Commun 13:969. https://doi.org/10.1038/s41467-022-28592-2

Zhang X, Xie N, Ding G, Ning D, Dai W, Xiong Z et al (2024) An evolutionarily conserved pathway mediated by neuroparsin-A regulates reproductive plasticity in ants. PLoS Biol 22: e3002763. https://doi.org/10.1371/journal.pbio.3002763

Zhu S, Liu F, Zeng H, Li N, Ren C, Su Y et al (2020) Insulin/IGF signaling and TORC1 promote vitellogenesis via inducing juvenile hormone biosynthesis in the American cockroach. Development 147:dev188805. https://doi.org/10.1242/dev.188805

Zitnan D, Kingan TG, Hermesman JL, Adams ME (1996) Identification of ecdysis-triggering hormone from an epitracheal endocrine system. Science 271:88–91. https://doi.org/10.1126/science.271.5245.88

Chapter 11
Regulation of Behavior

Shinji Nagata and Hitomi Seike

Abstract Insects exhibit a remarkable repertoire of innate behaviors—molting, feeding, mating, aggression, and migration—each governed by intricately coordinated neuroendocrine mechanisms. Neuropeptides and bioactive peptides are central to this regulation, which serve as integrative mediators between environmental cues and internal physiological states, enabling context-dependent behavioral modulation. These signaling systems establish a dynamic regulatory framework that aligns developmental processes, physiological demands, and ecological conditions, thereby facilitating behavioral plasticity and evolutionary resilience. This chapter highlights insect-specific behavioral phenomena through the aspects of neuropeptide-mediated regulation, offering insights into the molecular underpinnings of behavioral diversity and adaptability.

Keywords Ecdysis · Feeding behavior · Mating behavior · Aggression · Collective behavior

11.1 Introduction

Insect behaviors, including molting, feeding, mating, and other instinctive behaviors such as aggression and migration, are tightly regulated by a complex interplay of external and internal factors. Among these, neuropeptides and bioactive peptides play critical roles in orchestrating physiological and behavioral responses. Various neuropeptides govern distinct behavioral processes by modulating their motivational control. These hormones sometimes trigger cascades of gene expression that lead to specific behavioral outcomes. In recent years, advances in molecular biology have revealed the precise mechanisms through which hormones influence neural circuits

---

S. Nagata (✉) · H. Seike
Department of Integrated Biosciences, Graduate School of Frontier Sciences, The University of Tokyo, Kashiwa, Japan
e-mail: shinjin@edu.k.u-tokyo.ac.jp; hitomis@edu.k.u-tokyo.ac.jp

A. Mizoguchi, N. Okamoto (eds.), *Insect Peptide Hormones and Neuropeptides*, Entomology Monographs, https://doi.org/10.1007/978-981-95-7510-7_11

and gene regulation, shedding light on how insects adapt their behaviors to environmental cues. This chapter examined the modulatory systems mediated by neuropeptides and bioactive peptides that govern insect-specific and characteristic behaviors.

11.2 Ecdysis Behavior

Ecdysis is a stereotyped innate behavior essential for molting in insects and other Ecdysozoa (Aguinaldo et al. 1997). In insects, the ecdysis sequence is orchestrated by a neuropeptide cascade centered on ecdysis-triggering hormone (ETH; see Sect. 2.4.1), which is secreted from Inka cells just before ecdysis (Zitnan et al. 1996). ETH release is induced by eclosion hormone (EH; see Sect. 2.3.9), and ETH then activates ETHR-expressing neurons in the central nervous system (CNS), initiating sequential neuropeptidergic outputs that drive pre-ecdysis, ecdysis, and post-ecdysis behaviors (Kingan et al. 1997; Gammie and Truman 1999; Zitnan and Adams 2000). Stimulation of EH is accompanied by an increase in intracellular cGMP as a second messenger to secrete ETH from Inka cells. Two receptors, ETHR-A and ETHR-B, have been characterized in the fruit fly *Drosophila melanogaster* (Iversen et al. 2002; Park et al. 2003). In the tobacco hornworm *Manduca sexta*, neurons expressing the *ETHR-A* gene produce diverse peptides such as allatostatin-A (AST-A; see Sect. 2.3.2), crustacean cardioactive peptide (CCAP; see Sect. 2.3.7), calcitonin-like diuretic hormone (CT-DH or DH31; see Sect. 2.3.8), CRF-like diuretic hormone (CRF-DH or DH44; see Sect. 2.3.6), EH, kinin (see Sect. 2.4.3), myoinhibitory peptide (MIP; see Sect. 2.4.4), and short neuropeptide F (sNPF; see Sect. 2.3.15), which can generate the sequential ecdysis program (Kim et al. 2006a). In the red flour beetle *Tribolium castaneum*, bursicon also functions downstream of ETH (Arakane et al. 2008). Collectively, these neuropeptides seem to be involved in the regulation of ecdysis mechanisms (Fig. 11.1).

Among these neuropeptides, CCAP is pivotal for switching motor patterns from pre-ecdysis to ecdysis, thereby terminating preparatory movements and initiating cuticle-shedding contractions (Gammie and Truman 1997; Kim et al. 2006a). In *D. melanogaster*, almost all neuromeres, including subesophageal, thoracic, and abdominal neuromeres, express CCAP in ETHR-A-expressing neurons (Kim et al. 2006b). CCAP-producing cells co-express MIP in abdominal neuromeres, whereas CCAP-producing cells in several abdominal and thoracic neuromeres co-express bursicon (Kim et al. 2006b). FMRFamide and kinins are expressed in thoracic neurosecretory cells and lateral abdominal neurons, respectively. These FMRFamide and kinin-producing cells extend their axons to CNS or neurohemal organs, indicating that ETH stimulates the CCAP-producing cells, simultaneously facilitating the peripheral release of FMRFamide and kinins (Kim et al. 2006b). FMRFamide and its related peptide, FMRFamide-like peptides (FaLPs), function in muscular contractility and behavioral transitions (Chowański et al. 2025). ETH directly activates FMRFamide neurons, integrating their activity into the motor program of ecdysis (Mena et al. 2016).

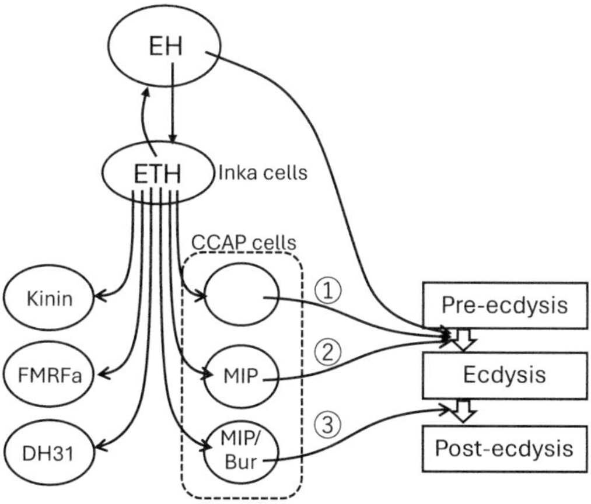

Fig. 11.1 Schematic representation of neuropeptide regulation in insect ecdysis. Eclosion hormone (EH) released from the brain stimulates ecdysis-triggering hormone (ETH) secretion from Inka cells, which in turn activates various neuropeptide-producing neurons involved in the ecdysis sequence. CCAP-producing cells contribute to the transition from pre-ecdysis to ecdysis behavior (①), partly through the effects of MIP expressed in these cells (②). Moreover, CCAP-producing cells co-expressing MIP and bursicon can modulate the subsequent transition from ecdysis to post-ecdysis behavior (③). CCAP Crustacean cardioactive peptide, MIP myoinhibitory peptide, Bur bursicon, DH31 diuretic hormone 31, CT-DH calcitonin-like diuretic hormone

DH31 contributes to ETH-mediated modulation of motor circuits, ensuring precise coordination of muscle activity (Kim et al. 2006a; Zitnan and Adams 2012). DH44, though primarily involved in osmoregulation and stress physiology, shows spatial and functional overlap with ETH-regulated circuits. Such coordination may buffer insects against stress during the vulnerable molting period (Zandawala et al. 2017; Malhotra and Basu 2023). Similarly, bursicon, released by CCAP/bursicon neurons targeted by ETH, governs cuticle tanning and wing expansion post-molt. ETH-driven activation of these neurons couples ecdysis behaviors with post-ecdysial processes (Arakane et al. 2008; Bai and Palli 2010; Loveall and Deitcher 2010; Kim et al. 2015). In the silkmoth *Bombyx mori*, sNPF, pigment-dispersing factor (PDF; see Sect. 2.4.6), and AST-CC also appear to be involved in the ETH ensemble, with stage-specific expression patterns (Daubnerová et al. 2021).

Real-time neuronal imaging has revealed a precise sequence of ETH-induced neuronal activations: FMRFamide neurons during pre-ecdysis, EH and CCAP/MIP ensembles at ecdysis onset, and CCAP/MIP/bursicon neurons during post-ecdysis. Disruption of specific ETHR-expressing cell populations causes distinct behavioral deficits, confirming the indispensability of stepwise neuropeptide recruitment for successful ecdysis (Kim et al. 2006b).

In summary, the sequential execution of ecdysis behaviors in insects critically depends on the stepwise activation of multiple neuropeptidergic ensembles downstream of ETH signaling. Kim et al. (2006b) demonstrated that ETH acts directly on the CNS, sequentially activating distinct groups of neurons that expressed a diversity of neuropeptides such as kinins, FMRFamides, EH, CCAP, MIPs, and bursicon.

11.3 Feeding Behavior

Insects exhibit finely tuned regulatory systems governing their feeding behavior, in which neuropeptides and bioactive peptides play pivotal roles as signaling messengers across the nervous and endocrine networks (Fadda et al. 2019; Nagata and Zhou 2019). These molecules orchestrate the initiation, modulation, and termination of feeding by integrating internal physiological status with external environmental cues, thereby enabling insects to precisely adapt their nutritional intake to varying metabolic demands (Nagata and Zhou 2019). Elucidating how neuropeptides interface with neural circuits and peripheral targets is fundamental to understanding the sophisticated mechanisms underlying behavioral plasticity and dietary adaptation in insects (Nagata and Zhou 2019; Lin et al. 2019).

11.3.1 Neuropeptide F (NPF)

One of the most well-studied peptide hormones in this context is neuropeptide F (NPF; see Sect. 2.3.11), which has been shown to play a pivotal role in regulating appetite. NPF is analogous to neuropeptide Y in vertebrates and promotes food intake (Fadda et al. 2019). The transcriptional level of NPF is driven by hunger state in *D. melanogaster* and the desert locust *Schistocerca gregaria* (Wu et al. 2003; Van Wielendaele et al. 2013a). In *D. melanogaster*, elevated levels of NPF significantly enhance feeding behavior, especially when flies are presented with rewarding food stimuli such as sugars and carbohydrates (Fadda et al. 2019). Enteroendocrine NPF signal in response to dietary sugar drives specific activation of lipid metabolism via suppression of adipokinetic hormone (AKH; see Sect. 2.3.1) and insulin-like peptides (ILPs; see Sect. 2.2.1) (Yoshinari et al. 2021; Malita et al. 2022). NPF is expressed not only in the enteroendocrine cells of the gut but also in the CNS, including the brain, and is known to regulate feeding behavior in both peripheral and central contexts. Behavioral assays have demonstrated that NPF neurons gate the preference for sucrose and other carbohydrate resources, directly influencing both the motivational and rewarding components of feeding (Wang et al. 2013).

11.3.2 Adipokinetic Hormone (AKH)

In addition to NPF, AKH also has a significant impact on feeding behavior. AKH is a functional counterpart of glucagon in vertebrates capable of mobilizing energy stores in response to starvation and acute energy demands. Indeed, this hormone has been identified and characterized as a factor that can mobilize lipid stores in response to long flights in *S. gregaria* (Stone et al. 1976). AKH has been definitively shown to regulate lipid metabolism in many insect species, including *D. melanogaster* and the two-spotted cricket *Gryllus bimaculatus*. Beyond its catabolic function, AKH intricately affects a diverse array of physiological functions, including the regulation of behavior and locomotion. AKH is produced by the corpora cardiaca (CC), and its receptor AKHR is expressed in the fat body and specific neurons. Deficiency of AKH signaling in *D. melanogaster* strengthens starvation resistance and reduces foraging-like behavior (Gáliková et al. 2015). AKH-abundant flies show lower lipid stores and heightened sensitivity to energy deprivation. The hormone accelerates lipid catabolism, as seen by reductions in triacylglycerol and diacylglycerol levels when AKH signaling is upregulated (Mochanová et al. 2018; Toprak 2020; Hughson 2021).

In *G. bimaculatus*, AKH's role in energy regulation is similarly well supported. Knockdown of the *AKHR* reduces hemolymph lipid levels and induces compensatory increases in feeding frequency, illustrating the hormone's critical function in lipid mobilization and homeostatic appetite suppression (Fukumura et al. 2018; Konuma et al. 2012). AKH administration also shifts the hemolymph fatty acid composition toward unsaturated fatty acids and inhibits acetate incorporation into newly synthesized lipids, emphasizing its rapid and reversible control of fat metabolism (Lorenz 2001). These findings affirm AKH's central role as a metabolic regulator that orchestrates both energy release and feeding behavior in insects. AKH/corazonin-related peptide (ACP; see Sect. 2.3.1), a peptide structurally related to AKH, has similar functions to AKH in *G. bimaculatus* (Zhou et al. 2018).

The interplay among ILP, AKH, and NPF forms a tightly regulated network that enables insects to coordinate their metabolic status and feeding behavior. Thus, alterations in the signaling of ILP, AKH, or NPF can strongly influence the propensity to feed or suppress appetite, ultimately impacting energy homeostasis and adaptive responses to changing nutritional environments (Chowański et al. 2021; Nässel and Zandawala 2019).

11.3.3 Feeding and Drinking Behavior Regulated by Other Neuropeptides

In addition to the three major neuropeptides—ILP, AKH, and NPF—a wide array of other neuropeptides and bioactive peptides are involved in the regulation of feeding behavior in insects. These neuropeptides function in response to internal metabolic states and external environmental cues, demonstrating the complexity of the regulatory systems that control feeding.

Neuropeptides such as AST-A, allatotropin (see Sect. 2.3.4), sNPF, corazonin (see Sect. 2.3.1), MIP, PDF, and RYamide (see Sect. 2.3.14) have all been implicated in modulating feeding behaviors. These peptides are involved not only in nutrient sensing but also in the regulation of appetite and food intake. Additionally, myosuppressin (see Sect. 2.4.5), CCHamide, and other myoinhibitory peptides contribute significantly to feeding behavior through their actions on gut motility and CNS signaling (Nagata and Zhou 2019; Zhu et al. 2022).

As mentioned above, myosuppressin inhibits gut motility, thereby inhibiting feeding behavior in *G. bimaculatus* (Zhou et al. 2019). Since feeding behavior is closely linked to gut motility, including peristaltic movements that facilitate digestion and food passage, neuropeptides and bioactive peptides play essential roles in modulating feeding via regulating these movements. In the cockroach *Blattella germanica*, sulfakinin acts as an inhibitory regulator of feeding behavior by suppressing gut contractions (Maestro et al. 2001). In *B. mori*, allatotropin and RYamide (also referred to as GSRYamides) similarly suppress feeding behavior by reducing intestinal contractile activity in conjunction with sNPF signaling (Matsumoto et al. 2019a; Matsumoto et al. 2019b).

Earlier studies have shown that the frontal ganglion plays an important role in regulating feeding behavior by making the pace of foregut contraction (Faivre 1863). Supporting this observation, peristalsis is also regulated by neuropeptides derived from the frontal ganglion. Neuropeptides including allatoregulatory peptides such as MIP (AST-B) in the frontal ganglion of lepidopteran species *M. sexta* and *Spodoptera litura* are regulatory peptides on the foregut contractive action for feeding (Audsley et al. 2005; Audsley and Weaver 2009). In *B. mori*, DH31 (CT-DH) in the frontal ganglion can inhibit feeding behavior (Nagata and Nagasawa 2017).

As water homeostasis is involved in drinking and feeding behavior, CAPA (capability; see Sect. 2.3.13) and DH44 (CRF-DH) are the crucial regulatory neuropeptides for drinking and thirst behavior in insects. Diuretic hormones (DHs), traditionally associated with water balance and excretion, also influence gut motility. Therefore, feeding behavior would be regulated by DHs in several insects. DH44, for example, is involved in feeding in the kissing bug *Rhodnius prolixus* (Mollayeva et al. 2018). This coordination between gut motility and water excretory function ensures that digestive and hydration processes remain in balance. CAPA, produced in the abdominal ganglia, functions as an anti-diuretic hormone in diverse insects, including mosquitoes and *D. melanogaster*, inhibiting Malpighian tubule secretion via signaling cascades involving nitric oxide (NO), cGMP, and protein kinase G (PKG) (Sajadi et al. 2020; Terhzaz et al. 2015). Functional ablation and pharmacological studies in the mosquito *Aedes aegypti* confirmed that CAPA neuropeptides act specifically through their cognate GPCRs to counteract diuretic hormone stimulation, ultimately protecting the insect from excessive fluid loss during desiccation or after blood feeding (Sajadi et al., 2020; Predel and Wegener 2006).

Complementing this, ion transport peptides (ITPs; see Sect. 2.5.1) represent a distinct neuropeptide class that directly drives thirst and represses excretion: in *D. melanogaster*, ITP is transcriptionally upregulated upon dehydration, increasing water-seeking behavior and reducing water loss, which is functionally analogous to vasopressin systems in vertebrates (Gáliková et al. 2018). Robust genetic evidence—including ITP receptor gene disruption in *B. mori*—demonstrates an essential role of this pathway in global water homeostasis, with the conserved mechanism via NO/cGMP signaling mediating anti-diuretic effects of ITP in insects (Sun et al. 2020).

11.4 Mating Behavior

Neuropeptides are central to the regulation in motivation of mating behavior in insects, operating at multiple stages from partner search and attraction to post-mating adaptation. The pheromone biosynthesis-activating neuropeptide (PBAN; see Sect. 2.3.13) is essential for female sex pheromone production in Lepidoptera such as *Helicoverpa zea* and *B. mori* (Raina et al. 1989; Kitamura et al. 1989). Gene knockdown and receptor disruption of PBAN have led to reduced pheromone emission and mating success (Kim et al. 2008; Lee et al. 2011). Meanwhile, NPF and its receptor have been shown to modulate courtship intensity and reproductive drive in *D. melanogaster* and *S. gregaria* (Liu et al. 2019; Van Wielendaele et al. 2013a, b).

A significant innovation in insects is the emergence of natalisin (see Sect. 2.7.3), a tachykinin (TK)-like neuropeptide, and its dedicated receptor, which are highly conserved and regulate both male and female reproductive behaviors (Jiang et al. 2013). RNAi and loss-of-function experiments in species such as *Spodoptera frugiperda* and *Bactrocera dorsalis* have demonstrated that disruption of either natalisin or its receptor strongly suppresses sexual receptivity, courtship, and mating frequency, indicating a crucial, pleiotropic role for this peptide in reproductive activity (Gong et al. 2024; Gui et al. 2018). These findings support the view that the TK peptide family, including natalisin, is essential for the orchestration of insect reproductive behaviors.

Classical TKs (tachykinin-related peptides or TRPs; see Sect. 2.3.18) themselves are primarily known for their roles in modulating aggression and other social behaviors in *D. melanogaster*, but they have also been implicated in the inhibition of specific aspects of courtship and sexual behavior through manipulation of tachykininergic neurons (Asahina et al. 2014; Wohl et al. 2023).

11.5 Other Behaviors

11.5.1 Orientation and Compassing Behaviors

Neuropeptides functioning in the neuronal circuits underlying compass orientation and navigation in models such as bees, monarch butterflies, and locusts exert their effects via a pivotal brain region that integrates sensory information about celestial cues (such as polarized skylight and sun azimuth) for heading computation. One of the best-supported neuropeptides involved in this system is PDF, which is closely associated with circadian clock neurons in the honeybee *Apis mellifera* (Zeller et al. 2015). PDF-immunoreactive fibers have been shown to project into sky compass pathways, especially where information from the dorsal rim area of the compound eye—essential for detecting polarized light—converges on neural processing centers such as the anterior optic tubercle and the central complex (Zeller et al. 2015). There,

PDF likely interacts with neurotransmitters such as serotonin and GABA to modulate time-of-day-dependent tuning of compass signals, thereby ensuring that the internal representation of heading remains aligned with changing solar azimuth throughout the day. Immunocytochemical and genetic studies have further revealed that neuropeptides act at multiple levels of compass orientation circuits—modulating input gain, altering neuronal excitability, or resetting the underlying oscillatory "ring attractor" code in the central complex's compass neurons. In honeybees and locusts, for instance, evidence suggests that PDF actively links the circadian clock to visual signals from the polarized sky during navigation.

11.5.2 Collective and Migratory Behaviors

Neuropeptides serve as crucial regulators of collective behavior and migratory events in insects by orchestrating transitions between solitary and gregarious phases; specifically, they achieve this by modulating neural circuits that govern behavioral responses, as well as influencing social plasticity that enables insects to flexibly adapt their social interaction and group dynamics according to environmental and physiological cues. A key example is NPF, which modulates locomotor activity, foraging dynamics, and aggregation in social insects such as ants, bees, and migratory locusts (Hou et al. 2023). In *S. gregaria*, NPF/nitric oxide (NO) signaling pathway functions as a critical molecular switch mediating density-dependent behavioral phase transitions (Hou et al. 2017). The transcriptional levels of both subtypes of NPF, NPF1a and NPF2, are downregulated in high-density conditions, while their receptor expression is elevated in isolated conditions. Both NPF1a and NPF2 act to suppress gregarious locomotor activity by inhibiting nitric oxide synthase (NOS) activity in the brain and thereby reducing NO production, illustrating that the NPF/NO axis is essential for controlling locomotor plasticity in response to the locust's social environment.

Corazonin (CRZ) is a neuropeptide originally identified as a factor associated with stress and pigment regulation. This peptide has emerged as a pivotal upstream regulator of gregarious behavior and swarm formation of locusts such as *Locusta migratoria* and *S. gregaria*. Experimental studies have demonstrated that injection of CRZ into solitary-phase nymphs of these locusts induces rapid shifts toward gregarious coloration, morphometric traits, and aggregation behaviors, closely replicating the effects of crowding (Sugahara et al. 2015; Verlinden et al. 2009; Maeno and Tanaka 2024).

CRZ's role as a gregarizing factor is well conserved among Acrididae locusts such as *S. gregaria* and *L. migratoria*, where it regulates phase polyphenism—shifting solitary individuals to gregarious swarming states. Comparable regulatory roles of CRZ have been revealed in social Hymenoptera, including ants such as *Harpegnathos saltator* and *Camponotus floridanus* and wasps such as *Polistes canadensis*: in these taxa, CRZ modulates behaviors related to caste identity and foraging activity, by maintaining high expression of CRZ in workers or forager castes along with suppressing reproductive functions in non-queen individuals (Gospocic et al. 2017).

Advanced mass spectrometry imaging, functional genomics, and behavioral assays in social insects have revealed that a broad array of neuropeptides, including sNPF, TK, and CRZ, regulate age-, caste-, and context-dependent collective behaviors at both central and peripheral targets (Hou et al. 2023). These peptides can modulate responsiveness to social signals, reproductive role shifts, and migratory activity, providing the flexibility required for complex group coordination and environmental adaptation (Gilbert et al. 2025). Intriguingly, neuropeptide cascades appear to integrate external cues (population density, social contacts) with internal physiological status to determine the emergence of mass movements and foraging swarms in diverse insect taxa, marking neuropeptidergic regulation as a core mechanism underlying the evolution of insect collective behavioral phenomena.

11.5.3 Aggressive and Related Behaviors

TK in *D. melanogaster* is expressed in a small cluster of male-specific neurons and powerfully promotes aggressive arousal and attack behavior (Asahina et al. 2014; Pavlou et al. 2014). Experimental activation of these neurons increases, and silencing decreases, inter-male aggression without affecting other social behaviors, indicating selective peptidergic control of agonistic drive. TK acts via TkR86C and TkR99D receptors, and its manipulation is sufficient to restore aggressive behavior, even if the environment is unfavorable for this behavior, highlighting the role of TK as a potent motivational signal for contest escalation (Asahina et al. 2014). Importantly, sexual dimorphism in the patterns of TK-producing neurons underlies male-biased aggression and parallels with the fact that substance P, a mammalian TK, mediates aggression in mammals, suggesting deep evolutionary conservation of endocrine control of aggression.

Other neuropeptides, including sNPF, NPF, and sulfakinin, contribute to the fine-tuning of aggressive behavior. NPF, the invertebrate homolog of mammalian NPY, generally suppresses high-intensity aggression, and genetic silencing of NPF or its receptors increases aggressive behaviors in flies and crustaceans (Dierick and Greenspan 2007; Bubak et al. 2019; Wu et al. 2020). Conversely, *Drosophila* sulfakinin (Dsk) acts as a tonic inhibitor of aggression, and its knockdown causes hyper-aggressive phenotypes. Social context, stress, and peptide level fluctuations modulate Dsk signaling to calibrate the threshold for aggressive action (Agrawal et al. 2020). Thus, insect aggression is steered by an integrated network of neuromodulators—primarily TK as positive and NPF/Dsk as negative regulators—whose dynamic interactions allow rapid, context-dependent adjustment of social conflict and dominance behavior (Sherer and Certel 2019).

11.6 Conclusion

Recent advances in insect neurobiology have revealed that multiple neuropeptides and bioactive peptides function in the precise regulation of diverse innate behaviors such as orientation, collective action, and feeding behavior. Importantly, these peptides display distinct, sometimes species-specific, functions, which likely result from differences in network harmonization, timing, and integration of sensory and physiological cues. For instance, NPF not only orchestrates hunger-driven feeding and food preferences but also plays critical roles in reward-related behaviors, social interaction, and sensory-driven behavioral flexibility in *Drosophila* and other insects. Such multifunctionality is also observed in the regulation of collective and orientation behaviors by NPF signaling.

Furthermore, the timing of peptide hormone secretion is crucial in developmental processes such as ecdysis, which are tightly controlled by orchestrated interactions between ETH, EH, and other neuropeptides, ensuring the correct sequencing of behavioral phases and adaptive developmental transitions. Variations in these neuropeptidergic cascades underlie species-specific molting and behavioral rhythms.

Altogether, the interplay among these neuropeptide networks enables precise, context-dependent regulation of innate behaviors, development, and growth in insects. This harmonization enables insects to flexibly respond to environmental change, optimize resource acquisition, and coordinate complex behavioral repertoires essential for insect survival.

References

Agrawal P, Kao D, Chung P, Looger LL (2020) The neuropeptide Drosulfakinin regulates social isolation-induced aggression in *Drosophila*. J Exp Biol 223(Pt 2):jeb207407. https://doi.org/10.1242/jeb.207407

Aguinaldo AM, Turbeville JM, Linford LS, Rivera MC, Garey JR, Raff RA et al (1997) Evidence for a clade of nematodes, arthropods and other moulting animals. Nature 387(6632):489–493. https://doi.org/10.1038/387489a0

Arakane Y, Li B, Muthukrishnan S, Beeman RW, Kramer KJ, Park Y (2008) Functional analysis of four neuropeptides, EH, ETH, CCAP and bursicon, and their receptors in adult ecdysis behavior of the red flour beetle, *Tribolium castaneum*. Mech Dev 125(11–12):984–995. https://doi.org/10.1016/j.mod.2008.09.002

Asahina K, Watanabe K, Duistermars BJ, Hoopfer E, González CR, Eyjólfsdóttir EA et al (2014) Tachykinin-expressing neurons control male-specific aggressive arousal in *Drosophila*. Cell 156(1–2):221–235. https://doi.org/10.1016/j.cell.2013.11.045

Audsley N, Weaver RJ (2009) Neuropeptides associated with the regulation of feeding in insects. Gen Comp Endocrinol 162(1):93–104. https://doi.org/10.1016/j.ygcen.2008.08.003

Audsley N, Matthews J, Weaver RJ (2005) Neuropeptides associated with the frontal ganglion of larval Lepidoptera. Peptides 26(1):11–21. https://doi.org/10.1016/j.peptides.2004.10.011

Bai H, Palli SR (2010) Functional characterization of bursicon receptor and genome-wide analysis for identification of genes affected by bursicon receptor RNAi. Dev Biol 344(1):248–258. https://doi.org/10.1016/j.ydbio.2010.05.003

Bubak AN, Watt MJ, Renner KJ, Luman AA, Costabile JD, Sanders EJ et al (2019) Sex differences in aggression: differential roles of 5-HT2, neuropeptide F and tachykinin. PLoS One 14(1): e0203980. https://doi.org/10.1371/journal.pone.0203980

Chowański S, Walkowiak-Nowicka K, Winkiel M, Marciniak P, Urbański A, Pacholska-Bogalska J (2021) Insulin-like peptides and cross-talk with other factors in the regulation of insect metabolism. Front Physiol 12:701203. https://doi.org/10.3389/fphys.2021.701203

Chowański S, Lubawy J, Pacholska-Bogalska J, Ragionieri L, Urbański A, Szymczak-Cendlak M et al (2025) FMRFamide-like peptides (FaLPs) – an overview of diverse physiological roles in insects and other arthropods. Int J Biol Sci 21(6):2725–2746. https://doi.org/10.7150/ijbs.106382

Daubnerová I, Roller L, Satake H, Zhang C, Kim YJ, Žitňan D (2021) Identification and function of ETH receptor networks in the silkworm *Bombyx mori*. Sci Rep 11(1):11693. https://doi.org/10.1038/s41598-021-91022-8

Dierick HA, Greenspan RJ (2007) Serotonin and neuropeptide F have opposite modulatory effects on fly aggression. Nat Genet 39(5):678–682. https://doi.org/10.1038/ng2029

Fadda M, Hasakiogullari I, Temmerman L, Beets I, Zels S, Schoofs L (2019) Regulation of feeding and metabolism by neuropeptide F and short neuropeptide F in invertebrates. Front Endocrinol (Lausanne) 10:64. https://doi.org/10.3389/fendo.2019.00064

Faivre E (1863) Experiences sur le role du cerveau dans L'ingestion des aliments chez les insectes, et sur les fonctions du ganglioan frontal. CR Soc Biol Paris 5:101–104

Fukumura K, Konuma T, Tsukamoto Y, Nagata S (2018) Adipokinetic hormone signaling determines dietary fatty acid preference through maintenance of hemolymph fatty acid composition in the cricket *Gryllus bimaculatus*. Sci Rep 8(1):4737. https://doi.org/10.1038/s41598-018-22987-2

Gáliková M, Diesner M, Klepsatel P, Hehlert P, Xu Y, Bickmeyer I et al (2015) Energy homeostasis control in *Drosophila* adipokinetic hormone mutants. Genetics 201(2):665–683. https://doi.org/10.1534/genetics.115.178897

Gáliková M, Dircksen H, Nässel DR (2018) The thirsty fly: ion transport peptide (ITP) is a novel endocrine regulator of water homeostasis in *Drosophila*. PLoS Genet 14(8):e1007618. https://doi.org/10.1371/journal.pgen.1007618

Gammie SC, Truman JW (1997) Neuropeptide hierarchies and the activation of sequential motor behaviors in the hawkmoth, *Manduca sexta*. J Neurosci 17(11):4389–4397. https://doi.org/10.1523/JNEUROSCI.17-11-04389.1997

Gammie SC, Truman JW (1999) Eclosion hormone provides a link between ecdysis-triggering hormone and crustacean cardioactive peptide in the neuroendocrine cascade that controls ecdysis behavior. J Exp Biol 202(Pt 4):343–352. https://doi.org/10.1242/jeb.202.4.343

Gilbert MB, Glastad KM, Fioriti M, Sorek M, Scarpa T, Purnell FS et al (2025) Neuropeptides specify and reprogram division of labor in the leafcutter ant *Atta cephalotes*. Cell 188(15): 3974–3991.e21. https://doi.org/10.1016/j.cell.2025.05.023

Gong W, Linghu JH, Xu HM, Luo LL, Smagghe G, Liu TX et al (2024) Neuropeptide natalisin regulates reproductive behaviors in *Spodoptera frugiperda*. Sci Rep 14(1):15122. https://doi.org/10.1038/s41598-024-66031-y

Gospocic J, Shields EJ, Glastad KM, Lin Y, Penick CA, Yan H et al (2017) The neuropeptide corazonin controls social behavior and caste identity in ants. Cell 170(4):748–759.e12. https://doi.org/10.1016/j.cell.2017.07.014

Gui SH, Pei YX, Xu L, Wang WP, Jiang HB, Nachman RJ et al (2018) Function of the natalisin receptor in mating of the oriental fruit fly, *Bactrocera dorsalis* (Hendel) and testing of peptidomimetics. PLoS One 13(2):e0193058. https://doi.org/10.1371/journal.pone.0193058

Hughson BN (2021) The glucagon-like adipokinetic hormone in *Drosophila melanogaster*–biosynthesis and secretion. Front Physiol 12:710652. https://doi.org/10.3389/fphys.2021.710652

Hou L, Yang P, Jiang F, Liu Q, Wang X, Kang L (2017) The neuropeptide F/nitric oxide pathway is essential for shaping locomotor plasticity underlying locust phase transition. elife 6:e22526. https://doi.org/10.7554/eLife.22526

Hou L, Wang N, Sun T, Wang X (2023) Neuropeptide regulations on behavioral plasticity in social insects. Curr Opin Insect Sci 60:101119. https://doi.org/10.1016/j.cois.2023.101119

Iversen A, Cazzamali G, Williamson M, Hauser F, Grimmelikhuijzen CJ (2002) Molecular identification of the first insect ecdysis triggering hormone receptors. Biochem Biophys Res Commun 299(5):924–931. https://doi.org/10.1016/s0006-291x(02)02798-5

Jiang H, Lkhagva A, Daubnerová I, Chae HS, Šimo L, Jung SH et al (2013) Natalisin, a tachykinin-like signaling system, regulates sexual activity and fecundity in insects. Proc Natl Acad Sci USA 110(37):E3526–E3534. https://doi.org/10.1073/pnas.1310676110

Kim YJ, Zitnan D, Cho KH, Schooley DA, Mizoguchi A, Adams ME (2006a) Central peptidergic ensembles associated with organization of an innate behavior. Proc Natl Acad Sci USA 103(38): 14211–14216. https://doi.org/10.1073/pnas.0603459103

Kim YJ, Zitnan D, Galizia CG, Cho KH, Adams ME (2006b) A command chemical triggers an innate behavior by sequential activation of multiple peptidergic ensembles. Curr Biol 16(14): 1395–1407. https://doi.org/10.1016/j.cub.2006.06.027

Kim YJ, Nachman RJ, Aimanova K, Gill S, Adams ME (2008) The pheromone biosynthesis activating neuropeptide (PBAN) receptor of *Heliothis virescens*: identification, functional expression, and structure-activity relationships of ligand analogs. Peptides 29(2):268–275. https://doi.org/10.1016/j.peptides.2007.12.001

Kim DH, Han MR, Lee G, Lee SS, Kim YJ, Adams ME (2015) Rescheduling behavioral subunits of a fixed action pattern by genetic manipulation of peptidergic signaling. PLoS Genet 11(9): e1005513. https://doi.org/10.1371/journal.pgen.1005513

Kingan TG, Gray W, Zitnan D, Adams ME (1997) Regulation of ecdysis-triggering hormone release by eclosion hormone. J Exp Biol 200(Pt 24):3245–3256. https://doi.org/10.1242/jeb.200.24.3245

Kitamura A, Nagasawa H, Kataoka H, Inoue T, Matsumoto S, Ando T et al (1989) Amino acid sequence of pheromone-biosynthesis-activating neuropeptide (PBAN) of the silkworm, *Bombyx mori*. Biochem Biophys Res Commun 163(1):520–526. https://doi.org/10.1016/0006-291x(89)92168-2

Konuma T, Morooka N, Nagasawa H, Nagata S (2012) Knockdown of the adipokinetic hormone receptor increases feeding frequency in the two-spotted cricket *Gryllus bimaculatus*. Endocrinology 153(7):3111–3122. https://doi.org/10.1210/en.2011-1533

Lee DW, Shrestha S, Kim AY, Park SJ, Yang CY, Kim Y et al (2011) RNA interference of pheromone biosynthesis-activating neuropeptide receptor suppresses mating behavior by inhibiting sex pheromone production in *Plutella xylostella* (L.). Insect Biochem Mol Biol 41(4):236–243. https://doi.org/10.1016/j.ibmb.2011.01.001

Lin S, Senapati B, Tsao CH (2019) Neural basis of hunger-driven behaviour in *drosophila*. Open Biol 9(3):180259. https://doi.org/10.1098/rsob.180259

Liu W, Ganguly A, Huang J, Wang Y, Ni JD, Gurav AS (2019) Neuropeptide F regulates courtship in *Drosophila* through a male-specific neuronal circuit. elife 8:e49574. https://doi.org/10.7554/eLife.49574

Lorenz MW (2001) Synthesis of lipids in the fat body of *Gryllus bimaculatus*: age-dependency and regulation by adipokinetic hormone. Arch Insect Biochem Physiol 47(4):198–214. https://doi.org/10.1002/arch.1052

Loveall BJ, Deitcher DL (2010) The essential role of bursicon during *Drosophila* development. BMC Dev Biol 10:92. https://doi.org/10.1186/1471-213X-10-92

Maeno K, Tanaka S (2024) Hormonal control of phase-related changes in the number of antennal sensilla in the desert locust, *Schistocerca gregaria*: possible involvement of [His^7]-corazonin. J Insect Physiol 50(9):855–865. https://doi.org/10.1016/j.jinsphys.2004.06.008

Maestro JL, Aguilar R, Pascual N, Valero ML, Piulachs MD, Andreu D et al (2001) Screening of antifeedant activity in brain extracts led to the identification of sulfakinin as a satiety promoter in the German cockroach. Are arthropod sulfakinins homologous to vertebrate gastrins-cholecystokinins? Eur J Biochem 268(22):5824–5830. https://doi.org/10.1046/j.0014-2956.2001.02527.x

Malhotra P, Basu S (2023) The intricate role of ecdysis triggering hormone signaling in insect development and reproductive regulation. Insects 14(8):711. https://doi.org/10.3390/insects14080711

Malita A, Kubrak O, Koyama T, Ahrentløv N, Texada MJ, Nagy S et al (2022) A gut-derived hormone suppresses sugar appetite and regulates food choice in Drosophila. Nat Metab 4(11): 1532–1550. https://doi.org/10.1038/s42255-022-00672-z

Matsumoto S, Kutsuna N, Daubnerová I, Roller L, Žitňan D, Nagasawa H et al (2019a) Enteroendocrine peptides regulate feeding behavior via controlling intestinal contraction of the silkworm *Bombyx mori*. PLoS One 14(7):e0219050. https://doi.org/10.1371/journal.pone.0219050

Matsumoto S, Ohara A, Nagai-Okatani C, Zhou YJ, Fujinaga D, Seike H et al (2019b) Antagonistic effect of short neuropeptide F on allatotropin-inhibited feeding motivation of the silkworm larva, *Bombyx mori*. Zoolog Sci 36(1):58–67. https://doi.org/10.2108/zs180119

Mena W, Diegelmann S, Wegener C, Ewer J (2016) Stereotyped responses of *drosophila* peptidergic neuronal ensemble depend on downstream neuromodulators. elife 5:e19686. https://doi.org/10.7554/eLife.19686

Mochanová M, Tomčala A, Svobodová Z, Kodrík D (2018) Role of adipokinetic hormone during starvation in *Drosophila*. Comp Biochem Physiol B Biochem Mol Biol 226:26–35. https://doi.org/10.1016/j.cbpb.2018.08.004

Mollayeva S, Orchard I, Lange AB (2018) The involvement of Rhopr-CRF/DH in feeding and reproduction in the blood-gorging insect *Rhodnius prolixus*. Gen Comp Endocrinol 258:79–90. https://doi.org/10.1016/j.ygcen.2017.07.005

Nagata S, Nagaswa H (2017) Calcitonin-like peptide hormone (CT/DH) in the frontal ganglia as a feeding regulatory peptide of the silkworm, Bombyx mori. Peptides 98:23–28. https://doi.org/10.1016/j.peptides.2016.05.002

Nagata S, Zhou YJ (2019) Feeding-modulating neuropeptides and peptide hormones in insects. Adv Insect Physiol 57:137–172. https://doi.org/10.1016/bs.aiip.2019.10.002

Nässel DR, Zandawala M (2019) Recent advances in neuropeptide signaling in *Drosophila*, from genes to physiology and behavior. Prog Neurobiol 179:101607. https://doi.org/10.1016/j.pneurobio.2019.02.003

Park Y, Kim YJ, Dupriez V, Adams ME (2003) Two subtypes of ecdysis-triggering hormone receptor in *Drosophila melanogaster*. J Biol Chem 278(20):17710–17715. https://doi.org/10.1074/jbc.M301119200

Pavlou HJ, Neville MC, Goodwin SF (2014) Aggression: tachykinin is all the rage. Curr Biol 24(6): R243–R244. https://doi.org/10.1016/j.cub.2014.02.017

Predel R, Wegener C (2006) Biology of the CAPA peptides in insects. Cell Mol Life Sci 63(21): 2477–2490. https://doi.org/10.1007/s00018-006-6187-3

Raina AK, Jaffe H, Kempe TG, Keim P, Blacher RW, Fales HM et al (1989) Identification of a neuropeptide hormone that regulates sex pheromone production in female moths. Science 244(4906):796–798. https://doi.org/10.1126/science.244.4906.796

Sajadi F, Uyuklu A, Paputsis C, Lajevardi A, Wahedi A, Ber LT (2020) CAPA neuropeptides and their receptor form an anti-diuretic hormone signaling system in the human disease vector, Aedes aegypti. Sci Rep 10(1):1755. https://doi.org/10.1038/s41598-020-58731-y

Sherer LM, Certel SJ (2019) The fight to understand fighting: neurogenetic approaches to the study of aggression in insects. Curr Opin Insect Sci 36:18–24. https://doi.org/10.1016/j.cois.2019.06.004

Stone JV, Mordue W, Batley KE, Morris HR (1976) Structure of locust adipokinetic hormone, a neurohormone that regulates lipid utilisation during flight. Nature 263(5574):207–211. https://doi.org/10.1038/263207a0

Sugahara R, Saeki S, Jouraku A, Shiotsuki T, Tanaka S (2015) Knockdown of the corazonin gene reveals its critical role in the control of gregarious characteristics in the desert locust. J Insect Physiol 79:80–87. https://doi.org/10.1016/j.jinsphys.2015.06.009

Sun L, Zhang Z, Zhang R, Yu Y, Yang F, Tan A (2020) Molecular disruption of ion transport peptide receptor results in impaired water homeostasis and developmental defects in *Bombyx mori*. Front Physiol 11:424. https://doi.org/10.3389/fphys.2020.00424

Terhzaz S, Teets NM, Cabrero P, Henderson L, Ritchie MG, Nachman RJ et al (2015) Insect capa neuropeptides impact desiccation and cold tolerance. Proc Natl Acad Sci USA 112(9): 2882–2887. https://doi.org/10.1073/pnas.1501518112

Toprak U (2020) The role of peptide hormones in insect lipid metabolism. Front Physiol 11:434. https://doi.org/10.3389/fphys.2020.00434

Van Wielendaele P, Dillen S, Zels S, Badisco L, Vanden Broeck J (2013a) Regulation of feeding by Neuropeptide F in the desert locust, *Schistocerca gregaria*. Insect Biochem Mol Biol 43(1): 102–114. https://doi.org/10.1016/j.ibmb.2012.10.002

Van Wielendaele P, Wynant N, Dillen S, Zels S, Badisco L, Vanden Broeck J (2013b) Neuropeptide F regulates male reproductive processes in the desert locust, *Schistocerca gregaria*. Insect Biochem Mol Biol 43(3):252–259. https://doi.org/10.1016/j.ibmb.2012.12.004

Verlinden H, Badisco L, Marchal E, Van Wielendaele P, Vanden Broeck J (2009) Endocrinology of reproduction and phase transition in locusts. Gen Comp Endocrinol 162(1):79–92. https://doi.org/10.1016/j.ygcen.2008.11.016

Wang Y, Pu Y, Shen P (2013) Neuropeptide-gated perception of appetitive olfactory inputs in *Drosophila* larvae. Cell Rep 3(3):820–830. https://doi.org/10.1016/j.celrep.2013.02.003

Wohl MP, Liu J, Asahina K (2023) *Drosophila* tachykininergic neurons modulate the activity of two groups of receptor—expressing neurons to regulate aggressive tone. J Neurosci 43(19): 3394–3420. https://doi.org/10.1523/JNEUROSCI.1734-22.2023

Wu Q, Wen T, Lee G, Park JH, Cai HN, Shen P (2003) Developmental control of foraging and social behavior by the *Drosophila* neuropeptide Y-like system. Neuron 39(1):147–161. https://doi.org/10.1016/s0896-6273(03)00396-9

Wu F, Deng B, Xiao N, Wang T, Li Y, Wang R et al (2020) A neuropeptide regulates fighting behavior in *Drosophila melanogaster*. elife 9:e54229. https://doi.org/10.7554/eLife.54229

Yoshinari Y, Kosakamoto H, Kamiyama T, Hoshino R, Matsuoka R, Kondo S et al (2021) The sugar-responsive enteroendocrine neuropeptide F regulates lipid metabolism through glucagon-like and insulin-like hormones in *Drosophila melanogaster*. Nat Commun 12(1):4818. https://doi.org/10.1038/s41467-021-25146-w

Zandawala M, Yurgel ME, Texada MJ, Liao S, Rewitz K, Keene AC et al (2017) Modulation of *Drosophila* post-feeding physiology and behavior by the neuropeptide leucokinin. Proc Natl Acad Sci USA 114(36):9709–9714. https://doi.org/10.1073/pnas.1716890115

Zeller M, Held M, Bender J, Berz A, Heinloth T, Hellfritz T et al (2015) Transmedulla neurons in the sky compass network of the honeybee (*Apis mellifera*) are a possible site of circadian input. PLoS One 10(12):e0143244. https://doi.org/10.1371/journal.pone.0143244

Zhou YJ, Fukumura K, Nagata S (2018) Effects of adipokinetic hormone and its related peptide on maintaining hemolymph carbohydrate and lipid levels in the two-spotted cricket, *Gryllus bimaculatus*. Biosci Biotechnol Biochem 82(2):274–284. https://doi.org/10.1080/09168451.2017.1422106

Zhou YJ, Seike H, Nagata S (2019) Function of myosuppressin in regulating digestive function in the two-spotted cricket, *Gryllus bimaculatus*. Gen Comp Endocrinol 280:185–191. https://doi.org/10.1016/j.ygcen.2019.05.001

Zhu Z, Tsuchimoto M, Nagata S (2022) CCHamide-2 signaling regulates food intake and metabolism in *Gryllus bimaculatus*. Insects 13(4):324. https://doi.org/10.3390/insects13040324

Zitnan D, Adams ME (2000) Excitatory and inhibitory roles of central ganglia in initiation of the insect ecdysis behavioural sequence. J Exp Biol 203(Pt 8):1329–1340. https://doi.org/10.1242/jeb.203.8.1329

Zitnan D, Kingan TG, Hermesman JL, Adams ME (1996) Identification of ecdysis-triggering hormone from an epitracheal endocrine system. Science 271(5245):88–91. https://doi.org/10.1126/science.271.5245.88

Zitnan D, Adams ME (2012) Neuroendocrine regulation of Ecdysis. In: Gilbert LI (ed) Insect endocrinology, Elsevier, pp 253–309

Chapter 12
Regulation of Tanning and Color Change

Hitomi Seike and Shinji Nagata

Abstract Insects exhibit a remarkable diversity in body coloration derived from structural properties and pigments, which is vital for their survival, communication, and reproduction. Coloration is highly adaptable and modulated by biotic and abiotic environmental cues (e.g., temperature, population density), with the endocrine system playing a critical role in regulating these changes. Melanin biosynthesis, the primary pigment system, begins de novo from tyrosine and proceeds via conserved core enzymes, yielding diverse melanin-derived colors. The levels of transcripts encoding these enzymes are highest immediately after molting, tightly linking melanization to cuticle tanning, accompanied by endocrine control. In locusts, neuropeptides such as corazonin (CRZ) induce density-dependent dark coloration. Elevenin signaling regulates melanization in the brown planthoppers by upregulating melanin-biosynthetic enzymes. Melanization and reddish coloration hormone (MRCH) induces larval dark body coloration in some moth larvae via a density-dependent pathway. Although previous reports dealing with the body coloration in insects are limited, we introduce the fact that insect body coloration is an adaptable phenomenon orchestrated by a complex and sophisticated network of hormonal, developmental, environmental, and anthropogenic factors in this chapter.

Keywords Body color · Melanin · Pigment · Density · Camouflage

12.1 Introduction

Insects exhibit a remarkable diversity in body coloration, which plays a vital role in their survival, communication, and reproduction. This coloration arises from both structural properties of the integument and pigments. The biosynthesis of pigments, particularly melanin and ommochromes, involves complex pathways regulated by various enzymes in insects (Shamim et al. 2014; Stavenga et al. 2014). Interestingly,

H. Seike (✉) · S. Nagata
Department of Integrated Biosciences, Graduate School of Frontier Sciences, The University of Tokyo, Kashiwa, Japan
e-mail: hitomis@edu.k.u-tokyo.ac.jp; shinjin@edu.k.u-tokyo.ac.jp

A. Mizoguchi, N. Okamoto (eds.), *Insect Peptide Hormones and Neuropeptides*, Entomology Monographs, https://doi.org/10.1007/978-981-95-7510-7_12

body color can be altered by biotic and abiotic conditions, for example, temperature, humidity, latitude, and UV as abiotic conditions and population density, growth and development, aging, and immunity as biotic conditions. It changes to accomplish their environmental adaptation, and sometimes to complete their camouflage against their surroundings. Also, sex differentiation in body color sometimes involving the pattern of body or wings can be observed in many species. Insect colorations seem to be directly established by pigments, cuticular surface structures, or their combination, which are induced by environmental cues including sunlight (Chapman et al. 2013). In addition, endocrine factors can contribute to decision-making that alters body color involving melanin biosynthesis and other coloration mechanisms to adapt to environmental cues in insects (Fig. 12.1a).

12.2 Melanin Biosynthesis in Insects' Coloration

Melanins are a major class of pigments commonly used in animals, plants, and fungi. They are synthesized de novo from the amino acid tyrosine, and the core pathway is conserved across insect species (Fig. 12.1b). Tyrosine is first converted to DOPA by tyrosine hydroxylase (TH, also known as pale in the fruit fly *Drosophila melanogaster*), and DOPA is subsequently decarboxylated to dopamine by dopa decarboxylase (DDC). Dopamine serves as a central precursor that can be further modified to form *N*-acetyldopamine (NADA) by arylalkylamine-*N*-acetyltransferase (aaNAT) or *N*-β-alanyldopamine (NBAD) by the enzyme "ebony" (*N*-β-alanyl dopamine synthase), which uses β-alanine supplied by an enzyme called "black." These dopamine derivatives, together with DOPA itself, are oxidized by enzymes such as phenoloxidase (PO), laccase2, multicopper oxidase 2 (MCO2), and products of members of the yellow gene family, producing different types of melanin pigments including dopa melanin, dopamine melanin, NADA melanin, and NBAD melanin (Whitten and Coates 2017; Popadić and Tsitlakidou 2021; Okude and Futahashi 2021).

The activities of these enzymes determine the range of colors observed in insect cuticles. DOPA and dopamine derivatives generally give rise to black and brown coloration, while NADA and NBAD contribute to lighter yellow and reddish hues. Expression of the melanin biosynthetic enzymes is tightly regulated at the transcriptional level, and their activities fluctuate in response to developmental and physiological conditions. Importantly, transcriptional levels of most enzyme genes are highest immediately after molting, implying a close association between melanin production, cuticle tanning, and the events of molting and metamorphosis. Melanization is suppressed by *ebony*, *black*, and *aaNAT* and promoted by *tan* and *yellow*. TH and DDC regulate the initial production of pigment, while other enzymes, such as MCO2, contribute to more specific pigmentation patterns, including black and reddish-brown hues in Odonata (Okude and Futahashi 2021).

Overall, the melanin biosynthetic pathway provides a versatile framework for generating diverse insect color patterns. A number of studies have demonstrated that melanin-based coloration is directly determined by the transcriptional levels of

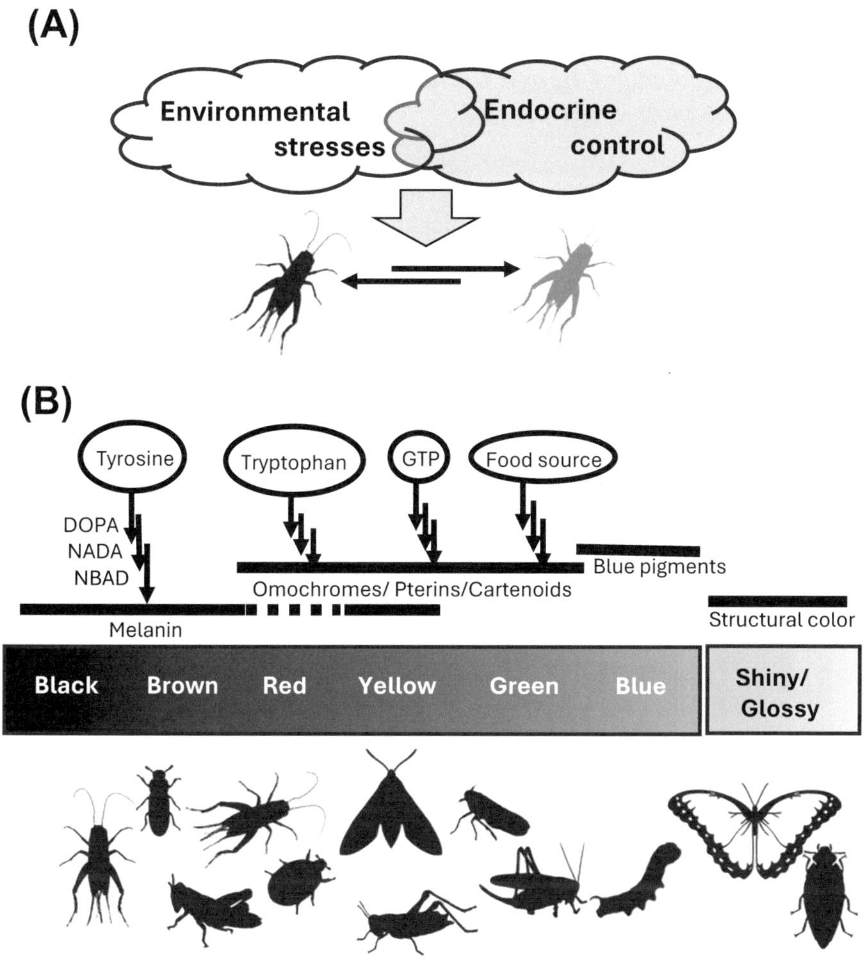

Fig. 12.1 Diversity of insect coloration. This figure is modified after Badejo et al. (2020). (**a**) Environmental stresses and endocrine signals coordinately regulate the modulation of body coloration in insects. (**b**) Major pigment and structural sources of insect colors. Tyrosine-derived melanins synthesized via DOPA, NADA, and NBAD induce black to brown coloration. Tryptophan serves as the precursor of ommochromes, contributing to brown to yellow coloration. Pteridine pigments, derived from guanosine, are responsible for green to yellow colors. Carotenoids generate yellow to orange coloration. In addition to pigment-based colors, structural coloration is produced by light interference through microstructured cuticular architecture

melanin-biosynthetic enzymes. Variations in the regulation and enzymatic activity of these conserved genes underlie interspecific and intraspecific diversity in pigmentation. Although other pigment classes, such as ommochromes and pteridines, also contribute to coloration, melanin represents the primary pigment system for structural and patterning functions in insects. The following sections outline the endocrine control of body color changes through melanin production.

12.3 Body Color Change by Endocrine Factors

12.3.1 Body Color Change Induced by Corazonin (CRZ) and Its Related Peptides

Among environmental cues influencing insect body color, it is intriguing that population density can affect body color through endocrine control. The desert locust *Schistocerca gregaria* is one of the most-investigated species for changing their body color according to their density, as observed by Uvarov (1921). Locusts exhibit phase transition from gregarious to solitarious state and vice versa, accompanied by their morphological and physiological changes. This polyphasic phenomenon is also involved in the body coloration in nymphs as well as adults of the migratory locust *Locusta migratoria*. In the case of locusts, the drastic alteration in green or black coloration occurs according to the phase transition. Changes in body coloration occur by removing the corpora allata (CA), application of juvenile hormone (JH), and transplantation of corpora cardiaca (CC) extracts. The regulatory role of the CA or JH in body coloration will be treated in Sect. 12.3.6.

In *L. migratoria*, dark coloration or the increased black pigmentation was induced by the extract from the CC (Staal 1961; Girardie and Cazel 1965). Also, in the albino nymphs of *L. migratoria*, implantation of the brain or CC-ablation induced dark body color. The effect of CC extract on body coloration was observed in closely related species *S. gregaria* (Tanaka and Yagi 1997). Meanwhile, the presence of [His^7]-CRZ was identified in the CC extract from *S. gregaria* as the dark-inducing neurohormone (Tawfik et al. 1999). CRZ is produced by the lateral neurosecretory cells of the brain and is secreted by the CC in several insect orders including Orthoptera *L. migratoria* and *G. bimaculatus* (Roller et al. 2003), consistent with the fact shown in the experiments of implantation and ablation. This peptide was homologous to [Arg^7]-CRZ, which had been already identified as a cardio-regulatory neuropeptide of the cockroach *Periplaneta americana* (Veenstra 1989; Veenstra 1991) (see Sect. 2.3.1). Subsequently, the presence of CRZ peptides was shown to be conserved among insect species, although the structural differences within their amino acid sequences are observed (Veenstra 1991; Predel et al. 2006).

Interestingly, CRZ-containing fractions extracted from ten other insect species led to dark patterns in locusts, although the CRZ-containing fraction or the extract from the brain-CC of Coleoptera did not affect body coloration of locusts (Tanaka 2000). Similarly, the black-patterned body color of the other orthopteran species, *Acrida cinerea*, *Gastrimargus marmoratus*, *Nomadacris succincta*, *Oxya yezoensis*, and *Atractomorpha lata*, is induced by [His^7]-CRZ (Tanaka 2000). However, in some insect species, CRZ does not alter the body color to a dark pattern. For example, the silkworm *Bombyx mori* and the two-spotted cricket *Gryllus bimaculatus* do not show coloration change by injection of their own CRZ. Body coloration of the katydid *Euconocephalus pallidus* is also not influenced by both [His^7]-CRZ and [Arg^7]-CRZ. These data observed in orthopterans illuminate that the regulatory mechanism of CRZ on body melanin-coloration is not a lineage-specific event. There are no reports that CRZ is involved in the regulation of biosynthesis of pigments other than melanin, such as ommochrome and pteridine, which directly determine insect body color.

12.3.2 Regulation of Pigment Synthetic Genes by Elevenin

Body color regulation by elevenin (see Sect. 2.7.1) has been well investigated in the brown planthopper *Nilaparvata lugens*. Elevenin contributes to melanization, which is disturbed by RNAi targeting on *elevenin* transcript in *N. lugens* (Uchiyama et al. 2018). Similarly, RNAi of *NlA42*, which is one of two elevenin receptor genes in *N. lugens*, results in cuticular melanization, while RNAi of another receptor gene *NlA44* does not (Uchiyama et al. 2018). Additionally, injection of elevenin peptide upregulates the expression levels of *ebony* and *adc* (aspartate 1-decarboxylase) rather than other melanin synthetic enzyme genes, *aaNAT* and *TH*, indicating the regulatory mechanisms via specific transcriptional targets by elevenin signaling (Wang et al. 2019). Dark coloration regulated by elevenin signaling, however, is not the only regulatory system for body coloration, because it cannot explain all coloration events occurring throughout *N. lugens* development. For example, the melanin biosynthetic pathway involved in body coloration of *N. lugens* is modulated by ion transport peptide (ITP) as described below (Yu et al. 2016).

Elevenin was a firstly identified from the L11 neuron of the gastropod mollusk *Aplysia californica* (Taussig et al. 1984). Homologs of elevenin also have been identified from other mollusks (snails), the nematode *Caenorhabditis elegans*, and some arthropods, including the blacklegged tick *Ixodes scapularis* (Veenstra 2010). cDNAs encoding elevenin have been identified in many arthropods, while no elevenin-coding genes have been found within the genomes of some insect groups, such as Diptera and Lepidoptera, which include two major model insects, *D. melanogaster* and *B. mori*. In insects, as mentioned above, two receptors for elevenin have been identified in *N. lugens* (Uchiyama et al. 2018), while only one elevenin receptor has been identified in the red flour beetle *Tribolium castaneum* (Veenstra 2014). The NlA42 receptor couples to the G protein subunits Gq and Gs, which activate protein kinase C (PKC) and protein kinase A (PKA) intracellular signalings, eventually modulating tyrosine-mediated body coloration (Wang et al. 2019).

RT-PCR and qRT-PCR (real-time PCR) revealed that *elevenin* is expressed predominantly in the brain, salivary glands, and integument in *N. lugens* and in the central nervous system, muscle, testes, epidermis, and wings in *G. bimaculatus* (Seike and Nagata 2021). The presence of elevenin in the salivary gland of *N. lugens* is also confirmed by immunolocalization. In *N. lugens*, the expression levels of elevenin receptors are highest in the brain and integument of the abdomen (Uchiyama et al. 2018). These data indicate that elevenin signaling contributes to multiple functions in addition to body coloration in insects.

12.3.3 Regulation of Body Coloration by MRCH

Melanization and reddish coloration hormone (MRCH) was first identified as pheromone biosynthesis activating neuropeptide (PBAN) in Lepidoptera (Matsumoto et al. 1985). MRCH is a hormone that induces density-dependent body color melanization in the common armyworm *Mythimna separata* (Ogura 1975; Matsumoto et al. 1985). The amino acid sequence of this peptide shares a C-terminal FXPRLamide that is conserved among members of the PBAN family (see Sect. 2.3.13). The precursor of MRCH/PBAN contains several FXPRLamides, including MRCH/PBAN, pyrokinin, diapause hormone, and other related peptides in *B. mori* and the corn earworm *Helicoverpa zea* (Kawano et al. 1992; Davis et al. 1992). To date, transcriptomic and genomic analyses have revealed that most insect species possess a gene encoding MRCH/PBAN. PBAN is produced and secreted by the brain and subesophageal ganglion and regulates the initiation of sex pheromone biosynthesis in the pheromone glands as reported in *H. zea* (Ma et al. 1994).

PBAN family peptides exhibit various functions, including the regulation of sex pheromone biosynthesis, cuticular melanization, initiation of embryonic diapause, termination of pupal diapause, contraction of visceral muscles, and acceleration of pupation (Rafaeli 2009). The effect of MRCH on body coloration involving cuticular melanization has been demonstrated by transplantation of the subesophageal ganglion, which contains MRCH-producing cells, in *M. separata* (Ogura 1975). As MRCH/PBAN knockout armyworm larvae did not exhibit density-dependent body coloration (Shirai et al. 2023), the larval dark body coloration in this species appears to be induced according to population density via an MRCH-dependent pathway. MRCH is also involved in cuticular melanization during the larval stage (Zheng et al. 2007). In *S. litura*, MRCH signaling activates MAP kinase via a PKC-dependent manner, resulting in melanization-mediated pigmentation (Zheng et al. 2007). However, although MRCH is conserved in most insect species, it depends on insect species whether its function can contribute to body coloration. In fact, body coloration in *L. migratoria* is not associated with the function of MRCH (Yerushalmi et al. 2002).

12.3.4 Regulation of Body Coloration by ITP/ITPL

Ion transport peptide (ITP) was identified as a factor that activates chloride transport through the ilea of *S. gregaria* (Audsley et al. 1992). The structure of ITP exhibits homology with crustacean hyperglycemic hormone (see Sect. 2.5.1). ITP-like (ITPL) was identified from the brain of *L. migratoria* as a peptide homologous to ITP (Macins et al. 1999). The structures of these peptides in *S. gregaria* were determined by cDNA cloning. ITP and ITPL are derived from alternative splicing products with similar structural features; both have three intramolecular disulfide bonds, and C-terminal amidation occurs in ITP but not in ITPL.

In *N. lugens*, knockdown of *ITP/ITPL* by RNAi caused incomplete molting, failed wing expansion, and dysfunction of melanization (Yu et al. 2016). The regulatory manner in melanization during body color change by ITP/ITPL seems to be different from that by elevenin signaling and CRZ-related signaling (Uchiyama et al. 2018). Other reports showing molting disorder by *ITP* silencing suggest that melanization is involved in the regulation of molting.

12.3.5 Regulation of Tanning by Bursicon

Bursicon was first discovered as a factor controlling cuticular hardening and tanning activities in the blowfly *Calliphora erythrocephala* (Cottrell 1962). Bursicon was partially purified from the grey flesh fly *Sarcophaga bullata* and the cockroach *Periplaneta americana*, and its molecular weight was estimated to be approximately 40,000 (Fraenkel et al. 1966). Later investigation revealed that bursicon is composed of two subunits, bursicon-α and bursicon-β (Luo et al. 2005; Mendive et al. 2005) (see Sect. 2.2.2). In many cases, these two subunits form a bioactive heterodimeric form of bursicon, but in some cases, monomers or homodimers of the subunits are also used (Dewey et al. 2004). Meanwhile, bursicon was demonstrated to act through the leucine-rich repeat-containing G protein-coupled receptor 2 (LGR2), which is encoded by the rickets gene (*rk*) in *D. melanogaster* (Luo et al. 2005). Another study has confirmed that LGR2 binds to heterodimeric bursicon (Mendive et al. 2005).

Bursicon RNAi using double-stranded RNA (dsRNA) encoding the β-subunit of bursicon in *B. mori* reduced the thickness and melanization of the exoskeletal chitin layer and caused a failure of wing expansion (Huang et al. 2007). Similarly, application of dsRNA targeting *bursicon* and *LGR2* reduces body coloration in *T. castaneum* (Arakane et al. 2008). Moreover, mutants for *bursicon* and *rk* in *D. melanogaster* exhibit the disturbance of wing expansion and tanning (Baker and Truman 2002). Since most body coloration occurs immediately after molting and metamorphosis, it is plausible that the hardening-related hormone bursicon can regulate body coloration that accompanies exoskeletal sclerotization.

12.3.6 Regulation of Body Coloration by JH

In addition to neuropeptides and bioactive peptides, JH plays important roles in regulating body coloration. We here summarize the key points of JH effects on body coloration in insects.

In the swallowtail butterfly *Papilo Xuthus* larvae, the decline of JH initiates a pre-patterning process during the fourth instar, which determines the spatial template for subsequent color patterning—specifically, the location where melanin-based markings develop in the final instar (Futahashi and Fujiwara 2008; Jin et al. 2019). During the molting period that follows, expression of melanin synthesis genes such as TH and DDC is spatially regulated according to this template, leading to precise deposition of melanin in black or brown markings (Futahashi and Fujiwara 2008; Jin et al. 2019; Futahashi et al. 2012). Thus, JH acts upstream of melanization, switching pattern fate by setting the boundaries for melanogenic enzyme expression, and the consequent cuticular melanization proceeds in the next instar (Futahashi and Fujiwara 2008; Futahashi et al. 2012).

In *S. gregaria*, JH application or CA transplantation to gregarious locusts induces green coloration by reducing melanin pigmentation (Maeno and Tanaka 2009). Similarly, ablation of the CA prevents this yellowing and delays maturation, while solitary-phase locusts treated with JH do not exhibit this change (Pener 1976). This is consistent with the observation that high JH titers induce solitary phase in locusts characterized by green coloration, whereas low JH titers result in gregarious phase characterized by darker coloration (Hasegawa and Tanaka 1994). Moreover, in gregarious-phase adults of *L. migratoria* and *S. gregaria*, JH is required for yellowish pigmentation during sexual maturation. Yellow coloration is mediated by the yellow protein of the takeout family (YPT), whose expression is regulated by JH, CRZ, environmental temperature, and background color (Sugahara and Tanaka 2018). These findings collectively support that the combination of JH and CRZ is essential for the endocrine control of body coloration in *L. migratoria* and related orthopteran species.

12.3.7 Other Possible Factors Regulating Insect Coloration

Pigment dispersing factor (PDF) in insects is a homolog of pigment dispersing hormone (PDH) of crustaceans, which was first identified from the shrimp *Pandalus borealis* as a factor that modulates crustacean chromatophore to modulate body color change (Rao et al. 1985) (see Sect. 2.4.6). PDF was first identified in the lubber grasshopper *Romalea microptera* as a homolog of PDH in insects (Rao et al. 1987). PDF is composed of 18 amino acid residues and is highly conserved among crustaceans and insects.

The functions of PDF in insects have been reported to be involved in the regulation of circadian rhythms (see Chap. 13), locomotor activity, courtship, and ecdysteroidogenesis (see Chap. 6). Although PDH is involved in pigmentation in crustaceans, such functions have not been reported in insects so far. This is not surprising given that insects lack chromatophores, unlike crustaceans that possess these pigment cells allowing rapid color changes. In contrast to the function of PDH in crustaceans, insect PDF differentially functions through visual mechanisms including circadian rhythms, which respond to environmental information though the optic lobe or the pan-shaped mushroom body in the brain (Dubowy and Sehgal 2017).

Furthermore, body coloration in insects can be influenced by anthropogenic factors. The classic example is industrial melanism in certain lepidopteran species, where darker wing morphs became prevalent due to air pollution during the

Industrial Revolution (Van't Hof et al. 2016). In urban environments, the oriental hornet *Vespa orientalis* shows changes in the yellow pigmentation of abdominal stripes, which are used for both thermoregulation and light energy conversion (Ishay et al. 1981).

12.4 Conclusion

Insect body coloration is a remarkably diverse and vital aspect of their survival, communication, and reproduction that is established through a complex interplay of pigments and structural properties and is profoundly influenced by environmental cues including population density. The endocrine system plays a critical role in modulating these color change processes. Basically, body color change in insects involves melanization and is intricately linked to molting processes. In conclusion, insect body coloration is a highly complex and adaptable phenomenon, orchestrated by a sophisticated network of diverse hormonal, developmental, environmental, and anthropogenic factors.

References

Arakane Y, Li B, Muthukrishnan S, Beeman RW, Kramer KJ, Park Y (2008) Functional analysis of four neuropeptides, EH, ETH, CCAP and bursicon, and their receptors in adult ecdysis behavior of the red flour beetle, *Tribolium castaneum*. Mech Dev 125(11–12):984–995. https://doi.org/10.1016/j.mod.2008.09.002

Audsley N, McIntosh C, Phillips JE (1992) Isolation of a neuropeptide from locust corpus cardiacum which influences ileal transport. J Exp Biol 173:261–274. https://doi.org/10.1242/jeb.173.1.261

Badejo O, Skaldina O, Gilev A, Sorvari J (2020) Benefits of insect colours: a review from social insect studies. Oecologia 194(1–2):27–40. https://doi.org/10.1007/s00442-020-04738-1

Baker JD, Truman JW (2002) Mutations in the *Drosophila* glycoprotein hormone receptor, rickets, eliminate neuropeptide-induced tanning and selectively block a stereotyped behavioral program. J Exp Biol 205(Pt 17):2555–2565. https://doi.org/10.1242/jeb.205.17.2555

Chapman RE, Simpson SJ, Douglas AE (2013) The insects: structure and function, 5th edn. Cambridge University Press, New York

Cottrell CB (1962) The imaginal ecdysis of blowflies. The control of cuticular hardening and darkening. J Exp Biol 39(3):395–412. https://doi.org/10.1242/jeb.39.3.395

Davis MT, Vakharia VN, Henry J, Kempe TG, Raina AK (1992) Molecular cloning of the pheromone biosynthesis-activating neuropeptide in *Helicoverpa zea*. Proc Natl Acad Sci USA 89(1):142–146. https://doi.org/10.1073/pnas.89.1.142

Dewey EM, McNabb SL, Ewer J, Kuo GR, Takanishi CL, Truman JW et al (2004) Identification of the gene encoding bursicon, an insect neuropeptide responsible for cuticle sclerotization and wing spreading. Curr Biol 14(13):1208–1213. https://doi.org/10.1016/j.cub.2004.06.051

Dubowy C, Sehgal A (2017) Circadian rhythms and sleep in *Drosophila melanogaster*. Genetics 205(4):1373–1397. https://doi.org/10.1534/genetics.115.185157

Fraenkel G, Hsiao C, Seligman M (1966) Properties of bursicon: an insect protein hormone that controls cuticular tanning. Science 151(3706):91–93. https://doi.org/10.1126/science.151.3706.91

Futahashi R, Fujiwara H (2008) Identification of stage-specific larval camouflage associated genes in the swallowtail butterfly, Papilio xuthus. Dev Genes Evol 218(9):491–504. https://doi.org/10.1007/s00427-008-0243-y

Futahashi R, Shirataki H, Narita T, Mita K, Fujiwara H (2012) Comprehensive microarray-based analysis for stage-specific larval camouflage pattern-associated genes in the swallowtail butterfly, Papilio xuthus. BMC Biol 10:46. https://doi.org/10.1186/1741-7007-10-46

Girardie A, Cazel M (1965) Role de la pars intercerebralis et des corpora cardiaca sur Laémelanization chez *Locusta migratoria* (L.). CR Hebd Seanc Acad Sci Paris 261:4525–4527

Hasegawa E, Tanaka S (1994) Genetic control of albinism and the role of juvenile hormone in pigmentation in *Locusta migratoria* (Orthoptera, Acrididae). Jpn J Entomol 62:315–324

Huang J, Zhang Y, Li M, Wang S, Liu W, Couble P et al (2007) RNA interference-mediated silencing of the bursicon gene induces defects in wing expansion of silkworm. FEBS Lett 581(4):697–701. https://doi.org/10.1016/j.febslet.2007.01.034

Ishay JS, Shimony TB, Schecter OS, Brown MB (1981) Effects of xanthines and colchicine on the longevity, photoconductive properties and yellow pigment structure of the Oriental hornet Vespa orientalis L. Toxicology 21(2):129–140. https://doi.org/10.1016/0300-483x(81)90123-2

Jin H, Seki T, Yamaguchi J, Fujiwara H (2019) Prepatterning of *Papilio xuthus* caterpillar camouflage is controlled by three homeobox genes: *clawless*, *abdominal-A*, and *Abdominal-B*. Sci Adv 5(4):eaav7569. https://doi.org/10.1126/sciadv.aav7569

Kawano T, Kataoka H, Nagasawa H, Isogai A, Suzuki A (1992) cDNA cloning and sequence determination of the pheromone biosynthesis activating neuropeptide of the silkworm, *Bombyx mori*. Biochem Biophys Res Commun 189(1):221–226. https://doi.org/10.1016/0006-291x(92)91547-4

Luo CW, Dewey EM, Sudo S, Ewer J, Hsu SY, Honegger HW et al (2005) Bursicon, the insect cuticle-hardening hormone, is a heterodimeric cystine knot protein that activates G protein-coupled receptor LGR2. Proc Natl Acad Sci USA 102(8):2820–2825. https://doi.org/10.1073/pnas.0409916102

Ma PW, Knipple DC, Roelofs WL (1994) Structural organization of the *Helicoverpa zea* gene encoding the precursor protein for pheromone biosynthesis-activating neuropeptide and other neuropeptides. Proc Natl Acad Sci USA 91(14):6506–6510. https://doi.org/10.1073/pnas.91.14.6506

Macins A, Meredith J, Zhao Y, Brock HW, Phillips JE (1999) Occurrence of ion transport peptide (ITP) and ion transport-like peptide (ITP-L) in orthopteroids. Arch Insect Biochem Physiol 40(2):107–118. https://doi.org/10.1002/(SICI)1520-6327

Maeno K, Tanaka S (2009) Artificial miniaturization causes eggs laid by crowd-reared (gregarious) desert locusts to produce green (solitarious) offspring in the desert locust, *Schistocerca gregaria*. J Insect Physiol 55(9):849–854. https://doi.org/10.1016/j.jinsphys.2009.05.012

Matsumoto S, Isogai A, Suzuki A (1985) N-terminal amino acid sequence of an insect neurohormone, melanization and reddish coloration hormone (MRCH): heterogeneity and sequence homology with human insulin-like growth factor II. FEBS Lett 189(1):115–118. https://doi.org/10.1016/0014-5793(85)80853-x

Mendive FM, Van Loy T, Claeysen S, Poels J, Williamson M, Hauser F et al (2005) Drosophila molting neurohormone bursicon is a heterodimer and the natural agonist of the orphan receptor DLGR2. FEBS Lett 579(10):2171–2176. https://doi.org/10.1016/j.febslet.2005.03.006

Ogura N (1975) Hormonal control of larval coloration in the armyworm, Leucania seperata. J Insect Physiol 21:559–576. https://doi.org/10.1303/aez.6.93

Okude G, Futahashi R (2021) Pigmentation and color pattern diversity in Odonata. Curr Opin Genet Dev 69:14–20. https://doi.org/10.1016/j.gde.2020.12.014

Pener MP (1976) The differential effect of the corpora allata on yellow coloration in crowded and isolated *Locusta migratoria migratorioides* (R. & F.) males. Acrida 5(4):269–285

Popadić A, Tsitlakidou D (2021) Regional patterning and regulation of melanin pigmentation in insects. Curr Opin Genet Dev 69:163–170. https://doi.org/10.1016/j.gde.2021.05.004
Predel R, Neupert S, Russell WK, Scheibner O, Nachman RJ (2006) Corazonin in insects. Peptides 28(1):3–10. https://doi.org/10.1016/j.peptides.2006.10.011
Rafaeli A (2009) Pheromone biosynthesis activating neuropeptide (PBAN): regulatory role and mode of action. Gen Comp Endocrinol 162(1):69–78. https://doi.org/10.1016/j.ygcen.2008.04.004
Rao KR, Riehm JP, Zahnow CA, Kleinholz LH, Tarr GE, Johnson L et al (1985) Characterization of a pigment-dispersing hormone in eyestalks of the fiddler crab Uca pugilator. Proc Natl Acad Sci USA 82(16):5319–5322. https://doi.org/10.1073/pnas.82.16.5319
Rao KR, Mohrherr CJ, Riehm JP, Zahnow CA, Norton S, Johnson L et al (1987) Primary structure of an analog of crustacean pigment-dispersing hormone from the lubber grasshopper *Romalea microptera*. J Biol Chem 262(6):2672–2675
Roller L, Tanaka Y, Tanaka S (2003) Corazonin and corazonin-like substances in the central nervous system of the Pterygote and Apterygote insects. Cell Tissue Res 312(3):393–406. https://doi.org/10.1007/s00441-003-0722-4
Seike H, Nagata S (2021) Different transcriptional levels of Corazonin, Elevenin, and PDF according to the body color of the two-spotted cricket, *Gryllus bimaculatus*. Biosci Biotechnol Biochem 86(1):23–30. https://doi.org/10.1093/bbb/zbab183
Shamim G, Ranjan SK, Pandey DM, Ramani R (2014) Biochemistry and biosynthesis of insect pigments. Eur J Entomol 111(2):149–164. https://doi.org/10.14411/eje.2014.021
Shirai Y, Ono H, Daimon T (2023) Redundant actions of neuropeptides encoded by the dh-pban gene for larval color pattern formation in the oriental armyworm *Mythimna separata*. Insect Biochem Mol Biol 157:103955. https://doi.org/10.1016/j.ibmb.2023.103955
Staal GB (1961) Studies on the physiology of phase induction in *Locusta migratoria migratorioides* R. & F. Doctoral thesis. Publ Fds Landb Expt Bur, 1916–1918, 40:1–125
Stavenga DG, Leertouwer HL, Wilts BD (2014) The colouration toolkit of the Pipevine Swallowtail butterfly, *Battus philenor*: thin films, papiliochromes, and melanin. J Comp Physiol A 200:547–561. https://doi.org/10.1007/s00359-014-0901-7
Sugahara R, Tanaka S (2018) Environmental and hormonal control of body color polyphenism in late-instar desert locust nymphs: role of the yellow protein. Insect Biochem Mol Biol 93:27–36. https://doi.org/10.1016/j.ibmb.2017.12.004
Tanaka S (2000) Induction of darkening by corazonins in several species of Orthoptera and their possible presence in ten insect orders. Appl Entomol Zool 35(4):509–517. https://doi.org/10.1303/aez.2000.509
Tanaka S, Yagi S (1997) Evidence for the involvement of a neuropeptide in the control of body color in the desert locust, *Schistocerca gregaria*. Jpn J Entomol 65:447–457
Taussig R, Kaldany RR, Scheller RH (1984) A cDNA clone encoding neuropeptides isolated from Aplysia neuron L11. Proc Natl Acad Sci USA 81(15):4988–4992. https://doi.org/10.1073/pnas.81.15.4988
Tawfik AI, Tanaka S, De Loof A, Schoofs L, Baggerman G, Waelkens E et al (1999) Identification of the gregarization-associated dark-pigmentotropin in locusts through an albino mutant. Proc Natl Acad Sci USA 96(12):7083–7087. https://doi.org/10.1073/pnas.96.12.7083
Uchiyama H, Maehara S, Ohta H, Seki T, Tanaka Y (2018) Elevenin regulates the body color through a G protein-coupled receptor NlA42 in the brown planthopper *Nilaparvata lugens*. Gen Comp Endocrinol 258:33–38. https://doi.org/10.1016/j.ygcen.2017.07.017
Uvarov BP (1921) A revision of the genus *Locusta*, L. (= Pachytylus, Fieb.), with a new theory as to the periodicity and migrations of Locusts. Bull Entomol Res 12(2):135–163. https://doi.org/10.1017/S0007485300044989
Van't Hof AE, Campagne P, Rigden DJ, Yung CJ, Lingley J, Quail MA et al (2016) The industrial melanism mutation in British peppered moths is a transposable element. Nature 534(7605):102–105. https://doi.org/10.1038/nature17951

Veenstra JA (1989) Isolation and structure of corazonin, a cardioactive peptide from the American cockroach. FEBS Lett 250(2):231–234. https://doi.org/10.1016/0014-5793(89)80727-6

Veenstra JA (1991) Presence of corazonin in three insect species, and isolation and identification of [His^7]corazonin from *Schistocerca americana*. Peptides 12(6):1285–1289. https://doi.org/10.1016/0196-9781(91)90208-7

Veenstra JA (2010) Neurohormones and neuropeptides encoded by the genome of *Lottia gigantea*, with reference to other mollusks and insects. Gen Comp Endocrinol 167(1):86–103. https://doi.org/10.1016/j.ygcen.2010.02.010

Veenstra JA (2014) The contribution of the genomes of a termite and a locust to our understanding of insect neuropeptides and neurohormones. Front Physiol 5. https://doi.org/10.3389/fphys.2014.00454

Wang SL, Wang WW, Ma Q, Shen ZF, Zhang MQ, Zhou NM et al (2019) Elevenin signaling modulates body color through the tyrosine-mediated cuticle melanism pathway. FASEB J 33(9):9731–9741. https://doi.org/10.1096/fj.201802786RR

Whitten MMA, Coates CJ (2017) Re-evaluation of insect melanogenesis research: views from the dark side. Pigment Cell Melanoma Res 30:386–401. https://doi.org/10.1111/pcmr.12590

Yerushalmi Y, Bhargava K, Gilon C, Pener MP (2002) Structure–activity relations of the dark-colour-inducing neurohormone of locusts. Insect Biochem Mol Biol 32(8):909–917. https://doi.org/10.1016/S0965-1748(01)00180-1

Yu B, Li DT, Wang SL, Xu HJ, Bao YY, Zhang CX (2016) Ion transport peptide (ITP) regulates wing expansion and cuticle melanism in the brown planthopper, *Nilaparvata lugens*. Insect Mol Biol 25(6):778–787. https://doi.org/10.1111/imb.12262

Zheng L, Lytle C, Njauw CN, Altstein M, Martins-Green M (2007) Cloning and characterization of the pheromone biosynthesis activating neuropeptide receptor gene in *Spodoptera littoralis* larvae. Gene 393(1–2):20–30. https://doi.org/10.1016/j.gene.2006.12.025

Chapter 13
Circadian Rhythm and Neuropeptides

Sakiko Shiga

Abstract Circadian rhythms are driven by the circadian clock in which a self-sustaining mechanism produces ca. 24-h rhythmicity. Genetic ablation or rescue experiments have shown that many kinds of neuropeptides are necessary for circadian rhythms. However, relatively few neuropeptides have been well characterized physiologically in relation to circadian rhythms. In *Rhyparobia maderae,* ten neuropeptides have been immunohistochemically identified in accessory medulla neurons, which constitute the circadian clock system. In *Drosophila melanogaster*, 12 neuropeptides have been reported of their presence in brain clock neurons. Pigment-dispersing factor (PDF) is the most well-characterized peptide regarding the activity rhythm. PDF is important to set an activity phase in *D. melanogaster*, *R. maderae*, and *Gryllus bimaculatus*, possibly by coordinating the clock neuron network. Diuretic hormone 31 is known as an output factor regulating fly sleep and temperature preference under the control of the circadian clock. Short neuropeptide F works for coupling of the central and peripheral clocks by inhibiting prothoracicotropic hormone-producing neurons, thereby controlling the circadian eclosion rhythm. Transcriptome and connectome studies on clock neurons in *D. melanogaster* suggest a variety of neuropeptides play important roles in organizing circadian rhythms.

Keywords Circadian clock · Pigment-dispersing factor · Eclosion rhythm · Activity rhythm · Temperature preference rhythm

13.1 Introduction

The rotating features of Earth have influenced the evolution of rhythms in living organisms (Dunlap et al. 2004). The most well-known is the circadian rhythm ubiquitously found from unicellular organisms to multicellular invertebrates and

---

S. Shiga (✉)
Department of Biological Sciences, Graduate School of Science, The University of Osaka, Osaka, Japan
e-mail: shiga.sakiko.sci@osaka-u.ac.jp

A. Mizoguchi, N. Okamoto (eds.), *Insect Peptide Hormones and Neuropeptides*, Entomology Monographs, https://doi.org/10.1007/978-981-95-7510-7_13

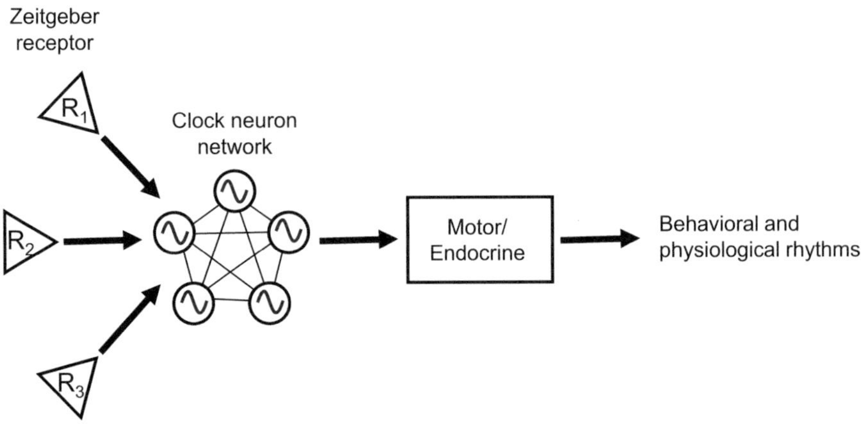

Fig. 13.1 Physiological components producing circadian rhythms in insects. Different receptors, R1, R2, and R3, receive Zeitgeber stimuli to inform environmental time cues to the circadian clock system composed of different clock neurons (sinewave circles), and then the rhythm is produced via motor or endocrine systems

vertebrates. The circadian rhythm is driven by the circadian clock in which a self-sustaining mechanism produces ca. 24-h rhythmicity. A common feature of transcription–translation feedback loops (TTFLs) of different clock genes has been demonstrated in both the fruit fly *Drosophila melanogaster* and the mouse *Mus musculus*, showing that they generate rhythms with a period of about 24 h (Patke et al. 2020). The TTFL machinery resides in a single clock neuron, and clock neurons form circuitries in the brain to drive different types of behavioral or physiological rhythms (Fig. 13.1).

It has been widely shown that circadian rhythms exhibit three characteristics: Endogenous sustainability of ca. 24-h oscillation by which the rhythm continues without any time cues from the environment, temperature compensation by which the rhythm period is compensated for environmental temperature (i.e., keeping nearly constant period in a certain range of temperature), and entrainability by which the cycle period can be adjusted to an environmental cycle (Dunlap et al. 2004). For rhythm entrainment, the clock responds to environmental stimuli to cause phase shifts, adjusting its phase to the environment. Such stimuli are called Zeitgeber in German (Dunlap et al. 2004). Light is the most effective Zeitgeber, although temperature and humidity are also known to cause phase shifts.

Organization of circadian rhythms in behavior or physiological functions includes receptor organs by which Zeitgeber stimuli are received, the circadian clock system composed of clock neurons in the brain or clock cells in peripheral organs, and clock-driven motor or neuroendocrine circuitries that produce each behavior or physiological function (Fig. 13.1). Genetic ablation or rescue experiments have shown that many kinds of neuropeptides are necessary for generating circadian rhythms. However, relatively few neuropeptides have been physiologically characterized for their roles in circadian rhythms. This chapter introduces neuropeptides involved in clock organization and clock outputs in circadian rhythm generation. A comprehensive survey of neuropeptides in *D. melanogaster* rhythm is found in Nässel and Zandawala (2019) and Reinhard et al. (2024).

13.2 General View of Neuropeptides Involved in Circadian Rhythms

Although a variety of neuropeptides are present in clock neurons of many insect species, their physiological roles in relation to circadian rhythms have only been examined in a few species, including the Madeira cockroach *Rhyparobia* (*Leucophaea*) *maderae* and the fruit fly *D. melanogaster*. The former has nocturnal and the latter diurnal locomotor activity rhythms, and both species commonly have clock neurons crucial for driving activity rhythms in a region close to the accessory medulla (aMe) (Reischig and Stengl 2003a; Frisch et al. 1994; Vosshall and Young 1995). The aMe is a small neuropil at the anterior base of the optic lobe medulla. In this chapter, I designate a region immediately beside the aMe "aMe region" and neurons with cell bodies located in the aMe region "aMe neurons." The importance of the aMe region including the aMe in the locomotor activity rhythm has been demonstrated by surgical ablation, implantation, or careful anatomy in *R. maderae* and the blowfly *Protophormia terraenovae* and by genetic mosaic studies in *D. melanogaster* (Reischig and Stengl 2003a; Grima et al. 2004; Stoleru et al. 2004, 2005, 2007; Rieger et al. 2006; Picot et al. 2007; Shiga and Numata 2009).

In *R. maderae*, eight groups of about 240 aMe neurons per hemisphere constitute the circadian clock network (Stengl and Arendt 2016). Apparent oscillation of the expression of circadian clock genes *period* (*per*), *timeless1*, and *cryptochrome2* has been shown in the aMe region under light-dark cycles (Werckenthin et al. 2012), and RNAi-based knockdown of these clock genes affected circadian activity rhythms (Werckenthin et al. 2020). These findings suggest that aMe neurons include clock neurons in *R. maderae*, although clock gene-expressing cells have not yet been identified. So far, ten neuropeptides have been immunohistochemically identified in the aMe neurons of *R. maderae* (Fig. 13.2).

Recent connectome analyses in *D. melanogaster* have revealed that ca. 240 clock neurons per brain form extensive connectivity within the clock neuron networks. The clock neurons are mainly classified into two super classes, dorsal neurons (DNs) and lateral neurons (LNs), and each of DNs and LNs is further classified into four classes (Reinhard et al. 2024). Careful analysis showed that DNs represent an important hub with high synaptic connectivity rates within the clock networks and also with clock output pathways. In the clock neuron networks, 12 neuropeptides are found, and all clock neuron classes express at least one neuropeptide. The majority expresses two neuropeptides, highlighting highly peptidergic pathways within the clock network. Pigment-dispersing factor (PDF; see Sect. 2.4.6) and short

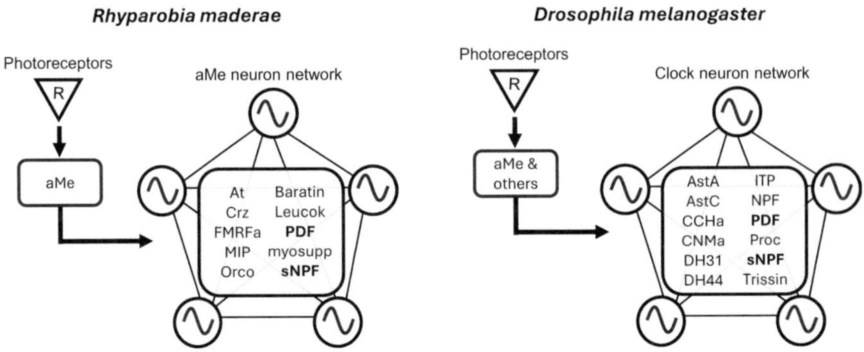

Fig. 13.2 Photic input pathway and neuropeptide lists in clock-neuron networks in the brain of *Rhyparobia maderae* and *Drosophila melanogaster.* Light signals are input to peptidergic clock neuron circuitry via accessory medulla (aMe). At Allatostatin, Crz corazonin, FMRFa FMRFamide, MIP myoinhibitory peptide, Orco orcokinin, Leucok leucokinin, PDF pigment-dispersing factor, myosupp myosuppressin, sNPF short neuropeptide F, Ast allatostatin, CCHa CCHamide, CNMa CNMamide, DH diuretic hormone, ITP ion transport peptide, NPF neuropeptide F, Proc proctolin. Adopted from Stengl and Arendt (2016), left) and Reinhard et al. 2024, right)

neuropeptide F (sNPF; see Sect. 2.3.15) are commonly found in both *R. maderae* and *D. melanogaster* (Fig. 13.2).

Light entrainment of circadian rhythms is mediated by visual neurons in addition to clock neurons themselves with a blue-light sensitive *Drosophila*-type cryptochrome (CRY) (Emery et al. 1998; Stanewsky et al. 1998; Helfrich-Förster et al. 2001; Klarsfeld et al. 2004). For light input pathways, light-responsive clock neurons make monosynaptic connections at the aMe with Hofbauer–Buchner eyelet photoreceptors and with interneurons conveying compound-eye visual signals via allatostatin C (AST-C; see Sect. 2.3.3) peptide (Li et al. 2018). AST-C mediates Zeitgeber information to clock neuron networks. Laser ablation of the aMe and eye removal both abolish light responses of clock neurons, revealing the aMe as a hub to channel eye inputs to the central circadian clock for light entrainment (Fig. 13.2, Li et al. 2018).

13.3 Neuropeptides for Clock Organizations

13.3.1 Pigment-Dispersing Factor

PDF is the most well-known neuropeptide in insect circadian clock systems. PDF has been originally discovered as pigment-dispersing hormone in the crustacean *Uca pugillator* to cause pigment dispersion in melanophores (Rao et al. 1985) (see Sect. 2.4.6). Later this peptide was immunohistochemically mapped to distinct sets of brain neurons in the aMe region in a range of insect species (e.g., Nässel et al. 1993; Ikeno et al. 2014; Hasebe et al. 2022). Since the colocalization of PDF with the circadian clock protein PER was shown in optic lobe neurons of *D. melanogaster* (Helfrich-Förster 1995), the physiological roles of PDF in circadian rhythms have been examined and well characterized mainly in the cricket *Gryllus bimaculatus*, *R. maderae*, and *D. melanogaster*.

In *D. melanogaster*, PDF-null mutant (*pdf01*) flies can anticipate light-to-dark transition in their locomotor activity. In constant darkness, about 70% of *pdf01* flies were deprived of locomotor rhythms, whereas the remaining flies displayed weak rhythms (Renn et al. 1999). PDF is expressed in eight pairs of ventral LNs (LNvs). Cellular circadian rhythms of PER content are detected as staining intensity rhythms of PER immunohistochemistry with a period of ca. 24 h in a small type of LNvs, named sLNvs. The PER intensity rhythm persists in the sLNv of *pdf01* mutants, indicating that PDF is not essential for the clock machinery (Lin et al. 2004). The phenotypes of *pdf01* mutants suggest that PDF-dependent hierarchical communication occurs from the LNvs to other clock neurons (Lin et al. 2004). Indeed, many classes of clock neurons express PDF receptors (Im and Taghert 2010; Abruzzi et al. 2017; Reinhard et al. 2024). Examination of behavioral rhythms and PER staining intensity rhythms in different clock neurons under LD cycles with longer cycle periods, such as LD 16:16, has suggested that PDF released from sLNvs to dorsal LNs has a phase-delaying effect on the evening peak component of the locomotor activity rhythm. This serves as the primary coordinating force under LD cycles, and PDF is important to set the phase relationship with environmental time (Vaze and Helfrich-Förster 2021). Most neurons expressing the PDF receptor co-express circadian photoreceptor CRY. This suggests that clock neuron-derived PDF signaling and environmental light signaling converge in the clock neuron network to influence the phase and amplitude of the rhythm (Im et al. 2011).

Direct effects of PDF on the clock neuron network have been demonstrated in a type of DNs called posterior DN1 (DN1p) by electrophysiological recordings. DN1ps respond to PDF by increasing membrane potential and producing bursting of action potentials (Seluzicki et al. 2014). These responses are missing in mutants of the PDF receptor; Membrane potential depolarization is induced by PDF under tetrodotoxin (TTX)-applied conditions in the brain. These observations strongly support an idea that PDF is directly received by DN1p, not by multi-synaptic pathways. This membrane potential depolarization and firing also occur in the presence of a protein kinase A (PKA) inhibitor (H89) or in a dominant negative PKA mutant (PKA-R1). However, an adenylate cyclase inhibitor blocks the PDF-induced excitation, and an adenylate cyclase activator or cAMP itself induces the PDF-evoked response, suggesting that cAMP is responsible for the PDF effect on excitability, but the effect is not mediated by PKA signaling (Seluzicki et al. 2014). However, in a long-term response, PDF has inhibitory effects on calcium levels in clock neurons (Liang et al. 2017). PDF is an important peptide for behavioral entrainment as a signaling molecule within the clock neuron network that excites other clock neurons. The roles of PDF in *D. melanogaster* circadian rhythms are detailed in several reviews (Mertens et al. 2007; Shafer and Yao 2014).

In *R. maderae*, PDF is found in three groups of neurons within the aMe neurons (Stengl and Arendt 2016). The electron microscopy of the aMe found both afferent and efferent synapses on PDF fibers (Reischig and Stengl 2003b; Wei et al. 2010), suggesting that extensive input and output information related to the circadian system comes to and goes from the PDF neurons at the aMe.

Taking advantage of the large brain size, aMe neurons were electrophysiologically and morphologically examined in *R. maderae* prior to studies in *D. melanogaster*. Intracellular recordings and neuroanatomy demonstrated that some of aMe neurons are responsive to Zeitgeber stimuli of light (Loesel and Homberg 2001). Several recorded neurons were morphologically suggested to be identical to PDF neurons due to their arborization patterns. These PDF-like neurons are not responsive to light, contrasting to *D. melanogaster* PDF-containing LNvs that are responsive to light (Loesel and Homberg 2001; Li et al. 2018). PDF-like neurons in *R. maderae* seem to serve as coupling pathways within and outputs from the clock neuronal circuitry (Loesel and Homberg 2001).

Later, Schneider and Stengl (2005) made extracellular recordings of the excised aMe region in *R. maderae*. aMe neurons in culture conditions produce regular interspike intervals (ultradian period) in their activities. Consequently, the cell membrane potential oscillated with ultradian periods. Most oscillating cells are coupled, forming different assemblies, whose neurons share the same ultradian period and the same phase (timing of spikes), which are apparently mediated by inhibitory GABAergic synaptic interactions. Application of PDF phase-locks and synchronizes different assemblies, causing aMe neurons to oscillate at the same frequency (Fig. 13.3). The authors suggest that the phase control of firing oscillations in an ultradian manner by PDF might control the circadian pacemaker network (Schneider and Stengl 2005).

Light as Zeitgeber induces a phase delay or advance of rhythms in a circadian time (CT)-dependent manner for rhythm entrainment to environmental cycles (Golombek and Rosenstein 2010). Injections of neuropeptides also cause phase shifts in a CT-dependent manner in constant dark conditions, and the phase-response curves allow the assignment of functional roles to the applied neuropeptides. The phase–response curve obtained by PDF injection into the *R. maderae* brain differs from that obtained with light pulses, suggesting that PDF-immunoreactive neurons are not part of the visual input pathway to the pacemaker but an integral part of it and/or part of a nonphotic input into the clock (Petri and Stengl 1997). In addition to PDF, injections of myoinhibitory peptide1 (see Sect. 2.4.4), orcokinin (see Sect. 2.3.12), and allatotropin (see Sect. 2.3.4) mainly cause phase delays at the beginning of the subjective night, whereas injections of corazonin (see Sect. 2.3.1) and myoinhibitory peptide2 induce phase advances at the late subjective night (Hofer and Homberg 2006; Stengl et al. 2015; Arendt et al. 2017).

In *G. bimaculatus*, 13–15 PDF-immunoreactive neurons exist with fibers in the aMe (Homberg et al. 1991). Intracellular recordings and immunostaining have shown that PDF-positive aMe neurons exhibit morphological variation regarding the presence or absence of fibers inside the aMe, but all are weakly responsive to light (Abdelsalam et al. 2008). PDF peptide content increases both in the optic and cerebral lobes during nighttime compared with daytime, and this change is also observed under constant dark conditions, suggesting that the PDF content is under

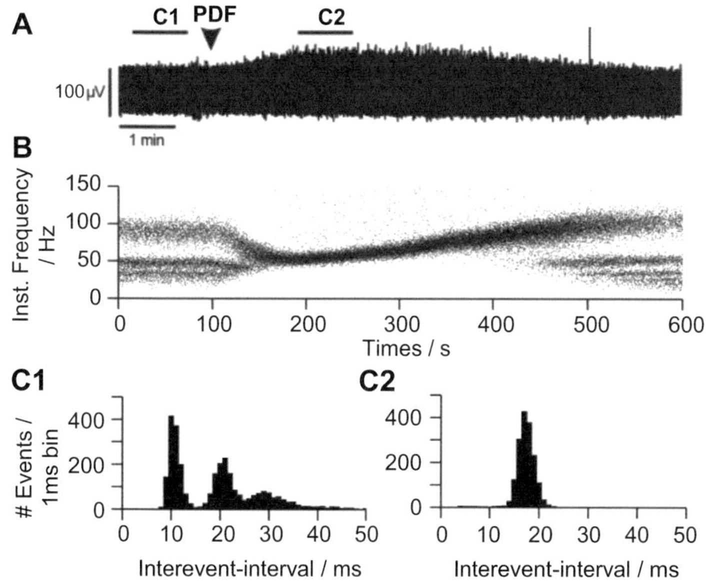

Fig. 13.3 Roles of pigment-dispersing factor (PDF) in electrical activities of accessory medulla (aMe) neurons in *Rhyparobia maderae*. PDF synchronizes aMe neurons. (**a**) Extracellular recording from an excised aMe. PDF was applied at arrowhead. (**b**) The parallel, distinct bands in the instantaneous (Inst.) frequency plot indicate that at least two assemblies of aMe neurons fire synchronously at the same inter-event interval or an integer multiple of it, but in different phases. Application of PDF synchronizes and phase locks the neurons transiently, causing all cells to fire at the same time. The interevent-interval distribution before (**c1**) and after PDF application (**c2**) show that at least two regularly firing synchronous cell assemblies became phase locked to the same phase. (Modified from Schneider and Stengl (2005). Copyright [2005] Society for Neuroscience)

the control of the circadian clock (Abdelsalam et al. 2008). The increase in PDF content during night period seems to fit well with the result from immunoreaction experiments using anti-PDF antibodies. It has been suggested that PDF is released during early night (Saifullah and Tomioka 2003). PDF injection in *G. bimaculatus* also causes phase shifts of the activity rhythm in a CT-dependent manner, but the phase-response curve is different from that to a light pulse, as in the case of *R. maderae* (Petri and Stengl 1997; Singaravel et al. 2003).

In terms of *PDF* expression or PDF content in *D. melanogaster*, neither the level of *PDF* mRNA in the head nor somata PDF-immunoreactivity exhibits circadian cycling (Park and Hall 1998; Park et al. 2000). However, circadian changes were observed in PDF immunoreactivity in terminal fibers in the dorsal protocerebrum of sLNv (Park et al. 2000). sLNv dorsal fibers exhibit circadian remodeling, with high fiber complexity during the day and low complexity at night, and this rhythm continues under constant conditions (Fernández et al. 2008). This remodeling causes different synaptic partners with sLNv at different phases along the day (Gorostiza et al. 2014) and is important for integrating Zeitgeber sensory inputs into circadian clock neuron networks (Fernandez et al. 2020).

PDF expression both in sLNvs and large type LNvs (lLNvs) is prerequisite for dorsal fiber plasticity of sLNvs, and PDF downregulation results in a configuration reminiscent of the nighttime pattern (less complex) of the fiber projection through the day (Herrero et al. 2020). Interestingly, depolarization of sLNv and lLNv using temperature-gated transient receptor potential ankyrin 1 (TRPA1) channel evokes increases in structural complexity and PDF levels in the terminal fibers of sLNv, and this seems to be mediated by PDF receptor and the cyclic-nucleotide-gated channels. This gives a possibility that sLNvs have a positive feedback loop in which the released PDF is received by sLNv and causes their depolarization to promote further PDF release, leading to higher fiber complexity during daytime (Herrero et al. 2020). Choi et al. (2012) have also suggested the involvement of an autoreceptor signaling system to activate sLNvs via PDF secretion.

13.4 Neuropeptides for Clock Outputs

13.4.1 Diuretic Hormone 31

Diuretic hormone 31 (DH31) and its receptor are homologous to vertebrate calcitonin and its receptor, and therefore DH31 is also called calcitonin-like diuretic hormone (CT-DH; see Sect. 2.3.8). A subset of DN1s expresses *DH31* (Kunst et al. 2014), and some DN1s and LNvs express *DH31 receptor* (Goda et al. 2018; Reinhard et al. 2024). DH31 signaling was characterized as an output factor regulating sleep in flies under the control of the circadian clock. Analysis of *DH31* loss- and gain-of function mutants has demonstrated that DH31 is a wake-promoting signal that acts late at night to arouse flies in anticipation of dawn. Electrical activity of *DH31*-expressing DN1s is high before dawn compared with that before dusk. Interestingly, it is suggested that DH31 secretion by DN1s is directly regulated by PDF signals from sLNvs (Kunst et al. 2014). As neither *DH31* nor *PDF* mutations disrupt sleep rhythms, the PDF to DH31 peptide cascade is thought to be specific to sleep itself and not involved in circadian rhythm generation. DH31 secretion by DN1s mediates the wake-promoting effect.

DH31 is also involved in temperature preference rhythms in *D. melanogaster*. As insects are small ectotherms, their body temperature is almost equal to the surrounding temperature. Adult flies select a comfortable temperature by avoiding warm and cold ones. Interestingly, they select relatively high temperatures before dusk, and the selected temperature goes down at the beginning of nighttime (temperature preference rhythm). Consequently, *D. melanogaster*'s temperature preference rhythm generates body temperature rhythms that in mammals are controlled by endogenous heat production (Goda et al. 2023). The fly's preferred temperature increases during the day, decreases with the onset of night, and then gradually increases during the night, and this rhythm continues under constant darkness. As the rhythm becomes arrhythmic in clock gene mutants, the temperature preference rhythm is thought to be driven by the circadian clock (Kaneko et al. 2012). *DH31 receptor* mutant flies prefer high temperatures constantly during daytime, causing disruption of the daytime increase in the preferred temperature. As the mutants maintain normal clock protein oscillations in the clock neurons, suggesting that DH31 signaling contributes to the output from the circadian clock in controlling temperature preference behavior during daytime (Goda et al. 2018). Unexpectedly, *DH31* mutant flies exhibited a normal daytime increase in preferred temperature, whereas the *DH31/PDF* double mutant flies additively disrupted the daytime increase in the preferred temperature, suggesting that DH31 and PDF compensate each other to control the preferred temperature in daytime (Goda et al. 2023).

13.4.2 Short Neuropeptide F

Locomotory activity or preference for a certain temperature occurs at a fixed time of the day, and therefore, their circadian rhythms are observable in a single individual. In contrast, we are not able to observe adult eclosion rhythm individually, because eclosion behavior is triggered at a fixed time on only 1 day in a lifetime. When a group of individuals with different ages is reared together, eclosion events are recorded daily, and this type of daily rhythm is called the circadian population rhythm. This eclosion rhythm is also a clock gene–dependent rhythm (Konopka and Benzer 1971).

It has been reported in *D. melanogaster* that the circadian eclosion rhythm is mediated through the coupling of a central clock composed of brain clock neurons and a peripheral clock residing in the prothoracic glands (Myers et al. 2003). It has been shown in several fly species that prothoracic gland cells express PER in a circadian manner (Myers et al. 2003; Yamamoto et al. 2017). In *D. melanogaster*, coupling of the central and peripheral clocks is mediated by a peptidergic signaling pathway in which sNPF released from sLNv inhibits prothoracicotropic hormone (PTTH; see Sect. 2.6.1)–producing neurons via the sNPF receptor (Selcho et al. 2017). Bath application of sNPF caused a rapid reduction in fluctuations in free calcium detected by GCaMP6 fluorescence in PTTH neurons, suppressing PTTH neuron activity (Fig. 13.4). Suppression of electrical activity in PTTH neurons by sNPF is also reported in larvae of the flesh fly *Sarcophaga similis* (Ohe et al. 2025). PTTH neurons secrete PTTH, which binds to a receptor tyrosine kinase encoded by *torso* expressed in prothoracic gland cells to control the production and release of ecdysteroids (Fig. 13.4) (see Sect. 2.6.1). Thus, the circadian clock system might be involved in determining the timing of ecdysteroid release.

Functional clocks both in the brain and prothoracic glands are required for producing a circadian eclosion rhythm, and they seem to make a hierarchical relationship in which the brain clock plays a dominant role as compared to the prothoracic gland clock (Selcho et al. 2017). In the flesh fly *Sarcophaga crassipalpis*, pharate adults are sensitive for eclosion timing to 20-hydroxyecdysone about 20 and 16 h before eclosion, whereas no effect of 20-hydroxyecdysone is observed about 12 and 8 h before eclosion. A low

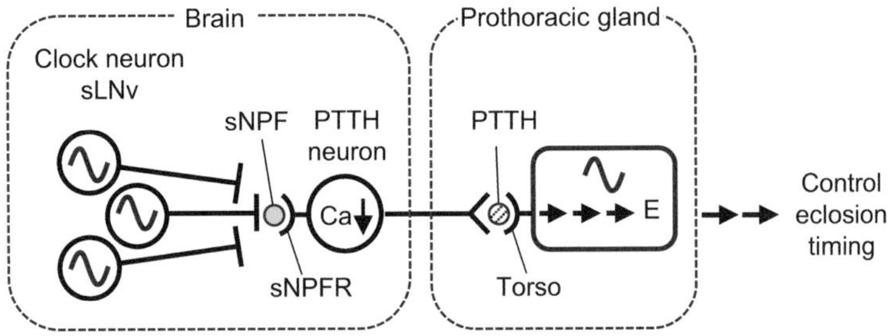

Fig. 13.4 Functional components required for producing the eclosion rhythm in *Drosophila melanogaster*. The central clock composed of small ventral lateral neurons (sLNvs) in the brain and the peripheral clock residing in the prothoracic gland are coupled via a peptidergic signaling pathway involving short neuropeptide F (sNPF) and prothoracicotropic hormone (PTTH). Sinewave clock neurons, E ecdysteroids, sNPFR sNPF receptor

temperature pulse as Zeitgeber causes a phase shift of eclosion rhythm in a CT-dependent manner until 4.5 h before eclosion, which is after ecdysteroid peak (Yamamoto et al. 2017). This suggests that the circadian clock control of eclosion timing is effective till just before eclosion, and ecdysteroids may not be the sole timing signal as clock output. The circadian clock in the prothoracic gland or other peripheral clocks may send a signal, distinct from ecdysteroids, from several hours before eclosion to time the onset of eclosion in *S. crassipalpis* (Yamamoto et al. 2017). The exact mechanism by which the timing of eclosion is determined remains unclear.

13.5 Future Perspectives

To generate circadian behavioral or physiological rhythms, a variety of neuropeptides are involved in the pathways of Zeitgeber inputs, clock organizations, or their outputs (Fig. 13.1). In the clock neuron network of *D. melanogaster*, a variety of neuropeptides have been localized by immunohistochemistry, imaging with a genetically expressed fluorescence probe, and neuropeptidome using single-cell RNA sequencing. Molecular genetic studies have revealed a requirement of these peptides in the circadian rhythm by distinguishing cells. Connectome and transcriptome studies of the clock neurons have greatly advanced our understanding of the spatial and temporal patterns of peptide expression in the clock neuron networks (Abruzzi et al. 2017; Shafer et al. 2022; Reinhard et al. 2024). Furthermore, there is a growing body of literature on how neuropeptides induce activity or molecular changes in the target cells. However, it is not fully understood how the resulting cellular changes affect the dynamics of the clock neuron networks.

In a single sLNv, PDF and sNPF are found with the conventional neurotransmitter glycine, and both *allatostatin C receptor* and *diuretic hormone 31 receptor* are co-expressed. Liang et al. (2017) have shown that PDF and sNPF share phase-setting mechanisms in a long-term effect by both suppressing Ca^{2+} activity in different clock neuron groups. This implies that both peptides are released at the same time. Frenkel et al. (2017) suggested that PDF and glycine released from sLNv play a similar role in other clock neurons. It remains unsolved why multiple peptide transmitters have to be employed in a single cell. It is also not answered very well about the significance of multiple peptide receptor types within a single neuron. In some clock neurons, four or more peptide receptors are co-expressed (Reinhard et al. 2024). Combined with physiological techniques and molecular genetics, circadian clock neuron networks provide a good study model for analyzing the functional roles of multiple peptidergic systems operating within the circuitry.

References

Abdelsalam S, Uemura H, Umezaki Y, Saifullah AS, Shimohigashi M, Tomioka K (2008) Characterization of PDF-immunoreactive neurons in the optic lobe and cerebral lobe of the cricket, *Gryllus bimaculatus*. J Insect Physiol 54(7):1205–1212. https://doi.org/10.1016/j.jinsphys.2008.05.001

Abruzzi KC, Zadina A, Luo W, Wiyanto E, Rahman R, Guo F, Shafer O, Rosbash M (2017) RNA-seq analysis of *Drosophila* clock and non-clock neurons reveals neuron-specific cycling and novel candidate neuropeptides. PLoS Genet 13(2):e1006613. https://doi.org/10.1371/journal.pgen.1006613

Arendt A, Baz ES, Stengl M (2017) Functions of corazonin and histamine in light entrainment of the circadian pacemaker in the Madeira cockroach, *Rhyparobia maderae*. J Comp Neurol 525(5):1250–1272. https://doi.org/10.1002/cne.24133

Choi C, Cao G, Tanenhaus AK, McCarthy EV, Jung M, Schleyer W, Shang Y, Rosbash M, Yin JC, Nitabach MN (2012) Autoreceptor control of peptide/neurotransmitter corelease from PDF neurons determines allocation of circadian activity in *Drosophila*. Cell Rep 2(2):332–344. https://doi.org/10.1016/j.celrep.2012.06.021

Dunlap JC, Loros JJ, DeCoursey PJ (2004) Chronobiology–biological timekeeping. Sinauer, Massachusetts

Emery P, So WV, Kaneko M, Hall JC, Rosbash M (1998) CRY, a *Drosophila* clock and light-regulated cryptochrome, is a major contributor to circadian rhythm resetting and photosensitivity. Cell 95:669–679. https://doi.org/10.1016/s0092-8674(00)81637-2

Fernández MP, Berni J, Ceriani MF (2008) Circadian remodeling of neuronal circuits involved in rhythmic behavior. PLoS Biol 6(3):e69. https://doi.org/10.1371/journal.pbio.0060069

Fernandez MP, Pettibone HL, Bogart JT, Roell CJ, Davey CE, Pranevicius A, Huynh KV, Lennox SM, Kostadinov BS, Shafer OT (2020) Sites of circadian clock neuron plasticity mediate sensory integration and entrainment. Curr Biol 30(12):2225–2237.e5. https://doi.org/10.1016/j.cub.2020.04.025

Frenkel L, Muraro NI, Beltrán González AN, Marcora MS, Bernabó G, Hermann-Luibl C, Romero JI, Helfrich-Förster C, Castaño EM, Marino-Busjle C, Calvo DJ, Ceriani MF (2017) Organization of circadian behavior relies on glycinergic transmission. Cell Rep 19:72–85. https://doi.org/10.1016/j.celrep.2017.03.034

Frisch B, Hardin PE, Hamblen-Coyle MJ, Rosbash M, Hall JC (1994) A promoterless *period* gene mediates behavioral rhythmicity and cyclical per expression in a restricted subset of the *Drosophila* nervous system. Neuron 12(3):555–570. https://doi.org/10.1016/0896-6273(94)90212-7

Goda T, Doi M, Umezaki Y, Murai I, Shimatani H, Chu ML, Nguyen VH, Okamura H, Hamada FN (2018) Calcitonin receptors are ancient modulators for rhythms of preferential temperature in

insects and body temperature in mammals. Genes Dev 32(2):140–155. https://doi.org/10.1101/gad.307884.117
Goda T, Umezaki Y, Hamada FN (2023) Molecular and neural mechanisms of temperature preference rhythm in *Drosophila melanogaster*. J Biol Rhythms 38(4):326–340. https://doi.org/10.1177/07487304231171624
Golombek DA, Rosenstein RE (2010) Physiology of circadian entrainment. Physiol Rev 90(3):1063–1102. https://doi.org/10.1152/physrev.00009.2009
Gorostiza EA, Depetris-Chauvin A, Frenkel L, Pírez N, Ceriani MF (2014) Circadian pacemaker neurons change synaptic contacts across the day. Curr Biol 24:1–7. https://doi.org/10.1016/j.cub.2014.07.063
Grima B, Chélot E, Xia R, Rouyer F (2004) Morning and evening peaks of activity rely on different clock neurons of the *Drosophila* brain. Nature 431(7010):869–873. https://doi.org/10.1038/nature02935. PMID: 15483616
Hasebe M, Kotaki T, Shiga S (2022) Pigment-dispersing factor is involved in photoperiodic control of reproduction in the brown-winged green bug, *Plautia stali*. J Insect Physiol 137:104359. https://doi.org/10.1016/j.jinsphys.2022.104359
Helfrich-Förster C (1995) The period clock gene is expressed in central nervous system neurons which also produce a neuropeptide that reveals the projections of circadian pacemaker cells within the brain of *Drosophila melanogaster*. Proc Natl Acad Sci USA 92(2):612–616. https://doi.org/10.1073/pnas.92.2.612. PMID: 7831339
Helfrich-Förster C, Winter C, Hofbauer A, Hall JC, Stanewsky R (2001) The circadian clock of fruit flies is blind after elimination of all known photoreceptors. Neuron 30:249–261. https://doi.org/10.1016/s0896-6273(01)00277-x
Herrero A, Yoshii T, Ispizua JI, Colque C, Veenstra JA, Muraro NI, Ceriani MF (2020) Coupling neuropeptide levels to structural plasticity in *Drosophila* clock neurons. Curr Biol 30:3154–3166.e4. https://doi.org/10.1016/j.cub.2020.06.009
Hofer S, Homberg U (2006) Evidence for a role of orcokinin-related peptides in the circadian clock controlling locomotor activity of the cockroach *Leucophaea maderae*. J Exp Biol 209(Pt 14):2794–2803. https://doi.org/10.1242/jeb.02307
Homberg U, Würden S, Dircksen H, Rao KR (1991) Comparative anatomy of pigment-dispersing hormone-immunoreactive neurons in the brain of orthopteroid insects. Cell Tissue Res 266(2):343–357. https://doi.org/10.1007/BF00318190
Ikeno T, Numata H, Goto SG, Shiga S (2014) Involvement of the brain region containing pigment-dispersing factor-immunoreactive neurons in the photoperiodic response of the bean bug, *Riptortus pedestris*. J Exp Biol 217(Pt 3):453–462. https://doi.org/10.1242/jeb.091801
Im SH, Taghert PH (2010) PDF receptor expression reveals direct interactions between circadian oscillators in *Drosophila*. J Comp Neurol 18(11):1925–1945. https://doi.org/10.1002/cne.22311
Im SH, Li W, Taghert PH (2011) PDFR and CRY signaling converge in a subset of clock neurons to modulate the amplitude and phase of circadian behavior in *Drosophila*. PLoS One 6(4):e18974. https://doi.org/10.1371/journal.pone.0018974
Kaneko H, Head LM, Ling J, Tang X, Liu Y, Hardin PE, Emery P, Hamada FN (2012) Circadian rhythm of temperature preference and its neural control in *Drosophila*. Curr Biol 22(19):1851–1857. https://doi.org/10.1016/j.cub.2012.08.006
Klarsfeld A, Malpel S, Michard-Vanhée C, Picot M, Chélot E, Rouyer F (2004) Novel features of cryptochrome-mediated photoreception in the brain circadian clock of *Drosophila*. J Neurosci 24(6):1468–1477. https://doi.org/10.1523/JNEUROSCI.3661-03.2004
Konopka RJ, Benzer S (1971) Clock mutants of *Drosophila melanogaster*. Proc Natl Acad Sci USA 68(9):2112–2116. https://doi.org/10.1073/pnas.68.9.2112
Kunst M, Hughes ME, Raccuglia D, Felix M, Li M, Barnett G, Duah J, Nitabach MN (2014) Calcitonin gene-related peptide neurons mediate sleep-specific circadian output in *Drosophila*. Curr Biol 24(22):2652–2664. https://doi.org/10.1016/j.cub.2014.09.077

Li MT, Cao LH, Xiao N, Tang M, Deng B, Yang T, Yoshii T, Luo DG (2018) Hub-organized parallel circuits of central circadian pacemaker neurons for visual photoentrainment in *Drosophila*. Nat Commun 9(1):4247. https://doi.org/10.1038/s41467-018-06506-5

Liang X, Holy TE, Taghert PH (2017) A series of suppressive signals within the *Drosophila* circadian neural circuit generates sequential daily outputs. Neuron 94(6):1173–1189.e4. https://doi.org/10.1016/j.neuron.2017.05.007

Lin Y, Stormo GD, Taghert PH (2004) The neuropeptide pigment-dispersing factor coordinates pacemaker interactions in the *Drosophila* circadian system. J Neurosci 24(36):7951–7957. https://doi.org/10.1523/JNEUROSCI.2370-04.2004

Loesel R, Homberg U (2001) Anatomy and physiology of neurons with processes in the accessory medulla of the cockroach *Leucophaea maderae*. J Comp Neurol 439(2):193–207. https://doi.org/10.1002/cne.1342

Mertens I, Husson SJ, Janssen T, Lindemans M, Schoofs L (2007) PACAP and PDF signaling in the regulation of mammalian and insect circadian rhythms. Peptides 28(9):1775–1783. https://doi.org/10.1016/j.peptides.2007.05.005

Myers EM, Yu J, Sehgal A (2003) Circadian control of eclosion: interaction between a central and peripheral clock in *Drosophila melanogaster*. Curr Biol 13:526–533

Nässel DR, Zandawala M (2019) Recent advances in neuropeptide signaling in *Drosophila*, from genes to physiology and behavior. Prog Neurobiol 179:101607. https://doi.org/10.1016/j.pneurobio.2019.02.003

Nässel DR, Shiga S, Mohrherr CJ, Rao KR (1993) Pigment-dispersing hormone-like peptide in the nervous system of the flies *Phormia* and *Drosophila*: immunocytochemistry and partial characterization. J Comp Neurol 331(2):183–198. https://doi.org/10.1002/cne.903310204

Ohe Y, Hasebe M, Hamanaka Y, Goto SG, Shiga S (2025) Photoperiodic plasticity of pigment-dispersing factor immunoreactive fibers projecting toward prothoracicotropic hormone neurons in flesh fly *Sarcophaga similis* larvae. J Comp Physiol A Neuroethol Sens Neural Behav Physiol. https://doi.org/10.1007/s00359-024-01729-y

Park JH, Hall JC (1998) Isolation and chronobiological analysis of a neuropeptide pigment-dispersing factor gene in *Drosophila melanogaster*. J Biol Rhythm 13(3):219–228. https://doi.org/10.1177/074873098129000066

Park JH, Helfrich-Förster C, Lee G, Liu L, Rosbash M, Hall JC (2000) Differential regulation of circadian pacemaker output by separate clock genes in *Drosophila*. Proc Natl Acad Sci USA 97(7):3608–3613. https://doi.org/10.1073/pnas.97.7.3608

Patke A, Young MW, Axelrod S (2020) Molecular mechanisms and physiological importance of circadian rhythms. Nat Rev Mol Cell Biol 21(2):67–84. https://doi.org/10.1038/s41580-019-0179-2

Petri B, Stengl M (1997) Pigment-dispersing hormone shifts the phase of the circadian pacemaker of the cockroach *Leucophaea maderae*. J Neurosci 17(11):4087–4093. https://doi.org/10.1523/JNEUROSCI.17-11-04087.1997

Picot M, Cusumano P, Klarsfeld A, Ueda R, Rouyer F (2007) Light activates output from evening neurons and inhibits output from morning neurons in the *Drosophila* circadian clock. PLoS Biol 5(11):e315. https://doi.org/10.1371/journal.pbio.0050315

Rao KR, Riehm JP, Zahnow CA, Kleinholz LH, Tarr GE, Johnson L, Norton S, Landau M, Semmes OJ, Sattelberg RM, Jorenby WH, Hintz MF (1985) Characterization of a pigment-dispersing hormone in eyestalks of the fiddler crab *Uca pugilator*. Proc Natl Acad Sci USA 82(16):5319–5322. https://doi.org/10.1073/pnas.82.16.5319

Reinhard N, Fukuda A, Manoli G, Derksen E, Saito A, Möller G, Sekiguchi M, Rieger D, Helfrich-Förster C, Yoshii T, Zandawala M (2024) Synaptic connectome of the *Drosophila* circadian clock. Nat Commun 15:10392. https://doi.org/10.1038/s41467-024-54694-0

Reischig T, Stengl M (2003a) Ectopic transplantation of the accessory medulla restores circadian locomotor rhythms in arrhythmic cockroaches (*Leucophaea maderae*). J Exp Biol 206(Pt 11):1877–1886. https://doi.org/10.1242/jeb.00373

Reischig T, Stengl M (2003b) Ultrastructure of pigment-dispersing hormone-immunoreactive neurons in a three-dimensional model of the accessory medulla of the cockroach *Leucophaea maderae*. Cell Tissue Res 314(3):421–435. https://doi.org/10.1007/s00441-003-0772-7

Renn SC, Park JH, Rosbash M, Hall JC, Taghert PH (1999) A *pdf* neuropeptide gene mutation and ablation of PDF neurons each cause severe abnormalities of behavioral circadian rhythms in *Drosophila*. Cell 99(7):791–802. https://doi.org/10.1016/s0092-8674(00)81676-1. Erratum in: Cell 2000 Mar 31;101(1):following 113

Rieger D, Shafer OT, Tomioka K, Helfrich-Förster C (2006) Functional analysis of circadian pacemaker neurons in *Drosophila melanogaster*. J Neurosci 26(9):2531–2543. https://doi.org/10.1523/JNEUROSCI.1234-05.2006

Saifullah AS, Tomioka K (2003) Pigment-dispersing factor sets the night state of the medulla bilateral neurons in the optic lobe of the cricket, *Gryllus bimaculatus*. J Insect Physiol 49(3): 231–239. https://doi.org/10.1016/s0022-1910(02)00270-6

Schneider NL, Stengl M (2005) Pigment-dispersing factor and GABA synchronize cells of the isolated circadian clock of the cockroach *Leucophaea maderae*. J Neurosci 25(21):5138–5147. https://doi.org/10.1523/JNEUROSCI.5138-A-04.2005. Erratum in: J Neurosci. 2005 Jun 8;25 (23):5666

Selcho M, Millán C, Palacios-Muñoz A, Ruf F, Ubillo L, Chen J, Bergmann G, Ito C, Silva V, Wegener C, Ewer J (2017) Central and peripheral clocks are coupled by a neuropeptide pathway in *Drosophila*. Nat Commun 8:15563. https://doi.org/10.1038/ncomms15563

Seluzicki A, Flourakis M, Kula-Eversole E, Zhang L, Kilman V, Allada R (2014) Dual PDF signaling pathways reset clocks via TIMELESS and acutely excite target neurons to control circadian behavior. PLoS Biol 12(3):e1001810. https://doi.org/10.1371/journal.pbio.1001810

Shafer OT, Yao Z (2014) Pigment-dispersing factor signaling and circadian rhythms in insect locomotor activity. Curr Opin Insect Sci 1:73–80. https://doi.org/10.1016/j.cois.2014.05.002

Shafer OT, Gutierrez GJ, Li K, Mildenhall A, Spira D, Marty J, Lazar AA, Fernandez MP (2022) Connectomic analysis of the *Drosophila* lateral neuron clock cells reveals the synaptic basis of functional pacemaker classes. Elife 11:e79139. https://doi.org/10.7554/eLife.79139

Shiga S, Numata H (2009) Roles of PER immunoreactive neurons in circadian rhythms and photoperiodism in the blow fly, Protophormia terraenovae. J Exp Biol. https://doi.org/10.1242/jeb.027003

Singaravel M, Fujisawa Y, Hisada M, Saifullah AS, Tomioka K (2003) Phase shifts of the circadian locomotor rhythm induced by pigment-dispersing factor in the cricket *Gryllus bimaculatus*. Zool Sci 20(11):1347–1354. https://doi.org/10.2108/zsj.20.1347

Stanewsky R, Kaneko M, Emery P, Beretta B, Wager-Smith K, Kay SA, Rosbash M, Hall JC (1998) The *cryb* mutation identifies cryptochrome as a circadian photoreceptor in *Drosophila*. Cell 95(5):681–692. https://doi.org/10.1016/s0092-8674(00)81638-4

Stengl M, Arendt A (2016) Peptidergic circadian clock circuits in the Madeira cockroach. Curr Opin Neurobiol 41:44–52. https://doi.org/10.1016/j.conb.2016.07.010

Stengl M, Werckenthin A, Wei HY (2015) How does the circadian clock tick in the Madeira cockroach? Curr Opin Neurobiol 12:38–45. https://doi.org/10.1016/j.cois.2015.09.007

Stoleru D, Peng Y, Agosto J, Rosbash M (2004) Coupled oscillators control morning and evening locomotor behaviour of *Drosophila*. Nature 431(7010):862–868. https://doi.org/10.1038/nature02926

Stoleru D, Peng Y, Nawathean P, Rosbash M (2005) A resetting signal between *Drosophila* pacemakers synchronizes morning and evening activity. Nature 438(7065):238–242. https://doi.org/10.1038/nature04192

Stoleru D, Nawathean P, Fernández MP, Menet JS, Ceriani MF, Rosbash M (2007) The *Drosophila* circadian network is a seasonal timer. Cell 129(1):207–219. https://doi.org/10.1016/j.cell.2007.02.038

Vaze KM, Helfrich-Förster C (2021) The neuropeptide PDF is crucial for delaying the phase of *Drosophila*'s evening neurons under long zeitgeber periods. J Biol Rhythm 36(5):442–460. https://doi.org/10.1177/07487304211032336

Vosshall LB, Young MW (1995) Circadian rhythms in *Drosophila* can be driven by period expression in a restricted group of central brain cells. Neuron 15:345–360. https://doi.org/10.1016/0896-6273(95)90039-x

Wei H, el Jundi B, Homberg U, Stengl M (2010) Implementation of pigment-dispersing factor-immunoreactive neurons in a standardized atlas of the brain of the cockroach *Leucophaea maderae*. J Comp Neurol 518(20):4113–4133. https://doi.org/10.1002/cne.22471

Werckenthin A, Derst C, Stengl M (2012) Sequence and expression of *per*, *tim1*, and *cry2* genes in the Madeira cockroach *Rhyparobia maderae*. J Biol Rhythm 27(6):453–466. https://doi.org/10.1177/0748730412462109

Werckenthin A, Huber J, Arnold T, Koziarek S, Plath MJA, Plath JA, Stursberg O, Herzel H, Stengl M (2020) Neither per, nor tim1, nor cry2 alone are essential components of the molecular circadian clockwork in the Madeira cockroach. PLoS One 15(8):e0235930. https://doi.org/10.1371/journal.pone.0235930

Yamamoto M, Nishimura K, Shiga S (2017) Clock and hormonal controls of an eclosion gate in the flesh fly *Sarcophaga crassipalpis*. Zool Sci 34(2):151–160. https://doi.org/10.2108/zs160153

Index

A. Mizoguchi, N. Okamoto (eds.), *Insect Peptide Hormones and Neuropeptides*, Entomology Monographs, https://doi.org/10.1007/978-981-95-7510-7